BREAKING THE
IRON BONDS

DEVELOPMENT OF WESTERN RESOURCES

The Development of Western Resources is an interdisciplinary series focusing on the use and misuse of resources in the American West. Written for a broad readership of humanists, social scientists, and resource specialists, the books in this series emphasize both historical and contemporary perspectives as they explore the interplay between resource exploitation and economic, social, and political experiences.

John G. Clark, University of Kansas, General Editor

BREAKING THE IRON BONDS
Indian Control of Energy Development

Marjane Ambler

 University Press of Kansas

© 1990 by the University Press of Kansas

Published by the University Press of Kansas (Lawrence, Kansas 66045), which was organized by the Kansas Board of Regents and is operated and funded by Emporia State University, Fort Hays State University, Kansas State University, Pittsburg State University, the University of Kansas, and Wichita State University

Library of Congress Cataloging-in-Publication Data

Ambler, Marjane, 1948–
    Breaking the iron bonds : Indian control of energy development /
Marjane Ambler.
       p.   cm. — (Development of western resources)
    Includes bibliographical references.
    ISBN 0-7006-0422-7
    1. Indians of North America—Land tenure.   2. Indians of North
America—Politics and government.   3. Indians of North America—
Government relations.   4. Energy development—United States.
5. Mines and mineral resources—United States.   I. Title.
II. Series.
E98.L3A43   1990
333.79 ′15 ′08997—dc20   89-27680
                                                                          CIP

British Library Cataloguing in Publication Data is available.

Printed in the United States of America
10 9 8 7 6 5 4 3 2 1

The paper used in this publication meets the minimum requirements of the American National Standard for Permanence of Paper for Printed Library Materials Z39.48-1984.

To the young warriors of the energy tribes, who will take up the battle with law books and tribal codes as their weapons.

If we fight for civil liberties for our side, we show that we believe not in civil liberties but in our side. But when those of us who never were Indians and never expect to be Indians fight for the Indian cause of self-government, we are fighting for something that is not limited by accidents of race and creed and birth; we are fighting for what Las Casas, Vitoria and Pope Paul III called the integrity or salvation of our own souls. We are fighting for what Jefferson called the basic rights of man. We are fighting for the last best hope of earth. And these are causes which should carry us through many defeats.

—Felix Cohen

CONTENTS

ILLUSTRATIONS, MAPS, AND TABLES

Illustrations

Maps

Tables

.

ACRONYMS

AIO	Americans for Indian Opportunity
AML	Abandoned Mines Land
ARCO	Atlantic Richfield Company
BIA	Bureau of Indian Affairs
BLM	Bureau of Land Management
BuRec	Bureau of Reclamation
CEP	Council of Economic Priorities
CERT	Council of Energy Resource Tribes
CONPASO	Consolidation Coal and El Paso Natural Gas Company
CUC	Charging Ute Corporation
DNA	Dinebeiina Nahiilna be Agaditahe (People's Legal Services)
DOE	Department of Energy
EIS	Environmental impact study
EPA	Environmental Protection Agency
ETSI	Energy Transportation Systems, Inc.
FBI	Federal Bureau of Investigation
FEA	Federal Energy Agency
FmHA	Farmers Home Administration
FOGRMA	Federal Oil and Gas Royalty Management Act
FTC	Federal Trade Commission
GAO	General Accounting Office
IMDA	Indian Mineral Development Act
IRA	Indian Reorganization Act
MMS	Minerals Management Service
NANRDF	Native American Natural Resources Development Federation
NARF	Native American Rights Fund
NEDA	Navajo Energy Development Authority
NEPA	National Environmental Policy Act
NIIP	Navajo Indian Irrigation Project
OPEC	Organization of Petroleum Exporting Countries
OSMRE	Office of Surface Mining Reclamation and Enforcement
PNM	Public Service Company of New Mexico
SMCRA	Surface Mining Control and Reclamation Act
STATSS	State and Tribal Support System
USGS	U.S. Geological Survey
WESCO	Western Gasification Company

PREFACE

In 1974 a retired schoolteacher in Ulysses, Nebraska, named Ida Ferrin convinced me to accompany her on a research trip to the Northern Cheyenne Reservation in Montana. Little did I know how that trip would influence my life. Since then several people have asked me why I chose to focus my writing on American Indians, a topic they assume to be terribly depressing. Other acquaintances assume that I am one of the romantics who glamorize the Indian; they cautiously ask me, "Haven't you ever driven the streets of Gallup on a Sunday morning?" Others, not so cautious, tell me that the time for the reservation has passed, that Indians would be better off abandoning both their attempts to be different and their "government handouts."

It is not surprising that there is so little understanding of contemporary Indian issues in this country. In school we are taught about Indians only in the past tense, either as innocent victims of western expansion or as brutal savages. Historians tend to ignore the twentieth century when writing about Indian issues. In neither our civics nor our history classes do we learn about concepts such as tribal sovereignty and self-determination or of the modern-day relevance of Indian treaties. Thus when the president of an energy company suddenly is told that he must abide by an Indian tribe's regulations, or when a fisherman finds that his neighbor, an Indian, can legally catch more fish than he, the reaction is often outrage and disbelief.

For ten years I lived in Lander, Wyoming, on the border of the Wind River Reservation. I, too, saw the drunks on Main Street and read the news accounts of violence. The Indian people I know at Wind River and on the other thirteen reservations where I have traveled—the dedicated tribal employees and officials working to improve their tribes' self-sufficiency while retaining important aspects of their culture and self-esteem—seem to be invisible to outsiders. Busy taking their children to basketball games and pow-wows, teaching language and traditional skills, and attending countless meetings, these Indians have no time to loaf on street corners in Lander or Gallup.

In this book I introduce such people in the context of their efforts to increase tribal control over energy development. I expose the external hurdles they face without glossing over the many internal, tribal barriers. I also cover efforts of the allottees, individual Indians who own minerals as a result of federal allotment policies at the turn of the century. As critics point out, many Indian tribes "want it both ways"—they want some federal protection and assistance as well as freedom to make their own decisions. We should remember that the American public also wants it both ways: We want the Indian tribes to become self-sufficient, but we don't want their efforts to compete with ours.

In this book I attempt to move beyond stereotypes and generalities about American Indians, and I try to broaden our understanding of the complex issues involved in their struggle for autonomy. In the first two chapters, I briefly introduce the history of federal Indian policies, economic development, and reservation energy development. In the rest of the book I focus on the efforts of tribes and allottees in the 1970s and 1980s to increase control over energy development using their power as mineral owners, as governments, and finally as partners in development. With federal policies as the backdrop, I study what is actually happening on the reservations and other Indian lands. I hope the information that I gathered from field research will lay to rest commonly held misconceptions founded primarily upon archival research—for example, that all Indian people oppose energy development and that extensive coal mining has occurred on the Northern Cheyenne Reservation. In reality, the Northern Cheyenne Reservation is the site of a valiant and successful effort to assert tribal control and prevent mining.

It is impossible to generalize about Indian people's attitudes toward energy development or about tribes' approaches toward gaining control: Each tribe and each allottee is different. I focus on the fifteen western reservations with the most coal, uranium, and oil and gas (excluding natives of Alaska, where policies are unique). Although there is little tribal mineral ownership in Oklahoma today, I look briefly at the Osages' mineral history and at modern-day allottees in that state because of the relevance to other tribes and allottees. I do not attempt to address other, nonenergy minerals, nor do I cover alternative energy sources such as solar and hydro power, which would warrant separate treatment.

I began work in earnest on this book in 1980 with a fellowship from the Alicia Patterson Foundation that allowed me to spend a year traveling to western energy reservations. The Nu Lambda Trust and the Fund for Investigative Journalism later helped with travel expenses, and I am grateful for their support. For research assistance I want to thank the Native American Research Information Service at the University of Oklahoma, Montana State University, and—for interlibrary loans—the Yellowstone National Park Research Library. I especially appreciate the assistance of Mary Mousseau at the National Indian Law Library in Boulder, Colorado, who ferreted out dozens of documents for me, sometimes using quite obscure clues. I also appreciate the early encouragement I received from those who thought a book on Indian energy was needed, especially Ron Therriault, Alvin M. Josephy, Jr., and Richard Hart. The enthusiasm of the University Press of

Kansas staff kept me going, and series editor John Clark helped show me where to go.

For sharing their thoughts and time I am grateful to the people in tribes, tribal organizations, industry, and government agencies who responded promptly to my many inquiries over the years, especially Hugh Baker, Rich Schilf, David Lester, Doug Richardson, Ahmed Kooros, Caleb Shields, Perry Baker, David Harrison, Carol Connor, Alan Taradash, Earl Old Person, Reid Chambers, John Echohawk, David Harrison, Lucille Echohawk, Robert Nordhaus, Bill Haltom, Allen Rowland, John Sledd, Tom Fredericks, Tom Acevedo, Orville St. Clair, Susan Williams, Steve Chestnut, Wes Martel, Gary Collins, Dave Allison, Don Aubertin, Eric Natwig, Ellis "Rabbit" Knows Gun, Scott McElroy, Leonard Robbins, Don Wharton, and Don Ami. The opinions and interpretations expressed in the book are, of course, my own responsibility.

Peter Iverson deserves special thanks for his continued encouragement and for his suggestions on improving a very rough manuscript. Sara Hunter-Wiles provided companionship and her invaluable perspective during several trips to reservations. Several diligent readers offered comments on parts of the text, including Jazmyn McDonald, Doug Richardson, Dave Geible, Bill Haltom, Tom Fredericks, Alan Taradash, Nancy Gregory, Lorna Wilkes, Steve Douglas, Eric Natwig, Tom Glenn, Reid Chambers, and Warner Reeser.

For his patience and his nurturing sense of humor I would like to thank my husband, Terry Wehrman. Thanks also to my father, Edward Ambler, for all that he has given me.

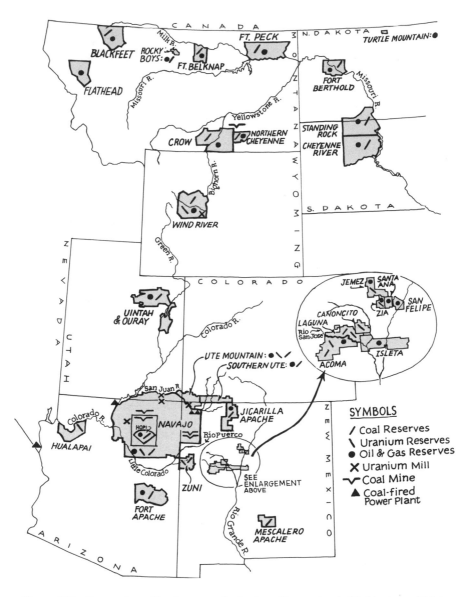

Energy Tribe Reservations. Not shown are the Spokane Reservation in Washington, which has uranium and one uranium mill, and the state of Oklahoma, where coal, oil, and gas are found under the former reservation lands of the Five Civilized Tribes. Oil and gas are also found under the former reservation lands of sixteen other tribes in the state. Some coal and oil in Oklahoma is still in Indian ownership, mostly individual allottees'. Other isolated reservations with possible oil or gas include the Isabella in Michigan, Big Cypress in Florida, Choctaw in Mississippi, Potawatomi and Kickapoo in Kansas, Quinault in Washington, Aqua Caliente in California, and native lands in Alaska. The coal resource information is from Douglas Richardson, The Control and Reclamation of Surface Mining on Indian Lands, for the Office of Surface Mining. Denver: CERT, September 1979, 3-1. The uranium and oil resource information is from Richard Nafziger, "Indian Uranium: Profits and Perils," in Americans for Indian Opportunity, You Don't Have to Be Poor to Be Indian, edited by Maggie Gover. Washington, D.C.: AIO, no date, 149, and from Department of Interior (DOI), "Indian Lands Map: Oil, Gas, and Minerals on Indian Reservations," 1978.   (Map by Hannah Hinchman.)

CHAPTER ONE

Out of the Mainstream:
The Importance of the Reservation

What do you people want anyway? we've heard
  the Mericano ask. . . .
Land and life, we have to say
for they are one and the same.
One cannot exist without the other.

—Simon Ortiz[1]

To many people in the United States the Indian reservation is an embarrassment. To them it represents the American version of apartheid—a prison without walls where the government confined Indians to keep them apart from other Americans, similar to the detention centers where Japanese Americans were kept during World War II. When they drive through reservations, they see only the signs of poverty: brightly colored, dilapidated frame houses; abandoned cars; dusty yards. They look for dancers dressed in feathers and bells to entertain them and vaguely wonder why such an archaic institution—the reservation—still exists.

The Indian people who live on reservations see it differently. To them the reservation is home, a locus for their political, social, economic, and cultural lives. Their families and their culture are tied to that patch of land. Young Indian people say they feel as if they are entering alien territory when they cross the reservation boundary to the outside world. The reservation is where they learned to hunt with their grandfathers or to herd sheep with their aunts. They know where to find the springs when the creeks dry up in August; where to cut the straightest lodgepole pines for powwow teepee poles; and where to pick buffalo berries, mint, and yarrow. Since they were young, they have known which places are sacred and where the powwows are held each summer. Most of their relatives still live there. When they drive back into the reservation, they come home.

Why have Indian people clung to their land and the reservation system despite various attempts by the U.S. government during the past century to assimilate them into the mainstream? The answer is critical to understanding why developing a reservation economy is important. In this chapter I briefly trace the history of the United States' Indian policies, with an emphasis upon the economic impact of those policies, especially the shrinking land base.

1

First graders at Wyoming Indian Elementary School practiced the Arapahoe language when they made this display in 1987 about a powwow celebration. Pictured (from left) are Tyrel White Buffalo, Christina Bull Calf, Edward Quiver, Lamar Spoonhunter, and Felicia Brown. (Photo by Mike McClure.)

Indian people have been bound by three iron chains: paternalism, exploitation, and dependency. These chains gained their crippling power through the decades as lawmakers vacillated about whether tribes should continue to exist as separate sovereigns or whether their separate rights should be terminated. Many of those who advocated "freeing the Indian" through each period of history were well-meaning but paternalistic. Others were merely greedy and wanted to exploit Indian resources. Whatever their motives the effects of the vacillating policies were the same: The tribes and individual Indians lost more and more of the land and other resources they needed to make them self-sufficient again. This exploitation left them increasingly dependent upon the federal government yet, ironically, resistant to change, because so often in the past change had brought disaster.

The constant conundrum throughout the history of Indian affairs in this country has been finding the best balance between dependency and independence; trust protection and self-determination; autocratic federal control and tribal control. Today it still underlies all disagreements between tribal and federal officials. For the energy tribes the conflict has been critical. In order to increase their autonomy and self-determination, they know they have to decrease their dependence upon federal funds and increase their self-sufficiency. But this does not mean that a tribe has to become an economic island, consuming only what it produces on the reservation. A tribe will have reached self-sufficiency when it can provide for the needs of community members and determine its own social and economic goals for the future without violating its cultural heritage.[2]

In the context of this book, "energy tribes" refers to tribes that receive a significant portion of their income from energy minerals or that own substantial undeveloped reserves. These include the Fort Peck Assiniboine and Sioux, Blackfeet, Northern Cheyenne, and Crow Tribes in Montana; Fort Berthold Three Affiliated Tribes in North Dakota; Wind River Arapahoe and Shoshone Tribes in Wyoming; Osage Tribe in Oklahoma; Spokane Tribe in Washington; Northern Ute Tribe in Utah; Southern Ute and Ute Mountain Ute Tribes in Colorado; Jicarilla Apache Tribe and Laguna Pueblo in New Mexico; Hopi Tribe in Arizona; and the Navajo Tribe in Arizona, New Mexico, and Utah (see map).

MISCONCEPTIONS ABOUT THE RESERVATION

In January 1983 Interior Secretary James Watt for a few days focused the nation's attention on the question of reservation versus assimilation by suggesting that Indians should be "freed" and integrated into the American system. The statement by Watt, a controversial and short-tenured appointee of President Ronald Reagan, infuriated most Indian people, many of whom called for his resignation. Nevertheless, Watt had verbalized sentiments shared by a few Indians and many non-Indians (that is, among those who thought about Indian reservations at all). Watt said, "Every social problem is exaggerated because of socialistic government politics on the Indian reservations. . . . The people have been trained through 100 years of government oppression to look to the government as . . . the provider, and they've not been trained to use the initiative to integrate into the American system.[3]

Apparently mystified by why Indians would stay on reservations, he blamed greedy tribal leaders who somehow coerced their members: "If you're the chief or the chairman, you're interested in keeping this group of people assembled in a desert environment where there are no jobs, no agriculture potential, no water, because if the Indians were allowed to be liberated, they'd go and get a job and that guy wouldn't have his government handout as a government paid Indian official."[4]

Although President Reagan at that time disavowed Watt's remarks, he later expressed the same misunderstanding about the reservation system. In 1988, responding to Soviet students who asked about American Indians, Reagan said he could not understand what their complaints could be, given that the American people had "humored" the Indians by providing them with reservations where their "primitive life style" was to blame for their poverty.[5]

Indians had heard such rhetoric before. Since the 1880s many people who considered themselves friends of the Indian had championed the cause

TABLE 1.1 Energy Tribes' Resources, Land, Population, and Form of Government

Reservation (Tribe)	Energy Mineral Potential	Trust Acreage (% allotted)	Resident Indian Population	Government
Blackfeet (Blackfeet)	coal, oil and gas[a]	937,701(68)	6,555	IRA
Crow (Crow)	coal[a], oil and gas[a]	1,516,005(73)	5,288	non-IRA constitution
Fort Berthold (Mandan, Hidatsa, Arikara)	coal, oil and gas[a]	419,198(83)	3,081	IRA
Fort Peck (Assiniboine and Sioux)	coal, oil and gas[a]	904,683(57)	5,022	non-IRA constitution
Hopi (Hopi)	coal[a], oil and gas	1,561,213(0)	8,945	IRA
Jicarilla Apache (Jicarilla Apache)	oil and gas[a], coal	823,580(0)	2,344	IRA
Laguna Pueblo (Keresan)	uranium[a], coal	461,099(0)	6,525	IRA
Navajo (Navajo)	coal[a], uranium[a], oil and gas[a]	11,289,959(1)	158,917	—
Northern Cheyenne (Northern Cheyenne)	coal, oil	436,947(27)	3,197	IRA constitution
Osage (Osage)	oil and gas[a]	168,794[b](100)	5,999	—
Southern Ute (Ute)	oil and gas[a], coal	309,970(1)	1,107	IRA
Spokane (Spokane)	uranium[a]	130,180(19)	1,963	—
Uintah and Ouray (Ute)	oil and gas[a], coal, oil shale	1,021,556(1)	2,244	IRA
Ute Mountain Ute (Ute)	oil and gas[a], coal, uranium	597,288(1)	1,554	IRA
Wind River (Arapahoe and Shoshone)	coal, uranium, oil and gas[a]	1,887,262(5)	5,385	—

[a]developed
[b]Osage surface acreage. The tribe owns 1.5 million acres of subsurface mineral rights.

Note: Acreage figures include only on-reservation, trust lands—not any privately owned or tribal fee lands. Allotted figures were rounded to the nearest percentage point. Population figures do not indicate the total number of enrolled tribal members, nor whether the resident Indians are members of the same tribe. IRA = Indian Reorganization Act.

Sources: U.S. Department of Interior (DOI), Bureau of Indian Affairs (BIA), *Annual Report of Indian Lands*, dated September 30, 1985 (Washington, D.C.: GPO, October 1988); DOI, BIA, *Indian Service Population and Labor Force Estimates* (Washington, D.C.: GPO, January 1983); Gilbert L. Hall, *The Federal-Indian Trust Relationship* (Vienna, Va.: Institute for the Development of Indian Law, 1979).

of assimilation with missionarylike zeal in language similar to Watt's when he said, "We ought to give them freedom, we ought to give them liberty, we ought to give them their rights, but we treat them as incompetent wards."[6] Most of those who had advocated terminating federal protection for tribes in earlier years were probably as bewildered as Watt and Reagan by the negative response.

The 1980 census reported that Indians on reservations had lower incomes and fewer modern conveniences than those who lived off the reservations and had dramatically lower incomes than non-Indians. True, in cultures in which subsistence agriculture and bartering are common, the standard socioeconomic indicators do not necessarily accurately reflect the quality of life. Nevertheless, the extent of the differences between reservation and off-reservation conditions did indicate a serious problem. More than half (58 percent) of the reservation males between 20 and 64 years old were not employed, compared with 26 percent of off-reservation Indian males and 18 percent of all races.[7]

Of reservation Indians, census takers found that 41 percent were living in households with incomes below the poverty level, compared with half as many (22 percent) off-reservation Indians and 12 percent of the U.S. population as a whole. About one-fourth (22 percent) of the reservation households received income from public assistance, compared with 15 percent for off-reservation Indian households and 8 percent for the U.S. population as a whole. Fewer than half (44 percent) of the reservation Indian households had telephones. Poverty and unemployment contribute to extremely high alcoholism and drug use rates as well as other social problems. The social and economic problems on reservations are real, but Watt's analysis of the causes was shallow and his solution—termination—too simple.[8]

So why would anyone stay on the reservations? Watt's theory notwithstanding, tribal leaders do not and cannot coerce members to stay. And, in fact, not everyone does stay. About half of the 1.5 million Indians in the country lived on or near reservations in 1980.[9] Many of the others had left for jobs or schooling, returning on weekends and tribal holidays to renew family and cultural ties. Others may never have lived on reservations and may feel more "Indian" than Crow or Assiniboine or Navajo. They take part in pan-Indian, urban festivities without regard for particular tribal affiliations or traditions. Still others have assimilated most of the values of the dominant culture and rarely think of themselves as Indian at all.

The Indians who choose to live on the reservations do so for many reasons. On some reservations the residents get more medical and economic benefits from the federal and tribal governments than members who live elsewhere, as Watt pointed out. Yet more intangible reasons encourage thousands of Indians to stay on reservations. Despite the "desert environment," many love their land, just as many non-Indians love special places of personal significance. They stay to be with their families and so their children can benefit from time spent with family elders. They participate in tribal religious and traditional activities. Many feel strongly that they want to serve their people.

On the tribal level the reservation land serves three critical, interrelated purposes: cultural, political, and economic. A contemporary reservation Indian does not fit the expectations of the non-Indian who believes the "real" Indian should wear skins and feathers and that his or her reservation should be a living museum depicting the 1700s. For example, one non-Indian visitor decided that he did not like telephone poles because they interrupted the Indians' psychic/spiritual connection to the earth, without asking the Indians whether they appreciated the convenience of finally being able to call friends, government agencies, and doctors. Others seem to think Indians should drop any pretense of being different if they are going to keep changing their life styles. One author ridiculed the contemporary Ute, saying, "What remains of their culture is a mongrelized version of tradition, lore, religion, and a quasi-language that has almost lost its purity." Anthropologist Loretta Fowler says that her colleagues sometimes encourage such misconceptions by focusing upon arbitrary measures of a tribe's traditional life and suggesting that successful adaptation is un-Indian.[10]

On most reservations many traditional cultural values persist. Although tribal members may be wearing three-piece suits with their beads and braids and sporting degrees from Harvard or the University of Oklahoma, they often have retained aspects of their culture that thrive within them, invisible to outsiders. The blue-eyed, outwardly assimilated tribal geologist may go to both a white doctor and a medicine man when she is sick. During his off-hours, a tribal councilman may go on a vision quest as he decides how best to serve his people. The supernatural is often part of everyday life.

Although non-Indians may believe that Indian cultures have not been viable for generations, a University of Chicago study in 1961 affirmed that Indian communities were increasing in population and were, as a whole, "distinct growing communities that still preserve the core of their native style of life." Since then the resurgence of old traditions has continued on many reservations as tribal members learn to tan hides the old way, using the animal's brains; make pemmican and chokecherry delicacies; and perform dances that once were nearly forgotten. On some reservations tribal languages are spoken extensively; children are taught their language by tribal elders, at the insistence of tribal councils. On the Jicarilla Apache Reservation in New Mexico, for example, a 1978 study found that half of the reservation residents spoke their language and one-third of the households used it regularly. A 1987 study on the Navajo Reservation found about 125,000 Indians who spoke Navajo fluently.[11]

Romanticized cultural expectations often cause serious problems. Non-Indians argue against fulfilling the requirements of "old treaties" by saying the people have assimilated and thus are not of the same culture as those

Evelyn Willow, a Northern Arapahoe, shows other tribal members how to make moccasins at a workshop on the Wind River Reservation in 1983. (Photo by Sara Hunter-Wiles.)

who signed the treaties. Members of Congress erect barriers to any kind of economic development that does not fit their expectations. For example, arts and crafts are acceptable, but bingo halls and factories are often restricted. In response, Indian leaders point to the value of adaptation—retaining important aspects of their own cultures while taking from the dominant culture. Adaptation has been key to the Indians' survival on this continent for the past hundred thousand years. Tribes adopted useful tools and trade materials from one another. When the Europeans arrived, tribes adapted to the horse and to the glass trade beads that today are typically identified with the "traditional" Plains Indian culture. No culture is static; it dies if it does not grow. For the contemporary Indian, the right to change is an important legal right, not just an academic assertion. Since 1908 the U.S. Supreme Court has recognized that the powers of tribal governments, including their water rights, can change to adapt to the twentieth century.[12]

The needs for a cultural, political, and economic base are intertwined. Every ethnic group has faced the challenge of change and its effect on traditional ways, but few in this country have kept their cultures as intact on such a wide scale as the American Indian. The explanation, as many Indians see it, is simple: the land. Reservation lands and the tribes' political authority help distinguish Indians from other religious or ethnic minorities. Although the land and its resources are more than just real estate to be exploited, their economic potential is important to the continued existence of the tribe as a political and cultural entity. Without land, tribal governments would find it difficult to practice self-government or to exercise the limited sovereignty guaranteed to them by the U.S. Constitution. Without economic development to nourish tribal coffers, self-government is meaningless.

A hundred years ago Chief Dull Knife told the Northern Cheyennes of Montana to send their children to school to prepare for the changing times. For that some of his people called him "the wife of a white man," a charge of treason at that time. Yet Dull Knife knew that through education young Cheyennes could help to develop their local economy and tribal government and to preserve Indian rights to land and resources. Just as their adaptation helped them retain their land, so, too, did the land help them retain their culture. In the 1980s the Northern Cheyennes revere Dull Knife and Little Wolf and the land they led them back to after their government-imposed exile to Oklahoma. Every gathering—from powwows to school board meetings—opens with a reference to the long journey back to Montana. The official tribal stationery, with pictures of Dull Knife and Little Wolf on the top, says, "Out of defeat and exile they led us back to Montana and won our Cheyenne homeland which we will keep forever."[13]

The invisible culture of the reservation is beside the point to many melting-pot missionaries (such as Watt) who, if they see it at all, see it as an obstacle to economic progress. To them a self-sufficient tribal government just perpetuates an anachronistic system that stands between the Indian and true equality. An example is another Wyomingite, former Governor Lester C. Hunt. In language remarkably similar to Watt's, Hunt argued forty years earlier against the government's continuing to be a "wet nurse" for the Indians. Saying the Indian had "lost his glamor as a showman," Hunt suggested taking away federal services and dividing the land among tribal members so that the "Indian as we know him today would soon lose his identity and would rapidly acquire the American way of living." He professed ignorance of why the Indian had not progressed: "The Indian reservations, probably all of them, are surrounded by the highest types of civilization, and how or why they have been retarded in their advancement as much as they have is a mystery to me."[14] Why have reservations been "retarded" in their economic advancement? Actually, in light of the history of federal Indian policy, the real mystery is how, despite the odds, tribes have been able to retain some of their resources and their culture and to progress economically as far as they have.

THE FORMATIVE YEARS: 1776–1871

The history of tribal resource development does not reflect a continuum of increasing tribal control. Before the coming of the white man, resource decisions were totally under tribal control, limited only by the environmental constraints of their territory. Through their strength as warriors and their adaptability, Indian people could expand that territory. After the reservations were established, however, tribal resource decisions were forever limited by both the lands' boundaries and by the federal government's laws and regulations.

Initially reservations developed as an alternative to trying to exterminate Indians who stood in the way of westward expansion. They were temporary holding places, halfway houses to civilize the Indian before he or she assimilated into the dominant culture. According to historian Francis Paul Prucha, the nation's leaders had "no notion" of the role reservations would later play in a pluralistic society—where land is occupied in part by European immigrants and their descendants and in part by American Indians adhering to their own customs.

As one commissioner of Indian affairs said bluntly in 1850, the reservations would civilize the Indians because they would be "compelled by sheer necessity to resort to agricultural labor or starve." Indian agents forbade the Indians to leave reservation boundaries to hunt or trade without a permit, thus forcing them to be dependent upon government rations and products from their farms. As happened again and again in later years, the implementation of Indian policy did not adhere to the theoretical goal behind it. Rather than good agricultural land, the Indians usually got lands that others did not, at the moment, want.[15]

President Reagan's comments notwithstanding, the United States did not "give" reservations to the Indian tribes. When the U.S. government made treaties with the tribes, the Indians gave up land in exchange for guarantees of recognized rights to other lands that the Indians either had reserved from their aboriginal domain or had obtained through purchase or exchange. From the beginning some of the non-Indians challenged treaties creating reservations, saying the roaming savages had no rights to land at all. When Congress in 1871 voted to end treaty making, the decision was based only partly upon the desire to end Indian tribal organization. The debate dealt more with the House's jealousy over the Senate's sole authority to ratify treaties and with charges that the Senate had abused its authority by giving former Indian lands to railroads and land companies.[16]

The vote did not affect rights under existing treaties or end the establishment of reservations. In fact, several of the energy tribes' reservations were established after 1871 by presidents' executive orders (until 1919) and/or acts of both houses of Congress. The executive order reservation tribes, however, faced frequent attacks in later years to their lands, minerals, and water by those who said executive-order tribes did not share the rights of treaty tribes.[17]

ALLOTMENT, LEASING, AND
ASSIMILATION ERA: 1871–1928

In deciding how to deal with the nation's original inhabitants, the federal government tended to apply its public land policies to Indian lands just as Europeans had when they first claimed Indian lands as their own. The desire to treat Indian lands identically with public lands was inappropriate. Nevertheless, it helps explain why Indians received such low per acre prices for their lands and how they lost some of their mineral lands during this period from 1871 to 1928. Public land policies, such as homestead and mineral laws, were designed to make public lands most accessible to exploitation for private gain. The privatization of western lands was an accepted public policy during this period. Western settlers were to get their little pieces of America just as eastern settlers had by claiming and improving them. Under the homestead acts and most mineral laws the settler and miner obtained not just the use but also the ownership of the resource. When applied to lands owned and occupied by Indian people, these laws had the impacts of colonization—enforced dependency and exploitation.[18]

In 1887 Congress passed the General Allotment Act, which generalized a policy included in several previous statutes and treaties. Tribal lands should be divided into equal parcels among tribal members and the "surplus" lands sold. The size of the allotments varied with each reservation's allotment act. Proponents believed strongly in the principle of private land ownership implemented for public lands by the various homestead acts. To them allotment was another means for developing the frontier as well as making the Indians more like the non-Indian settlers. Indians received their "homesteads," generally from communally owned tribal lands; non-Indians received theirs from communally owned public lands.[19]

Senator Henry Dawes deeply distrusted communal ownership, regardless of evidence of its benefits. Before sponsoring the Allotment Act in 1887, he visited the Five Civilized Tribes (Cherokee, Creek, Choctaw, Chickasaw, and Seminole). Later he reported:

> The head chief told us that there was not a family in that whole nation that had not a home of its own. There was not a pauper in that nation, and the nation did not owe a dollar. It built its own capital, and it built its own school and its hospitals. Yet the defect of the system was apparent. They have not got as far as they can go because they own their land in common, and under that [system] there is no enterprise to make your home any better than that of our neighbors. There is no selfishness, which is at the bottom of civilization.[20]

Like Watt, Dawes believed that only greedy tribal officials skimming coal royalties could oppose dissolving tribal governments. He and other allotment proponents, such as the Indian Rights Association, considered themselves strong advocates of Indian rights. They opposed non-Indian settlers' and miners' encroachment on Indian reservations, which the military was not preventing, and believed that individual ownership would offer the Indians more security than communal ownership.[21]

Nevertheless, a selfish desire to open "surplus" Indian lands lay beneath these rationales. Recognizing this, the minority report of the House Indian Affairs Committee said, "If this were done in the name of greed, it would be bad enough; but to do it in the name of humanity and under the cloak of an ardent desire to promote the Indian's welfare by making him like ourselves whether he will or not, is infinitely worse."[22]

The General Allotment Act of 1887 provided for mandatory allotment of Indian lands. Yet language of specific reservation allotment and cession acts prior to 1903 emphasized three points: the rationale behind the release of lands, the willingness of the Indians to part with them, and the compensation the tribes were intended to receive. The 1891 law providing for the Three Affiliated Tribes of Fort Berthold in North Dakota to relinquish their surplus lands, for example, said,

> Whereas it is the policy of the government to reduce to proper size existing reservations when entirely out of proportion to the number of Indians existing thereon with the consent of the Indians, and upon just and fair terms; and whereas the Indians of the several tribes, parties hereto, . . . are desirous of disposing of a portion thereof in order to obtain the means necessary to enable them to become wholly self-supporting . . . .[23]

Using whatever means were necessary, Indian Bureau representatives tried to convince tribes to *agree* to cessions—especially when a tribe's treaty specified that lands could not be taken without consent. The process of haggling over terms took time and sometimes substantially delayed opening the surplus lands. The Crow Tribe, for example, agreed in 1899 to sell over one million acres, but the sale was delayed for years by their insistence that they get one dollar per acre.[24]

Later, however, during the first two decades of the twentieth century, the U.S. Congress and Indian commissioners tried to terminate all special tribal rights. Congress said tribal treasuries should be divided among the members. The Office of Indian Affairs in many cases stripped tribal governments of most of their powers, forbidding leaders to meet and removing from leadership individuals who did not conform. Tribal courts were abolished. All cultural expression was suppressed.[25]

The U.S. Supreme Court contributed greatly to this termination effort in 1903 when it ruled in *Lone Wolf v. Hitchcock*. A Kiowa leader with the evocative name of Lone Wolf challenged Congress's decision to sell 2.4 million acres of "excess" tribal land in Indian Territory that belonged to the Kiowa, Comanche, and Apache Tribes without the adult male tribal members' approval, as required by an 1867 treaty. The high court said Congress had the power to abrogate the provisions of an Indian treaty, though "presumably" Congress would do so only when the interests of the country and the Indians justified it. Although subsequent cases established that Congress must compensate tribes when it broke treaties, the *Lone Wolf* decision struck a crippling blow to tribal sovereignty and independence.[26]

The *Lone Wolf* decision devastated energy tribes' land base, just as it did other tribes' land base. President Theodore Roosevelt took 2.5 million acres of Indian timber lands—including mineral rights—for national forests during his term. Although most energy tribes opposed allotment, after *Lone Wolf* few escaped its ravages. Indian Inspector James McLaughlin went to reservation after reservation with the same message. In 1904 he told the Arapahoe and Shoshone Tribes of the Shoshone Reservation (later known as the Wind River Reservation):

> Up until a little over a year ago, no Indian reservation has been open to settlement without first consulting the Indians, but it is not now deemed necessary by Congress that such negotiations should be conducted with Indians for opening their surplus lands. . . . In days gone by, years ago, when your reservation was set apart, large reservations were possible, because the whiteman did not desire the lands, but the tide of emigration is now pressing from both east and west . . . and . . . the department having charge of the Indians cannot prevent it.

McLaughlin told the Northern Ute Indians in Utah: "Indians have no right to any part of their reservations except what they may require for allotments in severalty or can make proper use of."[27]

Many of the energy tribes' lands were not allotted and opened to non-Indian settlement and mineral prospecting until after the *Lone Wolf* decision. The Uintah and Ouray Reservation allotment act was implemented March 3, 1903, two months after the *Lone Wolf* decision. In 1904 Congress authorized taking land from the Crows, unilaterally changing the previous agreement to say the tribe would not get paid one dollar an acre but would instead only get paid when lands were settled. The Wind River Reservation was allotted and opened in 1905. The Blackfeet Reservation was allotted in 1907, Fort Peck in 1908, and Fort Berthold in 1910.[28] Although the Northern Cheyenne Reservation allotment law was passed in 1926, the tribe had actually requested allotment, believing it was the only way to fight white ranchers' attempts to take their lands and to escape from the Indian agent's autocratic control of land use.[29]

Navajo allotments did not fit the pattern established elsewhere in either objectives or outcome. Thousands of tribal members lived on lands that had

This Northern Ute delegation went to Washington in 1905 for the final negotiations on opening the reservation to homesteaders. From left—first row: Appah, Arrive; second row: Red Cap, David Cooperfield, Charlie Shavanaux, Wee-che; third row: Wallace Stark, Charley Mack, John Duncan, Suckive, unknown, Boco White, unknown. (Courtesy of Utah State Historical Society.)

not been included in a reservation boundary. With the help of local Indian Bureau superintendents who realized the limitations of desert grazing lands, the Navajos succeeded in dramatically increasing their land base to provide for their people through both allotments on "public lands and new executive-order reservations.[30]

The Jicarilla Apache Tribe also increased its land holdings during this period. Because southwestern interests had forced earlier presidents to remove the Jicarillas from two executive-order reservations, their permanent home was not established until 1887, just a few months before the General Allotment Act was passed. Bungled surveys delayed allotment, and by the time it began, Congress had recognized the benefits of ranching as opposed to farming allotments. Thus the Jicarilla Apache allotments were larger, and President Theodore Roosevelt doubled the size of the reservation in 1907 to accommodate more livestock. That addition, which was never allotted, later produced most of the tribe's oil and gas.[31]

Continued Indian expansion incensed southwestern land interests, including Senator Albert B. Fall of New Mexico. They vigorously fought back by convincing Washington to restore some of the Navajo executive-order additions to the public domain between 1908 and 1922 and to curtail allotment periodically. As a result of these pressures Congress in 1917 voted to prohibit any more executive-order reservations in New Mexico or Arizona. The following year the prohibition was extended to all states. During this period of southwestern backlash and national termination sentiment, the Navajos lost mineral-rich executive-order additions as well as minerals under allotments, which were returned to the public domain through executive orders and administrative decisions. Two large pieces of land in the coal- and uranium-rich San Juan Basin that had been added to the Navajo Reservation by President Theodore Roosevelt in 1907 were subsequently returned to the public domain—one piece by President Roosevelt in 1908 and the other by President William H. Taft in 1911—leading in later years to a bitter dispute between the state of New Mexico and the Navajo Tribe involving jurisdiction over the McKinley coal mine.[32]

What became the most devastating aspects of the allotment law—leasing and forced patenting—were not added until later. In 1891 Congress allowed agricultural and mineral leasing, saying that Indians were not using allotted acres and that lease income could provide for minors, women, and the elderly, who could not work the land themselves. Not incidentally, leasing made Indian lands available to neighboring non-Indian farmers and ranchers at low prices. Many people objected to the idea, including ex-Senator Dawes, because it conflicted with the original concept behind allotment. Dawes wrote to Samuel Brosius of the Indian Rights Association in 1901 and said the change "has been of infinite damage to the whole undertaking of making a self-supporting citizen of the Indian" because Indians would not be working the land themselves. Congress expanded its authorized mineral leasing for allotments in 1909. Whether or not it affected the Indians' work ethic, leasing did take more resource control out of their hands.[33]

Allotment had been intended to do away with communal tribal property and thus with tribal governments so individual Indians would assimilate. The General Allotment Act said, however, that allotments would be held in trust for twenty-five years. The trust provisions prevented sale and taxation of the allotments; the legal title was held by the United States while the Indians held the beneficial title. These trust provisions initially thwarted the law's purpose by keeping Indian property separate and thus its Indian owners separate from mainstream society.

Between 1906 and 1920, however, Congress and the administration released more and more allotments from federal protection by liberalizing the definition of "competence": Indians of half or less Indian blood were automatically competent. Federal protection continued only for Indians "incapable of progress." After Indians were declared competent, they received fee simple titles to their land (which removed the federal trust restrictions) and could sell it. They then became "citizens" of the nation, in effect renouncing their tribal citizenship. "Freeing" Indians from federal protection led inevitably to freeing them from their property. As this became clear, Indians resisted being declared competent, yet the Indian Bureau forced patents upon unwilling allottees. At one point Commissioner Francis Leupp tried to get Congress to issue patents to allottees who broke state laws.[34] This blatant attempt to make a designation of competence a punishment was not successful. Yet the principle behind it thrived. Between 1913 and 1921 Commissioner Cato Sells said frequently that he would make the Indian self-supporting by the end of his term. In the process he presided over the sale of more than one million acres of trust land. The Indians learned that success breeds failure and gains lead to losses. Appearing to be incapable, incompetent, and illiterate has its rewards.[35]

The first of many non-Indian ideas for Indian economic development failed. Not only did it leave most Indian people destitute, but it also left a legacy that would undermine future development efforts. In 1976 tribal leaders of eleven of the sixteen allotted reservations surveyed told the American Indian Policy Review Commission staff that allotment was their major obstacle to development. The Bureau of Indian Affairs (BIA) estimated that 75 percent of the 10 million acres of allotted lands was held undivided by multiple heirs.[36]

Part of the legacy of the allotment era is continued non-Indian use of lands. The Bureau of the Census estimated in 1980 that 50.8 percent of the people living on Indian reservations were non-Indians. On many reservations, particularly in the Northern Plains, much of the land within reservation boundaries is owned by non-Indians as a result of the allotment period, often in a checkerboard pattern with tribal lands and allottees' lands. Such inholdings contribute to jurisdictional disputes that undermine tribal powers and hinder economic development.[37]

Even if non-Indians do not own the reservation agricultural land, they still lease a majority of it. With the continued division of allotted lands through inheritance, leasing often becomes the only viable use of them. Income can be divided when land cannot. BIA policies encourage continued leasing, and the rents it sets are often low. On allotted lands BIA was leasing 70 percent of the best farmland to non-Indian farmers in the early 1980s, whereas on nonallotted reservations the percentage was much lower. The number of Indians who were farming and ranching their lands continued dropping. In 1983 Indians farmed 38 percent of the Indian-owned farmlands and ranched on 95 percent of the grazing lands, but in 1987, those percentages had dropped to 35 percent and 85 percent, respectively. Economists say that leasing land cannot provide the income that using it can, nor does it provide meaningful employment.[38] By reducing the Indian voice and by treating Indian lands as part of the public domain, the decision makers of the allotment era contributed to the continued dependency of the energy tribes. By providing for non-Indian inholdings within reservations, they crippled the governments of these tribes for years to come.

THE INDIAN NEW DEAL: 1928–1945

Indian tribes could have lost all of their lands to allotment except that in the late 1920s the nation's political atmosphere began changing. John Collier of the American Indian Defense Association and others called for an end to assimilationist policies, a viewpoint that predominated a few years later. Assimilation was not working: Indians on reservations lived in dire poverty and, despite the odds, had not rejected tribal cultural practices and traditions. Recognizing the changing political atmosphere, Interior Secretary Hubert Work in 1926 hired the Institute for Government Research to conduct a study, headed by Lewis Meriam, of Indian policies. In 1928 Meriam's report, *The Problem of Indian Administration*, sparked a major reform of Indian policies.[39]

Coincidentally or not, interest in Indian lands and minerals began dropping at about the same time because of national economic changes. Just when the nation had less demand for Indian lands and resources, it adopted more humane and equitable policies. After a disastrous drop in oil prices caused by overproduction, Indian Commissioner Charles Burke in 1927 discontinued oil leasing on Indian reservations. In 1929 mineral prospecting permits for public domain and Indian lands were suspended. In 1933 Interior Secretary Harold L. Ickes called a moratorium on patenting allotments because of the poor land market. In 1934 Congress passed both the Indian Reorganization Act (IRA), which ended allotment, and the Taylor Grazing Act, which ended homesteading of public lands in most states.[40]

Whether the change in Indian policies was motivated by the reduced demand for Indian resources or by humanitarian reasons, it came just in time. The assimilation policy had not resulted in the disappearance of Indian people into mainstream society; however, it had resulted in the disappearance of many Indian resources. A growing Indian population was trying to live off fewer resources. By 1934 John Collier, who had been named Indian commissioner in 1933, told Congress that Indian land holdings had been cut by two-thirds.[41] Many of the lands remaining in Indian ownership had become virtually unusable except by leasing.

The Indian Reorganization Act repudiated some of the policies of the past. When Congress turned from the assimilation philosophy, it affected federal policies on Indian economic development and resources. As Indian historian D'Arcy McNickle (Salish-Kootenai) put it:

Until the third decade of the present century, Indian policy was rooted in the assumption that the Indians would disappear. Authorities responsible for policy continued to refer to a diminishing population long after the growth had turned upward. Given this premise, it seemed to not be a serious consequence that Indian land was shrinking or that the revenues credited to the tribes from land sales and treaty payments were dissipated in administrative costs and small doles, while nothing was invested in the development of tribal and individual resources. Then in 1933, the outset of the Roosevelt Administration, Congress undertook a radical revision of Indian policy.[42]

At the heart of the revision lay protection of Indian lands. Collier and Interior Secretary Ickes operated by an overriding principle that was largely adopted by Congress in the Indian Reorganization Act: "The land . . . is fundamental in any lifesaving program." To protect and increase Indian land ownership, the law provided for ending allotment, limiting sales of trust lands, and returning "surplus" lands.[43] In addition it recognized tribal governments as legitimate institutions with rights, including the right to reject the law itself and the right to control tribal assets. The legislation sought to limit the dictatorial powers of the Interior Department and to give Indians more control over their affairs. On the other hand, it perpetuated federal guardianship. Collier had said he wanted the Office of Indian Affairs, which he then headed, to become purely an advisory body, as the Department of Agriculture served American farmers. The final law, however, reserved to Interior the power to review many actions of tribal governments. Collier and other authors of the IRA recognized the need for economic development. The law set up a revolving fund to make loans to tribes and provided for tribes setting up chartered business corporations.[44]

The law succeeded in stopping allotment and in increasing the tribal land base. Between 1933 and 1950 tribes regained about 7.6 million acres, according to the American Indian Policy Review Commission's estimates. Whether or not they accepted the provisions of the IRA and organized under IRA constitutions, many of the energy tribes eventually got back the "surplus" lands that had not been settled. Several also received mineral rights under settled lands. Section 3 provided for restoring lands of *any* reservation opened, subject to valid rights or claims, without distinguishing between tribes that had been organized under the law and those that had not. Ironically, tribes benefited from Congress's decisions at the turn of the century not to immediately pay them for lands because the unpaid-for lands were returned after passage of the IRA. The secretary of Interior issued an order in September 1934 withdrawing lands from the public domain pending specific actions. His withdrawal order affected lands listed by John Collier as falling under Congress's intent, including those belonging to both organized and unorganized tribes. After the withdrawal order, neither miners nor homesteaders could claim the lands, no matter how long it took to actually restore them to tribal ownership.[45]

Only organized tribes could get lands restored by secretarial order; unorganized tribes required acts of Congress. Nevertheless, some of the unorganized energy tribes had their lands restored before several of the organized tribes. It often took many years for either Congress or the secretary to act, as indicated by the following land restoration dates: (1) Tribes organized under the IRA:[46] 1938—Southern Utes; 1938—Ute Mountain Utes; 1945—Northern Utes of the Uintah and Ouray Reservation; 1948, 1961, 1965, 1970—Three Affiliated Tribes of the Fort Berthold Reservation. (2) Tribes that did not organize under the IRA:[47] 1939—Arapahoes and Shoshones of the Wind River Reservation; 1958—Crows; 1958—Assiniboines and Sioux of the Fort Peck Reservation; 1958—Spokanes.

Most of the tribes in the nation accepted the IRA (181 accepted it; 77 did not). Of the energy tribes, however, many did not organize under the act, including the largest and those with the most energy minerals. In addition to the ones already mentioned, the Navajos and Osages did not organize either. Energy tribes that did organize under the act include the Hopis, Blackfeet, Northern Cheyennes, and Jicarilla Apaches.[48]

It is not clear if early energy development influenced tribes' decisions to organize under the IRA or not, but there might be some connection. The federal government stifled most tribes' traditional political and economic systems prior to 1934, especially the sophisticated systems of the coal- and oil-rich Five Civilized Tribes in Oklahoma.[49] On the other hand, the Interior Department allowed—and in some cases encouraged—the tribal governments of some energy reservations to continue functioning through those early years because it needed their action on energy leases. Perhaps because of this, the Osage, Navajo, and Shoshone and Arapahoe tribal governments were sufficiently ensconced that the tribal members saw no need to start over with different forms of government organization.

Anthropologist Loretta Fowler found that tribal leaders on the Wind River Reservation were suspicious of the IRA, partially because of their dissatisfaction with the federal government's past performance on Indian oil and gas leases. Former Arapahoe Councilman Yellow Calf said, "I am a little bit afraid of this idea for this reason—that there has been nothing said from that bill directing the disposal of our valuable oil, mines, or anything of that kind. The white people are smart and they can foresee things a long ways ahead and there might be a time when I will be lifted clear off this reservation and put somewhere else."[50]

Later, traditional leaders of some tribes, including the Hopis, criticized the IRA, saying it imposed the white man's system of governing. Yet the differences between organized and unorganized tribes are not that clear cut. Most of both types have hierarchical, representative governments, and some have constitutions. Some tribes (both organized and unorganized) integrated their old systems with the new, electing hereditary chiefs or religious leaders or relying upon elders' counsel. Although outsiders came to believe that all tribal councils were imposed upon unwilling people to meet the needs of the dominant society, the IRA generally benefited tribal governments because it restored powers that had long been dormant.[51]

Collier was disappointed in later years by Congress's failure to implement the promises of the IRA. In 1947 he said that "policies established by the legislation in 1934 have withstood every attack except the attack through appropriations." Congress failed to provide the money necessary for a revolving credit fund. It also failed to provide the means for achieving Collier's main goal—restoring a communally owned land base. Although the law provided funds for buying land, Congress never appropriated the money. Where money was allocated, such as on the Jicarilla Apache Reservation, and where tribes chose land acquisition as a high priority, such as on the Northern Cheyenne Reservation, the tribes were at a distinct advantage for future development.[52]

The study by the American Indian Policy Review Commission showed that in later years both BIA and Congress erected other roadblocks to consolidating lands under tribal ownership. BIA often did not notify tribes when allotments were for sale. Congress blocked some energy tribes from using their own mineral income to purchase land both outside and inside reservation boundaries.[53] Despite the odds, many tribes succeeded in buying back land from allottees and fee owners.[54]

Implementation of IRA policies continued to be hindered by the lack of a consolidated land base. As the Meriam study team had recognized in 1928, the "economic underpinnings of the old culture had been destroyed" by the reservation and the allotment era. Although the IRA stopped allotment, it did not restore those underpinnings, which had once made the Indians self-sufficient. Historian Kenneth Philp believes that as a result of the IRA's failure to provide economic progress, the IRA era led necessarily into another era of termination.[55]

SECOND TERMINATION ERA: 1946–1960

By 1946 the mood of the nation had swung back toward termination of special Indian rights and protection. The goals were remarkably similar: divide the tribal lands and treasuries and free the Indian from federal interference. Once again the change was justified as a means for helping the Indians. Once again it resulted in large losses to the Indian land and resource base. Yet this time, because of increasing activism by Indian tribes and intertribal organizations, some of the proposals were eventually subverted.[56]

In 1953 Congress adopted by unanimous vote House Concurrent Resolution 108. The preamble language emphasized helping Indians:

Whereas it is the policy of Congress, as rapidly as possible, to make the Indians within the territorial limits of the United States subject to the same laws and entitled to the same privileges and responsibilities as are applicable to other citizens of the United States, to end their status as wards of the U.S., and to grant them all the rights and prerogatives pertaining to American citizenship.[57]

The resolution stated that at the earliest possible time all tribes and individual members in certain states should be "freed from federal supervision and control and from all disabilities and limitations specially applicable to Indians" and that BIA offices in those places would be abolished. Under the termination policy the federal government tried to limit tribal governmental authority and terminate the federal government's special relationship with the tribes. Once again energy tribes were targeted.

During World War II land increased in value again. Soon after the war, federal policies made Indian lands more accessible to non-Indians. In 1948 the secretary of Interior lifted constraints on selling allotments to non-Indians that had been imposed during the 1930s. Restrictions on leasing and mortgaging trust lands were also relaxed, opening the door to foreclosure. In the next ten years, 2.6 million acres of allotments were removed from trust status, mostly in the Northern Plains and Oklahoma—areas of the country that contained considerable oil and gas resources.[58]

During this period the government used its power of eminent domain to seize 500,000 acres for military use and in the Dakotas, hundreds of thousands of acres for federal dams. The Corps of Engineers and Bureau of Reclamation started construction on some of the dams before tribal officials were even consulted. Along with 152,360 acres of mostly rich bottomlands, the Three Affiliated Tribes of the Fort Berthold Reservation lost their mineral rights, which were not restored to tribal ownership for fifty years. Alarmed at the trend, Senator James E. Murray, chairman of the Senate Committee on Interior and Insular Affairs, finally in 1958 convinced the Interior secretary to suspend sales of Indian lands while the committee investigated the land losses.[59]

Nothing revealed the hypocrisy of the termination movement more than the taking of Indian resources during this period. Termination proponents emphasized that they wanted Indian tribes to be self-sufficient, but while saying this, they took away the means for the tribes to be so. Congress used four criteria to determine which tribes should be terminated; one criterion was the "economic resources upon which the tribes could rely after termination." Thus Congress chose resource-rich tribes for termination planning, including several timber tribes and the mineral tribes of Osage, the Arapahoe and Shoshone of the Wind River Reservation, Northern Ute of the Uintah and Ouray Reservation, and Spokane. Of these, only the mixed blood Utes were actually terminated. The Osage and Wind River Tribes were never terminated, but they were forced to pay much of the costs for their continuing federal supervision—without having any control over how the agencies were run.[60]

During the termination era and for years afterward the federal government denied tribal rights to thousands of acres of mineral-rich lands that an earlier Congress had purchased for the Indians. In the 1930s, when farmers were going bankrupt, the administration of Franklin D. Roosevelt had created the Submarginal Land Retirement Program. Congress had authorized BIA to acquire "submarginal lands" within reservation boundaries from the non-Indian farmers and ranchers rather than allowing the banks to foreclose on the lands. The lands were intended to be used for the "readjustment and rehabilitation of the Indian population." Submarginal lands purchased for cities, state governments, and national parks were turned over to them immediately. Yet only a few of the purchased lands were given to tribes. When it was discovered that some of the remaining submarginal lands had oil and gas potential, Congress hesitated to transfer the minerals to the tribes, calling it a giveaway and saying tribes would not use the lands well. It was not until 1975—after the termination era ended—that Congress turned over 480,000 acres of submarginal lands with mineral rights to some eighteen tribes, including three energy tribes (Blackfeet, Fort Peck Assiniboine and Sioux, and Navajo).[61]

During this period when the federal government was taking away some tribal resources, BIA was also emphasizing resource development and job placement. At that time job placement meant relocating tribal members to urban areas rather than developing reservation jobs. BIA discouraged reservation economic development just as it had during the earlier termination period of the 1910s when the Indian Bureau had discontinued agricultural assistance and had forced the Plains tribes to sell the breeding stock for their flourishing cattle enterprises. Between 1952 and 1960 over thirty-five thousand Indians moved to cities with federal encouragement and limited assistance—usually one-way bus tickets. For some energy tribes the relocation effort coincided with an influx of energy companies seeking Indian minerals. Thus the tribes' planning capacities were limited by the loss of some of their more able and ambitious members.[62]

In a frequently cited dissent, Supreme Court Justice Black in 1960 repudiated the philosophy behind termination:

> It may be hard for us to understand why these Indians cling so tenaciously to their lands and traditional tribal way of life. The record does not leave the impression that the lands of their reservation are the most fertile, the landscape the most beautiful, or their homes the most splendid specimen of architecture. But this is their home—their ancestral home. There, they, their children, and their forebears were born. They, too, have their memories and their loves. Some things are worth more

than money and the costs of a new enterprise. . . . I regret that this Court is to be the governmental agency that breaks faith with this dependent people. Great nations, like great men, should keep their word.[63]

Because of Indian activism, few tribes were actually permanently terminated. Yet many were psychologically crippled. The termination rationale reinforced the idea that tribal efforts at self-sufficiency would be punished. Although Congress set a new direction later when it passed the Indian Self-Determination Act (1975), and although it did not pass termination legislation in the 1970s and 1980s, it did not specifically repeal House Concurrent Resolution 108 until 1988. Both Presidents Nixon and Reagan said the largely symbolic gesture was important to erase the continuing threat of termination. Decades after the termination resolution, tribal leaders said their self-determination efforts were continually hampered by tribal members— often older ones—who begged them not to mess with the trust relationship. Such members saw only a couple of letters' difference between termination and self-determination. They warned their leaders that if a tribe contracted to provide services then provided by BIA, the federal government would once again cut off federal responsibility and terminate the tribe as a separate, sovereign governmental body. Some went so far as to call younger Indians Communists for suggesting increasing tribal powers.[64]

Many Indians continued to lose land in the 1970s and 1980s as a result of changes in mortgaging policies made during the termination period. In 1976 the American Indian Policy Review Commission said that securing loans with trust land was a "risky method for attempting land consolidation." The truth of those words was recognized too late, ten years later, when the Senate Select Committee on Indian Affairs in 1987 called for an investigation of Farmers Home Administration (FmHA) foreclosures on fourteen reservations in the Great Plains. Termination policies also exacerbated jurisdictional problems years later by extending state jurisdiction over some reservations and by increasing nontribal land and mineral ownership within reservation boundaries.[65]

SELF-DETERMINATION ERA: 1960 TO PRESENT

As a result of growing activism by Indian tribes and intertribal organizations, both major political parties rejected termination in their 1960 party platforms.[66] In the following years the nation's leaders adopted official policies that recognized the continuing importance of reservations and tribal governments. The actual role taken by the federal government did not, however, always reflect those policies. Despite several defeats suffered by tribes in the courts, Congress, and government agencies, the era would be remembered as one of remarkable progress toward tribes' determining their own goals. To benefit from their political progress, however, tribes needed much more on-reservation economic development than had occurred by 1988.

During the administration of John F. Kennedy and during President Lyndon B. Johnson's War on Poverty, money began pouring into the reservations. With unemployment rates as high as 80 percent and health problems worse than anywhere else in the nation, the reservations were logical targets for LBJ's war. The job training and administrative positions created during this period produced many talented and politically astute administrators who later became tribal leaders and reservation entrepreneurs. The antipoverty funds supported a resurgence of tribal governments. College enrollment among Indian students began increasing, and the Red Power movement gained momentum. Educated Indians could hope to find jobs on reservations, and many returned to their own or other reservations. Peter MacDonald, for example, returned to the Navajo Reservation in 1963 to serve as director of the Office of Navajo Economic Opportunity before deciding in 1970 to run for tribal chairman.[67]

The administration of Richard M. Nixon offered a political context for continued efforts to revitalize reservation economies. In a message sent to Congress in 1970, Nixon sought a middle ground that would provide for self-determination without termination:

> This . . . must be the goal of any new national policy toward the Indian People: to strengthen the Indian's sense of autonomy without threatening his sense of community. We must assure the Indian that he can assume control of his own life without being separated involuntarily from the tribal group. And we must make it clear that Indians can become independent of Federal control without being cut off from Federal concern and Federal support.[68]

Nixon recognized the connection between sound reservation economies and self-government. He suggested several initiatives to increase self-determination and self-sufficiency on the reservations. In 1974 Congress acted on one of those suggestions and passed the Indian Financing Act, which established an Indian revolving loan fund to promote economic development of both individual reservation Indians and Indian organizations. Congress also provided for interest subsidies, Indian business grants, and authorized some federal agencies to give special emphasis to Indian economic needs.[69]

In 1975 Congress acted on another of Nixon's suggestions and passed the Indian Self-Determination and Educational Assistance Act.[70] The law signaled a significant reversal of public policy. It directed BIA and the Indian Health Service to contract with tribes that wanted to provide programs and services previously administered by those agencies. The law was critical for both the energy and other tribes because it recognized tribal governments as institutions with long-term viability rather than just transitional structures. It meant they could take over many of the functions that the federal government had performed inadequately, such as monitoring oil fields for thefts, planning land use, and protecting air and water quality and cultural resources.

Stunning court victories accompanied these administrative and legislative recognitions of tribal self-determination. Through the late 1970s and 1980s, however, a backlash plagued Indian tribes, especially concerning their hunting and fishing rights. The public still was not aware of such concepts as "self-determination," "tribal self-government," and "sovereignty"—that is, until they felt their own rights being threatened by these strange, special rights of the Indians. By 1989 the backlash had not won wide favor in Congress as it had during the termination movements of the 1910s and 1950s. Partly as a result of the termination era, tribes and tribal organizations were much more vigilant than they had been in 1953 when House Concurrent Resolution 108 passed unanimously without objection from any member of either house.[71]

The administration of President Jimmy Carter strongly opposed the 1977 backlash legislation, the Native Americans Equal Opportunity Act, which would have abrogated all treaties and terminated federal supervision of Indian people and property. Testifying before a House subcommittee, Interior Secretary Cecil Andrus argued that treaties were legal contracts that were "every bit as legitimate" as, for example, home mortgages. The Indians let go of their property in exchange for the U.S. government's guarantees that it would protect the land, water, and other resources. Just because energy minerals were found under the land or because the government had allowed someone else to use the water did not relieve the United States of its legal obligations, he said.[72]

Instead of adopting the antitreaty legislation Congress in the 1980s adopted laws that recognized greater tribal governmental powers, including several that reinforced energy tribes' powers. In 1982 it passed the Indian Mineral Development Act (IMDA), which allowed tribes to become energy producers instead of just royalty holders, and the Indian Tribal Governmental Tax Status Act, which enabled tribes to issue certain tax-exempt revenue bonds, among other powers. In 1986 Congress amended three environmental laws that recognized tribes' authority to be similar to that of states. At the same time, Congress restricted tribes' efforts to establish gambling casinos and to use tax-exempt bonds to build off-reservation industries.[73]

President Ronald Reagan said in his Indian policy message on January 14, 1983, that he recognized the importance of reservation economic development to tribal governments. He set up the Presidential Advisory Commission on Indian Reservation Economies, with Ross Swimmer (Cherokee) as chairman. Although Indian leaders generally agreed with Reagan that tribes should become self-supporting and reduce their dependence upon federal funds, they were alarmed at Reagan's abrupt and severe budget cuts, which they considered economic termination. In 1981 the Reagan administration had called for a drastic reduction in federal assistance to tribes, including an 82 percent cut in economic development funds. Although Indians represented 0.6 percent of the country's population, they absorbed 2.5 percent of the budget cuts. Unlike other places in the nation, reservations had few businesses to make up the differences. Not only economic development but also health suffered as a result of Reagan's policies.[74]

During Reagan's second term, the National Congress of American Indians (NCAI) also called for the resignation of Reagan's assistant Interior secretary for Indian affairs, Ross Swimmer. A former tribal leader himself (principal chief of the Cherokee Nation), Swimmer understood the role of tribal governments and reservations better than Watt and also took a stronger stance defending tribal natural resources. Nevertheless, when he attempted to reduce the involvement of BIA in tribal affairs, many Indian tribes believed that Swimmer was trying to terminate their relationship with the federal government. Swimmer advocated phasing out BIA, and although some Indians agreed that it should be abolished, most reservation Indians believed the bureau was the only thing standing between them and termination of their reservations' special status.[75]

SURMOUNTING ECONOMIC DEVELOPMENT HURDLES

Despite the political support for Indian self-determination in the 1970s and 1980s and the concomitant recognition for the need for reservation economic development, such development continued to lag. Although some tribes had found that carefully planned programs could bring their members success and productivity, most still were dependent upon federal funds. Reagan acknowledged that "federal policies have . . . inhibited the political and economic development of the tribes . . . and promoted dependency rather than self-sufficiency."[76]

Various studies agreed and pointed to many other ways in which reservation economic development had been inhibited and dependency promoted during the self-determination era. Although they differed radically in political-economic theory and in some of their conclusions, the studies all said that the hurdles were not impossible to overcome. Tribes could become more "self-sufficient," even without the resources that they held prior to the coming of the white man, so long as self-sufficiency did not mean total independence from outside funding, which, after all, neither states nor local governments had yet achieved. Instead tribes could strive to provide for the

Inez Oldman, a Northern Arapahoe, receives a traditional shawl at a giveaway at the Northern Arapahoe Powwow in 1979. (Photo by Sara Hunter-Wiles.)

needs of their people in accordance with their own goals rather than federally imposed goals.[77]

From the 1960s through the early 1980s, federal monies poured into the reservations. As one Indian commentator put it, the government became the Indians' new buffalo. Federal spending for Indian programs rose significantly from $1.08 billion in 1973 to $2.75 billion in 1981 before it began declining again. Yet increasing federal funding alone would not solve the problems, the studies said. Although new social programs created jobs and decreased the suffering of poverty-stricken families, they rarely helped economic development. Thus they increased dependency upon federal government funding.[78]

The studies also agreed that the greatest obstacle was the lack of businesses on reservations. This deficiency caused two serious problems with far-reaching ramifications: First, when money flowed into the reservations (from the government or from mining jobs, welfare, or resource payments), it leaked out again into the coffers of off-reservation banks and grocery, clothing, and liquor stores. A study conducted by Anne Seip and Ahmed Kooros of the Council of Energy Resource Tribes (CERT) staff on such financial leakages showed that money hardly turned over at all on reservations.[79] The second problem created by the lack of businesses on reservations was that because the reservation economy was so underdeveloped, the tribal governments could not support themselves by taxation. Therefore federal funds had to be directed toward basic services—social, educational, and health—instead of economic development.[80]

The studies' conclusions about hurdles and possible solutions help in understanding mineral-based economic development because so many of the concepts relate to both. President Reagan's Commission on Indian Reservation Economies identified several obstacles that tribes erected to economic development, including weak business management, rapid turnover of tribal governments, and an unskilled and unreliable Indian labor force. Although not problems on all reservations, these were serious concerns on many. Tribally owned businesses on many reservations failed to produce profits and had to be continually subsidized by tribal governments. Such enterprises tended to be oriented toward providing jobs and sharing benefits rather than toward profits, and many were hamstrung by tribal politics. The commission said that tribes needed to separate business more from politics, to make their governments more efficient, and to create a more stable work environment for private business.[81]

The commission accused BIA of incompetence and of imposing trust restrictions overzealously and said BIA was incapable of reforming itself to remedy such problems. Consequently, Swimmer and his commission recommended dismantling the agency and privatizing many of its former functions. Not surprisingly, the Department of Interior's task force on economic development did not agree that BIA should be dismantled. In its rebuttal, the task force concluded that the BIA problems could be solved internally.[82]

Nevertheless, the DOI task force agreed with most of the changes the commission recommended in federal labor and economic stimulation laws and regulations. Tribes should be able to issue industrial revenue bonds; regulations for Small Business Administration incentives should be changed so tribes and more Indian-owned businesses could qualify; more federal agencies should be required to give preference under the Buy Indian Act; and national labor and wage laws should be changed so they would not continue to encourage importing non-Indians for reservation jobs.[83]

All of the studies agreed that increased tribal control was key to future success. All said this could be done without forsaking either the trust relationship or tribal cultural values. The CERT study said it most forcefully. Kooros and Seip looked at six reservations where they analyzed the impacts of federal expenditures upon social and economic problems. The study said increasing federal expenditures had "perpetuated tribal dependency" because projects were usually federally managed and delivered, that is, done for the tribes. The study found more success in tribally managed projects, which were funded by the government but administered by the tribes. The projects done by tribes (1) transferred skills and information to the tribe, (2) transferred control to the tribe, (3) enhanced local and tribal employment, and (4) reduced tribal financial leakages. In the success stories, economic, social, and cultural development proceeded together, and tribal infrastructures improved. However, economic efforts that ignored social and cultural factors failed. An earlier government study of light-industrial firms had reached a similar conclusion. Plants that had the lowest failure rate were Indian owned and resource based.[84]

Although some previous studies concluded that in order to succeed, Indian people had to reject their heritage and adopt the dominant culture's values, the CERT study and Swimmer's commission said assimilation was not necessary. Quoting from a World Bank study of tribal issues overseas, the Swimmer commission said cultural and economic self-determination can be combined. Despite its endorsement of traditional values, the commission throughout its report emphasized the need to "modernize" tribal governments and to get away from communal ownership. It said tribal enterprises should be turned over to individual Indian entrepreneurs, either by selling stock or the businesses themselves.[85]

Two commentators writing in a University of New Mexico study had a much different concept of economic solutions for tribes. Roxanne Dunbar Ortiz and Lorraine Turner Ruffing suggested more agricultural cooperatives and more tribal subsidies for small, labor-intensive, traditional occupations. Ortiz said, in fact, that tribal government should intervene—in the form of taxation, service contracts, supervision of outside companies, and internal integration of the economy—to "break the power of monopolistic outside interests." To her, economic development under the power of such outside interests had represented "neocolonialism." Although recognizing that industrialization was not necessarily undesirable for Indian reservations, Ortiz said mineral development under corporate control had deepened the gap between traditional and modern tribal sectors.[86]

In general the studies indicated that without economic development, reservation residents remained completely dependent upon federal funds for jobs and social services and thus subject to the paternalism and exploitation of outsiders. With development, they could become more self-supporting. The tribal governments could provide more services for reservation residents, both Indians and non-Indians. Thus the reservation could become a more viable place to live.

MINERALS FOR ECONOMIC DEVELOPMENT

Some tribes look to their minerals as a possible way of freeing themselves from complete dependency on the whims of Washington, D.C. Mineral development cannot, however, be the answer for every tribe, as illustrated by the conclusions of the economic studies. In the early 1970s an oil embargo by the Organization of Petroleum Exporting Countries (OPEC) led to high gasoline prices, long lines at gas stations, and a reevaluation of national energy policies in this country. During that time the media and the tribes themselves called so much attention to Indian mineral wealth that the general public was left with the mistaken impression that all tribes were rich. It is true that some Indians own incredibly large amounts of the nation's energy minerals, given that they own only 2 percent (53 million acres in trust) of the nation's land base. Government agencies say the Indians own 30 percent of the coal west of the Mississippi River, 37 percent of potential uranium resources, and 3 percent of the nation's known oil and gas reserves.[87]

Yet only a small percentage of the tribes could possibly benefit from those resources. Of the three hundred federally recognized Indian reservations in the nation, only about forty of them have energy resources, according to the government figures just cited. On some of those reservations, individual Indians—as opposed to tribes—own much of the potential wealth. Indian-owned energy resources were still largely untapped in the late 1980s, compared with the non-Indian resources surrounding them. Where energy development had occurred, it had been limited largely to oil and gas extraction. Uranium and coal had been mined on only a handful of other reservations, including the Navajo (coal and uranium), Hopi (coal), Crow (coal), Cheyenne River Sioux (coal), Laguna Pueblo (uranium), and Spokane (uranium).[88]

Where energy development had occurred prior to the mid-1970s it had failed to provide as many jobs or as much income as could be possible. Mineral development had tended to promote dependency rather than self-sufficiency for many of the same reasons that other forms of reservation economic development had.[89] Mineral prices fluctuated wildly, creating boom/bust cycles, just as the federal commitment to providing funds for reservations fluctuated from one administration to the next.

In the early years, increasing mineral revenue did not help reservation economies any more than increasing federal revenue, and for the same basic reason: Neither the tribe nor the companies invested the revenue in broadening the reservation economy; thus Indians did not gain many jobs. In 1975, for example, during the height of the energy boom on the Navajo Reservation, mineral development provided jobs for less than 5 percent of the Navajo labor force; 67 percent remained unemployed or underemployed. Because of BIA's inadequate managing of trust resources, including water and minerals, environmental and social costs were not compensated.[90]

Leaders of energy tribes became increasingly aware of such problems during the 1960s and 1970s. But rather than rejecting the idea of mineral exploitation, they began taking over the reins in the early 1970s and even earlier for some tribes. With the help of astute Indian and non-Indian attorneys, some leaders filed lawsuits to stop uncontrolled development and to force the Interior Department to abide by its own regulations. They took their battle for control beyond the courtrooms to the halls of Congress and to corporate board rooms.

By the late 1980s these Indian leaders were seeing the fruits of their work not only in energy development but also in more generalized economic development. The U.S. Supreme Court had recognized their authority to impose various taxes upon energy producers. Indian royalties and tax revenues helped them ride out the budget cuts during the Reagan years. The major energy tribes began using their energy income for economic development— investing in real estate and building stores, gasoline service stations, banks, and resorts.[91]

The tribes found that control over energy development entailed more than just boosting the royalty rate. Meaningful control for tribes and (where appropriate) for individual Indian mineral owners required the following:

1. information about the extent of tribal resources (land, minerals, and water), their value in the marketplace, alternative contractual arrangements, development options, and other tribes' successes and failures;

2. power to negotiate mineral contract terms, including monetary return, employment quotas, environmental and cultural safeguards, and access to financial and geologic information specific to a certain project;

3. regulatory authority to collect taxes and protect the reservation environment, including cultural sites;

4. involvement in off-reservation energy decisions affecting Indians;

5. financial control, including involvement in accounting for and field monitoring of energy development; and

6. access to development capital.

Before learning the importance of such issues the tribes endured many decades in which they had to direct all of their attention to preserving their rights of ownership and consent for their energy minerals.

CHAPTER TWO

The Rubber-Stamp Era: Early History of Indian Mineral Leasing

There was a time when we sat and listened and followed whatever was brought to us.

—Earl Old Person[1]

The discovery of millions of dollars in energy minerals beneath several Indian reservations in the United States presents one of the great ironies of history. From the pioneer farmers' perspective the sites chosen for Indian reservations in the late nineteenth century generally were wastelands that safely could be promised as "permanent" homelands. No one predicted the extent of the underground wealth or how much it would change the lives of the Indian people living over it.

Once the minerals were discovered, those tribes' fates diverged from others on a path noted for its paradoxes. Although the energy tribes were targets for exploitation, they often had more opportunity early on to practice self-government as they made energy decisions. Mineral income made them less vulnerable to federal Indian agents' coercion, such as cutting food rations. Some tribal governments—for example, that of the Osages—survived the allotment era largely because of their mineral holdings. Yet later, Congress threatened the survival of some energy tribes when in the 1950s it tried to terminate their federal protection because of their "wealth."

In this chapter I look at the energy tribes' history from the establishment of the reservations until the tribal renaissance of the 1970s. That period is crucial to understanding contemporary Indian problems. Within their historical context some of the most criticized federal laws were at one time progressive steps toward improving tribal authority and protecting resources. Between the taking of the Black Hills in the 1870s and the strip mining of Black Mesa in the 1960s lie many less known but equally important decisions by tribes and the federal government—decisions that to some extent prescribe the limits of tribal authority today. Those decisions have clouded the title to minerals and the jurisdiction over those minerals for decades.

The energy tribes' history revolves around three main questions: Who owned the minerals? Who decided whether they were developed? Who received the proceeds and, equally important, who decided how the money

31

Blackfeet Chairman Earl Old Person (right) speaks with Ouray McCook of the Northern Ute Tribe at the CERT meeting in 1980. Old Person served as chairman of the Blackfeet Tribe for more than twenty years and simultaneously served as a traditional chief. (Photo by John Martin.)

was to be spent? When the federal government answered these questions in paternalistic and exploitative ways, it planted the seeds of the tribes' dependency. For the most part during those early years, the tribal governments served as rubber stamps for mineral decisions made in corporate board rooms and DOI offices. Sometimes the tribes were less than rubber stamps—they were not even asked. Other times they served in more active, although usually not more effective, roles, negotiating their own contracts in the 1950s and 1960s. It was not until the tribal renaissance of the early 1970s that the major energy tribes stopped merely "listening and following whatever was brought to them." They began taking control.

SETTING THE BOUNDARIES:
RESERVATION ESTABLISHMENT THROUGH 1903

"There is nothing more dangerous to an Indian reservation than a rich mine," Interior Secretary Carl Schurz wrote in 1881. Before the turn of the century, Indian lands were often purchased or traded in the quest for precious metals. Georgians' lust for gold led to the removal of the Cherokees from that state in the 1830s. During the late 1800s gold fever reshaped the boundaries of several western reservations, lands that previously had been set apart for "the absolute and undisturbed use of the Indians," including the Shoshone (later called the Wind River) Reservation in Wyoming, the Ute Reservation in Colorado, the Fort Belknap Reservation in Montana, and the Sioux reservations in South Dakota. The taking of the Black Hills after General George Armstrong Custer confirmed the existence of gold in 1874 is the best known example.[2]

The philosophy of the U.S. government was rationalized in 1872 when gold miners, already illegally settled on the southern portion of the Wind River Reservation, clamored for releasing the area from Indian ownership. Commissioner of Indian Affairs Francis A. Walker wrote:

> It is the policy of the government to segregate such [mineral] lands from Indian reservations as far as may be consistent with the faith of the United States and throw them open to entry and settlement in order that the Indians may not be annoyed and distressed by the cupidity of miners and settlers who in large numbers, in spite of the efforts of the government to the contrary, flock to such regions of the country on the first report of the gold discovery.[3]

The Indian agent had reported in 1866, two years before the reservation was established, that numerous oil springs had been discovered in the same portion of the Wind River Reservation, but it was obvious from Commissioner Walker's statement that gold was a much more powerful incentive for taking the Indian lands. This was generally the case in the West. Although the federal government routinely justified taking gold lands from Indians for a pittance and for their own good, few examples have been documented outside of Oklahoma of non-Indians taking Indian lands just for the energy minerals. The Jicarilla Apache Reservation represents one of the exceptions. When President Grover Cleveland established the tribe's permanent home in 1887, it looked as if a bite had been taken out of its northeast corner. In a way it had. The boundary excluded the area where coal had been found and where a coal mine had been established illegally.[4]

The ignorance of the pioneers generally protected the energy tribes' lands except for those of the Five Civilized Tribes in Oklahoma. Before the twentieth century little was known about either the existence or the potential value of the energy fuels hidden beneath western lands, although Indians were known to have used coal in small quantities to fire pottery in the eleventh century. Later, explorers reported Indians using coal both as fuel and barter. Unless one could see coal outcroppings or oil seeps on the surface, the black gold beneath reservation lands went unrecognized. A Sioux allottee discovered coal under his land in South Dakota only after prairie dogs had brought coal dust to the surface. Between 1874 and World War II the average price of oil rarely rose above a dollar a barrel. Uranium did not become important until World War II for defense purposes and not until later for energy production.[5]

The 1907 addition to the Jicarilla Apache Reservation demonstrates the fortuity for the tribe of the pioneers' ignorance. Although the Indians lost land where coal had been recognized, they received land that much later produced millions of dollars in oil and gas revenue for them. The Ute Indians' experiences even more dramatically illustrate the advantages of such ignorance. They lost lands that held precious metals and asphalt but retained lands that much later became recognized for their energy minerals. Following illegal intrusions into their territory by gold and silver miners, Ute bands reluctantly agreed in nine different treaties and agreements to give up more and more of western Colorado. The federal government moved two of the Ute bands away from the others to a new reservation in Utah, where no gold or silver—and little farming potential—existed. In fact, President Abraham Lincoln set aside the Uintah Reservation in 1861 only after the Mormon leader Brigham Young said it was not suitable for Mormons; Young wondered why God had created the area unless it was to hold the other parts of the world together. Prospectors soon found Gilsonite (a form of asphalt) there, however, and the Indians lost even more land.[6]

But fortunately for the Utes, neither Young nor the early prospectors had seismograph equipment. In 1984 the Northern Ute Tribe estimated that its mineral reserves included 500 million recoverable barrels of oil, 100,000 acres of oil shale, as well as tar sands, coal, and other nonenergy minerals, including Gilsonite. The Southern Ute and Ute Mountain Ute reservations in Colorado also contain significant deposits of oil, natural gas, and coal. If the non-Indian residents of Colorado and Utah had recognized the energy wealth, they might have arranged to relieve the Utes of those lands, too.[7]

Whether or not energy wealth was recognized at the time, it was often ceded away as incidental to agricultural, forested, or railroad right-of-way land cessions. The Utes lost minerals when President Theodore Roosevelt established national forests in Utah and Colorado.[8] The Crow Reservation established by the 1851 Fort Laramie Treaty encompassed 39 million acres, including most of the Powder River Basin coal fields in Montana and Wyoming, part of the Fort Union coal formation, part of the Gas Hills uranium district in Wyoming, and what were later to become lucrative oil and natural gas fields. The tribe ceded away 30 million acres in 1868. In 1882 Congress ceded a railroad right of way through the reservation, which included a grant of mineral rights to the railroad. Yet another cession of lands in 1891 retroactively validated illegal, non-Indian mining claims on the ceded portion. After more treaties, cessions, and restorations, the Crow Reservation now encompasses only 2.2 million acres plus the minerals under some ceded lands. Despite millions of tons of coal, the reduced reservation obviously has much less mineral value than the original 39 million acres had. In later years Indians filed many claims based upon such losses of mineral lands without just compensation.[9]

BEGINNING OF INDIAN LEASING

In his 1942 *Handbook of Federal Indian Law*, Indian law expert Felix Cohen said that even during early years, Congress usually assumed that "in the absence of a clear expression to the contrary, tribal possession extends to the center of the earth." In a landmark case involving the Wind River Reservation, the U.S. Supreme Court ruled in 1938 that the minerals were "constituent elements of the land itself" and thus owned by the Shoshone Tribe, despite the federal government's arguments to the contrary. Cohen pointed to certain cases in which minerals were reserved to the United States, most of which—with the exception of Navajo allottees—did not involve energy minerals.[10]

In the cases cited above, tribal possession meant that the tribe was to be compensated when its minerals were taken. It did not mean that the minerals could not be purchased. Yet more and more often, mineral lands remained in Indian ownership and were leased under the authority of specific, authorizing statutes years before reforms provided for leasing energy minerals on public lands. For example, although coal, oil, and gas development began at the turn of the century at Wind River, those minerals were leased—not taken out of Indian ownership as the gold had been. Unlike public lands, Indian lands were occupied. As early as 1880 Indian Commissioner Hiram Price recommended a leasing system, explaining that "the Indians involved [in Arizona Territory] could not be prevailed upon to remove again [and] that the government could not undertake to work the mines."[11]

In contrast to energy minerals on western reservations, those in Oklahoma had been developed extensively by the turn of the century: In 1894, 130 billion barrels of oil were produced in Oklahoma Territory, and in 1901, thirty-nine coal corporations were operating in the Choctaw Nation alone, producing 1.5 million tons a year. Because non-Indians were stealing coal, the Cherokees, Creeks, and Choctaws adopted their own laws in 1869 to regulate leasing it. The federal government in 1898 passed leasing laws for Indian coal, oil, and gas and terminated tribal leasing. Ironically, because of dishonest federal administration of royalty revenues and because they feared, with cause, that they would lose their minerals without any compensation, the Choctaws and Chickasaws opposed leasing during the early 1900s and wanted to sell their minerals instead. In 1918 Congress agreed to allow those minerals to be sold. The early recognition of minerals contributed to the nearly complete destruction of communally owned lands in Oklahoma.[12]

One of the earliest western reservations to experience coal development was the Fort Belknap Reservation, which was established in 1888 near the Canadian border in Montana for the Assiniboine and Gros Ventre Tribes. Although the federal government sold the tribes' gold in 1895, it kept in tribal ownership the coal that Indian Agent William R. Logan found in

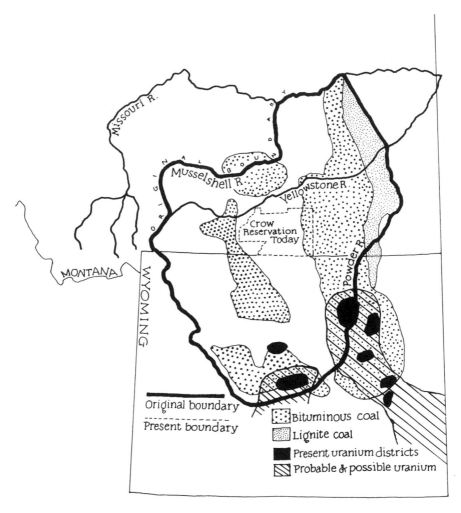

Crow Tribe's Loss of Energy Resources. Under the Treaty of Fort Laramie (1851), the Crow Tribe retained 39 million acres, which included vast coal and uranium resources as shown by the approximate boundaries above. (The northern boundary line was not stipulated between the Yellowstone and Missouri rivers, according to Joe Medicine Crow and Daniel S. Press, A Handbook of Crow Indian Laws and Treaties *[Crow Tribe: Crow Agency, Mont., 1966].) In later years the tribe lost most of those minerals in various transactions that concerned primarily the surface, not the mineral, ownership.* (Map by Hannah Hinchman.)

1903. Unlike the gold, coal was useful to his agency, which burned it for heating and industrial processing on the reservation until 1920, when another agent rented the coal to a white man, Charles N. Damon. Although Indians had to buy coal from Damon, he allowed local white settlers to use it for free.[13]

The Indian agent justified renting the coal to Damon by saying that the Indians did not know enough about Anglo practices to mine it. This rationale was used often, sometimes to justify taking Indian lands and sometimes leasing them. The nation believed fiercely in the sanctity of using the land and imposed this cultural bias upon Indians. During a congressional debate in 1880 over opening Ute lands in Colorado, one of the arguments was that "the Indians had too much land which they did not use for worthy purposes, i.e. farming and mining. Surplus lands could be used by whites properly." Commissioner of Indian Affairs Cato Sells told the Indian superintendents in 1914 that "I hold it to be an economic and social crime . . . to permit thousands of acres of fertile land belonging to the Indians and capable of great industrial development to lie in unproductive idleness."[14]

Until 1891 Congress closed Indian lands to mining access by various nonintercourse and allotment acts unless federal lawmakers had specifically authorized mineral development on a particular reservation. The nonintercourse laws, which dated back to the 1790s, were intended to protect tribes from trespass and from unauthorized trading. Because the minerals were going to waste from the dominant society's perspective, Congress authorized mineral leasing of tribal and allotted lands in 1891, four years after the General Allotment Act. Although designed primarily to provide access, three positive aspects of the 1891 law should not be ignored: It provided for leasing rather than whole-scale cessions of mineral lands; it represented a significant departure from existing policy on public lands whereby prospectors who found minerals could obtain the ownership of the minerals and the surface land under claim/patent laws (with a few exceptions, Congress did not authorize leasing public energy minerals until 1920); and it required tribal consent.[15]

The 1891 law said minerals could be leased only with the authority of the council speaking for the Indians, with terms to be approved by the secretary of Interior. But this consent authority provided little tribal control because tribes did not have the information they needed to make decisions and because various means of coercion, including economic necessity, influenced their decisions. Later, in the 1910s and 1920s, most Indian leasing laws did not provide for even this token tribal consent. The 1891 law applied to tribal lands that had been "bought and paid for," which was later interpreted to exclude most executive-order reservations and some treaty reservations. Although the law authorized leasing allotments, most allotted mineral leasing came under the authority of the 1909 Indian Appropriations Act.[16]

TRIBAL OWNERSHIP AND LEASING
AFTER *LONE WOLF*

Despite these problems, the early years were marked by some recognition of Indian authority in decision making and of Indian ownership of the land. But, as historian Frederick Hoxie noted in his book *A Final Promise*, this trend changed. During the decade after 1894 the federal government gradually reduced the Indian role in decision making. That gradual shift became abrupt with the Supreme Court's *Lone Wolf* ruling in 1903, which stated that Congress could take Indian lands without tribal consent. The attention of tribes and tribal advocates focused necessarily on the attacks on tribal ownership rights because tribal lands were being ceded away for non-Indians' agricultural use. The federal government extended the *Lone Wolf* decision beyond its original scope to say that tribal involvement was also not needed in other Indian resource decisions.[17]

When it was suggested in 1914 that tribal councils should have more authority to approve the sale of their resources, Commissioner Cato Sells said they should not: "I am confident that a successful administration of Indian affairs is dependent upon a maintenance of the principle that the United States has jurisdiction over tribal property." This patronizing attitude typified the times and contributed to tribes' enforced dependency. Hoxie believed that through such leasing policies the government after the turn of the century began treating Indian reservations more like colonies.[18]

In the 1910s and 1920s, during the period of increasing termination sentiment, Congress acted several times to open more Indian lands to mineral development. Tribal advocates were able in most cases to maintain tribal rights of mineral ownership—sometimes with the help of the Indian Bureau and sometimes despite it—but in the process they sacrificed several other aspects of control, giving up the right of consent and allowing state taxing authority.

Although the 1891 leasing act had provided for mineral leasing on most treaty reservations, controversy arose in the 1910s over mineral access to the ceded portion of the Wind River Reservation in Wyoming; the tribes had ceded this land to the United States for homesteading in 1905. Oil had been discovered in that area and was being produced on both privately owned and allotted lands. Non-Indians began pressuring the Department of Interior to open the other lands, which had not been homesteaded and thus were in limbo between public and Indian ownership. Consequently, it was not clear whether they could be leased under the 1891 leasing act or located under public land law. An earlier withdrawal order also impeded access to these lands.[19]

Fortunately for the Indians, the land had not proven as attractive for agricultural pioneers as had been expected. Less than a tenth of the 1.5 million acres opened to entry in 1905 had been sold to homesteaders, and DOI had not sold the rest because the local superintendent said the tribes needed the income from grazing and, potentially, from oil. Much of the remaining lands could have been lost to Indian ownership because of legal ambiguity about whether the minerals could be leased or located. Under location mining laws, miners could obtain fee-patent ownership of lands where they established mining claims. The situation created tremendous confusion among would-be developers. Locators, who assumed the land was federally owned, tried to establish claims on lands that had already been leased by others, such as the Carter Oil Company. Pressure from the developers forced Congress to act. Despite his responsibility to protect Indian interests, Interior Secretary Franklin K. Lane told Congress that developers could locate claims on the remaining lands because, in his opinion, the Indians had no interest in them except for proceeds. This conformed with a 1911 departmental policy that said Indians retained a right to use of and proceeds from such ceded lands. If his advice had been followed, the tribes would have lost ownership of patented oil lands and received $2.50 per acre for them, which was more than the average $1.50 they received under the homestead law but much less than their potential value.[20]

Fortunately Congress ignored Lane's advice and in 1916 it adopted a leasing law for the lands that provided for continued tribal ownership, royalties of at least 10 percent, and renegotiation every ten years. The law did not provide for tribal consent; instead, the secretary of Interior would set the terms, thus continuing the policies of the era. Yet it protected tribal ownership until after the Indian Reorganization Act (1934), when Congress restored the lands to the tribes; it prevented the tribes' oil from being drained by developers on private and allotted lands; and it provided an opening for renegotiating contract terms, which later became important.[21]

The Wind River tribes apparently benefited from the timing of the legislation. A debate over leasing versus locating minerals on public lands raged in Congress between 1910 and 1920. The East favored leasing, but to westerners, leasing "smacked of socialism." Various coal acts, beginning in 1864, had authorized selling public coal lands for $10 to $20 an acre. Oil was also considered "locatable." The western lawmakers favored continuing locating laws (under which the "little man" had an opportunity to own land) instead of leasing laws, which kept public lands under federal ownership where they could not be taxed. Leasing advocates were gaining momentum when the Wind River act was passed; they finally won their battle for leasing energy minerals on public lands with passage of the General Mineral Lands Leasing Act of 1920.[22]

Senator Henry Ashurst of Arizona, chairman of the Senate Committee on Indian Affairs, advocated opening the Wind River ceded lands as well as more Indian mineral lands in the Southwest. The latter effort culminated in a critical battle over ownership of Indian mineral lands. Under the 1891 act and the 1909 appropriations act, Congress had authorized mineral development only on allotments and on lands for which Indians had paid—mostly reservations created by treaty. Congress had not authorized developing minerals on reservations created by executive order. In 1919 Congress passed the metalliferous minerals leasing act, which opened up development of precious metals and later of coal on executive-order reservations in nine states. It extended to Indian reservations the policy of selling rather than leasing certain minerals. The law did not provide for tribal consent, and it gave Congress the authority to appropriate the proceeds for purposes that the lawmakers decided would benefit the Indians, including per capita payments.[23]

No one arguing for the precious metal legislation tried to argue that the tribes did not own the minerals, according to historian Lawrence Kelly. When Congress later deliberated over oil and gas development on executive-order reservations, however, that changed. In 1922—the year that oil and gas were discovered on an executive-order portion of the Navajo Reservation—another southwesterner, Secretary of Interior Albert B. Fall, testified before Congress that executive-order reservations were "merely public lands temporarily withdrawn by executive order." He believed that because the Indians did not really own them, such reservations could be developed under public land laws. If Indian minerals were developed under those public laws, the tribes would not have the power of consent, but that was a minor issue at the time given that Congress was questioning the Indians' title to the lands. It was one of the most far-reaching threats to the American Indian's land base because executive-order reservations represented two-thirds of it, including the Hopi, Jicarilla Apache, Northern Cheyenne, Uintah and Ouray reservations and parts of the Fort Berthold and the Navajo.[24]

When Congress would not act on executive-order reservation oil and gas legislation, Fall took matters into his own hands and issued a fiat in 1922 that placed the executive-order Indian reservations under the law for public lands, the General Mineral Lands Leasing Act, explaining that presidents (especially Theodore Roosevelt) often had restored executive-order reservations to public access. According to Indian advocate John Collier (then with the American Indian Defense Association), the effect of Fall's decision was to immediately withdraw the Indian claim to title from some sixteen million acres of land. Fall and the Indian Bureau also proposed an Indian omnibus bill that would have terminated all tribes and divided their assets among members, but Congress narrowly defeated it. A rancher and miner in territorial New Mexico, Fall believed in exploiting the natural resources throughout the West, not just on Indian lands. Fortunately for the Indians, Fall resigned in March 1923 prior to his indictment in the Teapot Dome scandal.[25]

Attorney General Harlan Stone issued a favorable opinion in 1924 that said the public lands leasing act could not apply to executive-order reservations. Stone included citations that proved Congress and the courts in the past had treated executive-order property rights the same as treaty rights.

Nevertheless, the belief that Indians did not own executive-order lands was held by many besides Fall. Congress seriously considered oil and gas bills for executive-order reservations under which the Indians would have gotten as little as one-third of the royalties from their minerals while both the state and the federal government would also have received one-third. Collier and Congressman James Frear of Wisconsin were the only ones to testify against a bill to provide the states with 37.5 percent of the royalties (the same as the state's share for publicly owned minerals). Although southwestern law-makers previously had argued that tribes needed mineral income to relieve the government of its burden, in this case they offered the astounding theory that the Navajos were rich and did not need all of the expected royalties.[26]

The Indian oil and gas leasing laws passed in 1924 and 1927 seemed innocuous in the context of the times because the tribes retained their ownership and received all the royalties. In fact, Kelly called the 1927 act a "great victory" for the tribes. The Indian Oil Leasing Act of May 29, 1924, amended the 1891 law for tribal lands on treaty reservations, requiring auctions and extending the term of leases from ten years to as long as the land produced oil and gas in paying quantities. It did not change the consent provision. The 1927 act, which applied to executive-order reservations, was more like the 1924 act than Fall and his cohorts had wished. It provided the Indians with 100 percent of the royalties and said that when the Indian oil monies were being appropriated by Congress, tribal councils should be consulted. No per capita payments were to be made with the royalties without an act of Congress. The act did not say outright that tribes owned their executive-order reservations, as Collier had wished, but it did say that there would be no further changes in the boundaries of Indian reservations without congressional action, thereby ending the practice of taking valuable mineral lands by presidential fiat.[27]

In opposing the attack on mineral ownership, however, tribal advocates opened the door to state taxation of Indian minerals under both the 1924 and the 1927 laws. Testifying against giving Indians only a portion of royalties from executive-order reservations, Indian Commissioner Charles H. Burke said, "I would be quite willing for a small percent to be paid to the state because of the fact that Indian reservation lands are not taxable." Collier went so far as to say he preferred providing for a state tax, just as on private lands, because he believed it would emphasize that the title to the lands was vested in the Indians.[28]

The debate over Fall's proposal bolstered the concept that Indian lands should be treated the same as public lands, a concept that still plagues Indian resource policies. It also reinforced the misconception that energy tribes were rich. By being forced to focus on ownership, tribes and their advocates neglected other critical issues. The decision to allow state taxation of Indian minerals, for example, seriously eroded tribal revenues and sovereignty in years to come. Yet loss of title to executive-order reservations or

This oil well pumped in the Two Medicine field on the Blackfeet Reservation in 1981. (Photo by Marjane Ambler.)

even loss of two-thirds of the royalties from those lands could have jeopardized the tribes' futures much more.

SPLIT ESTATES

As these battles over leasing Indian minerals took place, the federal government was also making critical decisions regarding the ownership of minerals under allotments and homesteads. The General Allotment Act of 1887 had broken up many tribes' communal estates into tracts owned by individual Indians and had opened the "surplus" lands to homesteading. The allotment act said the allottee had full rights to the timber and minerals, presumably so that he or she could obtain economic independence. Other allotment acts and subsequent legislation, however, often gave the mineral rights to either the tribe or the federal government.

The complicated land and mineral ownership pattern of the energy reservations today results from such decisions made over the years about whether a tribe, an allottee, a non-Indian settler, or the U.S. government should get the minerals under a parcel of land. Those decisions seemed to depend upon four factors: the national sentiment about leasing versus selling public land minerals; the national sentiment about whether tribal governments should continue to exist; demand for land; and what was then known about mineral potential. In most cases the federal government seemed to adopt Indian policies as afterthoughts to public land policies, whether or not they were appropriate.

These mineral ownership decisions predetermined the income flow and political dilemmas faced on the energy reservations in the second half of the twentieth century. When the tribe got the minerals, that decision virtually guaranteed the continued existence of the tribe, even where no tribal ownership of the surface continued, such as in Osage County, Oklahoma. It also facilitated consolidated administration. A tribal government could plan how to use the resources and could protect the reservation environment as a whole; an individual could not. Where the allottees got the minerals, the decision resulted in a few rich Indians and many headaches. Because of the split ownership, allottees ended up in bitter battles against their tribal governments and the federal government over mineral ownership and surface owners' rights for years after.

When most energy tribes' reservations were allotted and opened to homesteading just after the *Lone Wolf* decision, the laws did not specify the fate of the minerals under either the allottee's or the settler's agricultural tracts. Indian agents in some cases assumed that allottees should not be allowed to select mineral lands just as homesteaders could not select mineral lands under public land laws. The allottees and the settlers were expected to farm, not mine. In some cases, such as on the Fort Peck Reservation, Congress specifically reserved the coal from allotment. For nine years the agent on the Blackfeet Reservation held up approvals of 2,400 allotments on oil and gas lands, and Indian Commissioner Cato Sells asked Department of Interior solicitors if those minerals could be reserved for the tribe. In 1913 and 1917, however, Interior attorneys told agents on the Wind River and Blackfeet reservations that Congress had not prohibited allottees from receiving agricultural lands that happened to also have minerals, especially when the mineral potential was not recognized until later. Without specific instructions from Congress to the contrary, therefore, allottees were assumed to own the minerals under their lands.[29]

In 1906 the Osage Tribe convinced Congress to split the ownership of its land and minerals, reserving the minerals to the tribe while the surface was allotted. This legislation marked the first time on either public or Indian lands when mineral and land ownership were split. Although Congress had provided in the Curtis Act of 1898 for reserving minerals to the Five Civilized Tribes, the tribes had insisted upon disposing of both the minerals and the surface instead. In the same year President Theodore Roosevelt began withdrawing from entry vast acreages of mineral lands in the West and suggested splitting mineral and surface ownership. Roosevelt did not want miners to continue acquiring public mineral lands for a pittance, and he feared private corporations would acquire monopolies over coal supplies. He preferred that the energy minerals be leased rather than sold. While waiting for Congress to act, Roosevelt wanted to assure federal ownership of minerals. He and his successor, President William Howard Taft, withdrew mineral lands from entry so they could be classified; if they were classified as mineral lands, they could be sold for more money than unclassified lands.[30]

At Roosevelt's suggestion Congress passed coal lands acts in 1909 and 1910, which allowed settlers to claim and patent homesteads on coal lands if they agreed to reserve the coal to the United States. This policy was extended to other minerals, including oil and gas, in 1914. Congress passed the three laws partially because of a growing demand for western land; the number of homestead entries reached its second highest peak in 1909–1910, and coal prohibition placed thousands of acres off limits to homesteaders. Thus split estates, and all their inherent conflicts, became commonplace on public lands in the West.[31]

In 1917 Congress passed a coal lands act for opened Indian lands that split estates on those lands. Much like the law for public coal lands, it allowed homesteaders to claim the surface of coal lands and reserved the coal to the United States. Like the public coal lands laws, it was designed to open more land to agricultural settlers. Unlike the public coal lands laws, it said the proceeds would be credited to the tribes as the proceeds from surface lands sales had been.[32] Under the authority of the 1917 law and a similar one passed for Fort Berthold in 1914, homesteaders acquired thousands of acres of Indian lands over coal on several energy reservations, including the Crow, Uintah-Ouray, Fort Peck, and Fort Berthold reservations. On the Crow Reservation, patents were issued to non-Indians for approximately seventy thousand acres of surface lands. Questions about the status of minerals on opened Indian lands continued to plague federal officials, Indians, and miners for decades. In the 1920s various federal agencies insisted they had no jurisdiction to lease minerals on those lands.[33]

As the nation decided to split minerals under homesteads and reserve them to the federal government, it adopted a similar policy for allotments. Beginning at Fort Berthold in 1912 Congress reserved the minerals to the tribes. Then for several decades Congress reserved minerals to tribes under subsequent allotments: on the Blackfeet Reservation by an allotment law passed in 1919; Crow in 1920; Fort Peck in 1920 and 1927; Fort Belknap in 1921; Northern Cheyenne in 1926; and Wind River (under grazing allotments) in 1928. Some reservations had already been allotted, and this mineral policy applied only to new allotments. In several cases, to keep later options open, it provided for tribal mineral ownership for a limited period.[34]

Why reserve minerals to the tribes? If Congress intended to divide the land among individual Indians, why not provide the allottees with the resources under that land to help them succeed? Was it to help the tribes succeed instead? Congress responded to different objectives during different periods. In some cases the decision to reserve minerals to tribes reflected the national policy for public lands, which was designed to avoid monopolistic ownership. Angie Debo in her book about the Five Civilized Tribes, *And Still the Waters Run*, says Congress was adamant in 1906 that the Choctaw and Chickasaw coal should be kept out of the hands of a monopoly. Interior Secretary Ethan Allen Hitchcock was equally adamant about protecting non-Indian leaseholders. Thus plans to sell the coal lands were changed and despite the tribes' wishes, the coal was kept in tribal ownership. When the Osages demanded a tribal mineral reservation in 1906, they and their congressional supporters no doubt were influenced by the horrifying crimes recently committed to obtain the neighboring Five Civilized Tribes allottees' oil and gas rights.[35]

Less laudable motives were also at work. During this assimilation period the federal government exerted considerable control over development and revenues from tribally owned minerals. It was assumed there would ultimately be no tribes to hold these minerals. Indian Commissioner Cato Sells, who had told Interior's attorneys he wanted to reserve Blackfeet oil and gas "for the tribal benefit" in 1917, was an effective agent of national policy at the time, which was to destroy the reservations, eliminate communal property, and take care of the Indian problem by "freeing" Indians from federal protection. On the Blackfeet Reservation as elsewhere, Sells had forced fee patents upon many unwilling Blackfeet allottees, terminating the trust protection on those lands and minerals. The 1907 Blackfeet allotment act said any opened lands that were not settled within five years would be sold, and two-thirds of the tribal treasury would be divided among tribal members. Thus it would seem that Sells hoped to assimilate the Blackfeet minerals into the national pool by reserving them to the tribe, just as he hoped to assimilate the Blackfeet people.[36]

Partly as a result of Sells's fervor and the department's liquidation frenzy, allotment began to lose favor in Washington. The administration began to curtail allotment in 1920 before Congress abandoned it altogether with the Indian Reorganization Act in 1934. Both the Northern Cheyenne and the Crow Tribes sought communal mineral ownership when their reservations were to be allotted in the 1920s, but the change in national sentiment toward allotment probably had more to do with Congress's decisions reserving minerals than with the tribes' wishes. In an analysis of the 1926 rationale half a century later, the U.S. Supreme Court decided that Congress had provided tribal mineral rights for the Northern Cheyenne Tribe because the assimilation policy was losing its appeal. Because the IRA reaffirmed the role of tribal governments, Sells's and other assimilationists' efforts to preserve tribal mineral estates backfired. Most tribal minerals remained in Indian ownership instead of being lost to avaricious outsiders, as hundreds of thousands of acres of the allotted minerals had been.[37]

Although the reasons were not always clear, most of the decisions regarding mineral ownership under allotments fit a general pattern during the early 1900s, with the trend toward tribal mineral reservations increasing. Navajo allotments in New Mexico, however, did not fit this pattern. In 1919, the same year that Congress provided a tribal mineral reservation for Blackfeet allotments, the General Land Office (without any specific mineral reservation instructions from Congress) began issuing allotments to Navajos that reserved the minerals to the United States and not to the tribe. To justify this decision to take the mineral rights, the government used the authority of laws enacted to allow homesteaders to claim public lands.[38]

The reason for the government agency's decision no doubt rests upon the unique history of the area. As the Navajos were choosing their allotments between 1907 and 1918, various executive orders changed the lands' status between reservation and off-reservation several times in response to local pressure. At the time of allotment, 2,000 allotments were on reservation lands and 1,700 were on public lands. The perplexing decision to use public-lands law when awarding the mineral rights seemed to rest upon the rationale that most of the Navajo allotments were on public lands at some point. Except when the lands were within reservation boundaries, no tribal ownership was at issue. Congress by then had indicated its preference not to award the mineral rights to allottees.[39]

The question of mineral rights became critical to 14,000 Navajo allottees in the 1980s when they were told that only the federal government had the right to decide whether their lands would be strip mined for coal and had the right to receive the proceeds. To challenge the decisions reserving minerals, the allottees sued the Department of Interior in 1983 in *Etcitty v. Watt*. By 1989 the case still had not been heard, and the government had not explained its rationale.[40]

Regardless of Congress's or the tribes' intentions, the Indians' attempts to reserve minerals often failed because of a lack of information on both sides. Minerals could be reserved to the tribes only if the lands were first classified as mineral lands. The Fort Berthold Reservation provides a vivid illustration of this problem. When it surveyed the reservation in the 1910s, the U.S. Geological Survey (USGS) discovered only about half of the actual coal potential. It found no oil and gas potential, which is not surprising because oil and gas were not discovered in the state until 1951. As was true on other reservations, DOI attorneys later determined that the early mineral classifications were final, and the tribes had no claim on minerals under allotments or homesteads that were not recognized during that classification. That early ignorance had lasting financial and political impacts at Fort Berthold. Instead of the tribes' owning all minerals on the reservation, non-Indian descendants of the settlers owned coal and oil and gas in the homestead area and allottees owned most of the oil and gas on the rest of the reservation. The non-Indians developed two rather large coal mines in the homestead area in the 1930s and a few oil and gas properties. The tribes had neither approval authority over nor revenue from those developments.[41]

When opened lands in various states were returned to tribal ownership after the Indian Reorganization Act, they were subject to existing rights and thus bore the scars of the earlier policies, often in the form of split estates. The non-Indian who acquired the mineral or surface rights "in good faith" was allowed to keep them, regardless of the impact upon the tribal land base. Ultimately both Indians and non-Indians paid the price for the vacillating federal policies in energy decision conflicts that continue today.[42]

Only a few energy tribes held their communal mineral estate intact into the 1980s—notably the Ute Mountain Utes, Rocky Boys' Chippewa-Crees, and Osages. The surface/mineral ownership was split on reservations in many different ways: tribal over federal (Navajo); tribal over allotted (Fort Belknap, Fort Peck, Southern Ute, Fort Berthold); allotted over tribal (Northern Cheyenne, Osage, Crow, Fort Berthold); allotted over federal (Navajo); private over allotted (Fort Peck, Southern Ute, Fort Berthold); private over tribal (Crow ceded strip, Blackfeet); private over federal (Crow ceded strip); tribal over joint tribal (Navajo, Hopi); and state over tribal (Uintah and Ouray). Some, such as the Navajo and the Crow Tribes, held minerals outside their reservation boundaries, further complicating their status.[43]

TRIBAL ROLE IN RESOURCE DECISIONS

Behind the Five Civilized Tribes' desire to allot their mineral rights was a deep distrust of the federal administration of their income from minerals and land. They preferred totally liquidating the tribal estate to having the government skim off the tribes' funds. A study conducted for Senator Robert L. Owen (Cherokee) in 1909 confirmed their worst fears, and disputes over control of various energy tribes' funds persisted for decades despite various legislative attempts to correct them.[44]

Relying upon the authority of the *Lone Wolf* decision, the Curtis Act of 1898 (which abolished tribal laws and courts in Oklahoma), and various other measures, the federal government decided how Indian resource income should be spent "for the benefit of the Indians." When deciding how to use the money, government agents and Congress frequently ignored tribal priorities, used money for projects that tribes specifically opposed, and vacillated in their own priorities. At the turn of the century in Oklahoma, the Department of Interior did not even distinguish between tribal and federal funds in its records; tribal funds were siphoned off—for example, selling tribal property to pay for assimilation costs. Arguing for tribal control one congressman in 1934 said that since 1900 the government had spent $500 million of tribal money in per capita payments and administrative costs, including irrigation projects, highways, and bridges that benefited the whites rather than the Indians. While using tribal funds for itself or other non-Indians, the Indian Bureau would ask Congress for "gratuities" to assist the Indians. Sometimes the Indian office invested trust funds, directly or indirectly, into railroads, which were among the more notorious exploiters of Indian lands. On the Jicarilla Apache and the Wind River reservations tribal members literally starved while their own timber or mineral monies sat in the federal treasury or subsidized Indian agencies. Mineral tribes, often perceived as "rich," were especially vulnerable.[45]

Between 1908 and 1947 the elected leaders of the Shoshone and Arapahoe Tribes of the Wind River Reservation in Wyoming sent frequent delegations to Washington to argue for control over their money and to seek cancellation of an oil lease in the Maverick Springs area. The tribes objected to their oil and gas revenues being used for agency telephone lines and agency salaries when the Indian people were living in tents; they wanted per capita payments. Although Congress several times during the same years insisted that other tribes' funds be divided into per capita payments in preparation for tribal termination, the government refused the Wind River pleas. The tribal leaders supported energy development; at the turn of the century they had survived on mineral lease income when the government cut rations. In 1905 they wrote to the Indian commissioner, "Please send out a good friend of the Indians to find them mines. We give them 15 cents for every dollar." Yet the federal government's intransigence on the per capita issue drove the tribes to vote at one point not to allow any more leasing unless the proceeds were divided. In his appeal in 1911 Arapahoe Councilman Yellow Calf said, "We made these leases for our children, for the old and helpless and for those who can't work. . . . Probably God has made the Indians just like the white man, and I do not understand why the government did not ask us if they could use this money." Congress finally in 1947 passed legislation providing for per capita payments.[46]

The Wind River Tribes' battle against the capping of the Maverick Springs oil wells lasted nearly as long as their battle for per capita payments. After several lease holders capped the Maverick Springs wells because the oil companies said prices were too low, the Wind River Tribes no longer received desperately needed money, even though the original lease holders had made nearly $1 million by selling the properties. In 1938 Shoshone Councilwoman Nell Scott told the superintendent that many strong tribal leaders who had failed to get the Maverick Springs issue resolved had lost the people's respect. Despite opposition from Interior and weak lease production terms, the joint council delegation to Washington in 1939 succeeded in getting the oil fields opened. This victory established the council members as effective advocates for their people, according to anthropologist Loretta

Shoshone and Arapahoe Business Council members took Assistant Secretary of Interior Oscar Chapman to the Maverick Springs oil field in June 1938 to explain their objections to the oil field's operation. Arapahoe officials pictured on the left are John Goggles, unidentified, Charlie Whiteman, Robert Friday, and Nell Scott. The man with the white hat in the center is probably Chapman. To his left are Shoshone officials Gilbert Day, Ben Perry, Bill McAdams, Frank Enos, Charlie Washakie, and (in front) Robert Harris and Lynn St. Clair. Harris served on the council until the 1980s. (Courtesy Buffalo Bill Historical Center, Cody, Wyo.)

Fowler, who studied 120 years of the Arapahoe Tribe's decision making for her book *Arapahoe Politics.*[47]

The Wind River Tribes' persistence and ultimate success on these two issues—per capita payments and the uncapping of the oil wells—points out that some energy tribes played a more active role before and just after the Indian Reorganization Act than is commonly thought. Studies of mineral decision making during this period on the Wind River, Osage, and Navajo reservations repudiate some misconceptions, showing the early leaders' strengths and how they were undermined. Rather than reluctantly accepting mineral development, they sometimes sought it. At the same time, their control was limited by their lack of information, the interference of the federal government, their economic desperation, the terms of leases, and the lack of enforcement of federal regulations and lease terms.

One of the first tribes to lease its minerals, the Osages in Oklahoma became the most notorious. Stories of their extravagant spending and of their victimization were used to justify "protecting" all Indians from spending their own money. Their story, well told by Terry Wilson in his book *The Underground Reservation*, illustrates the conflicting expectations of the non-Indians. The Osage Tribal Council approved the first oil and gas lease in 1896 by a narrow margin. By 1906 tribal members came to be considered the "richest people on earth." Although the white man wanted the Indian to use his resources, he was not happy with the ramifications. In 1900 Commissioner William A. Jones told superintendents across the country to withhold food rations to force Indians to abandon their tribal ceremonies and dress; but the Osage Tribe, like those of Wind River, had too much income to be vulnerable to coercion. Some Indian agents referred derisively to their "unearned wealth" and were pleased when oil prices fell, saying the reduced circumstances would "finally instill in them the idea that they should *make* their livelihood" (emphasis added).[48]

The Osage Tribe agreed to a blanket lease arranged by the Department of Interior, covering the whole reservation. The tribe suffered from both the terms of the lease and the lack of enforcement. This leasing policy caused problems that haunted Indian people for years to come—for example, lack of surface owner and environmental protection; lack of competition; speculation; and bribery of both Interior and tribal officials. Because the blanket lease covered the entire reservation, it did not allow the competition necessary to improve income for the Osages. Oil crews could use as much of the surface—and the wood and water—as they needed for constructing derricks and housing. The lessee did not produce oil within eighteen months as his contract required so the Osage Tribe and the Interior secretary agreed his contract should be canceled. After the lessee appealed, however, the blanket lease was mysteriously reinstated.[49]

Although the Interior Department tried to impose its will upon the Osage Tribe by dissolving the tribal government at one point and later removing several uncooperative tribal council members from office, the Osages did not quietly acquiesce to federal control. They took a relatively active role in land and resource decisions. The Osages delayed but did not succeed in blocking the division of surface lands into allotments, which occurred in 1906. As a condition of its allotment act, however, the tribe made a critical decision—it insisted upon retaining tribal rights to the subsurface, thus forming an "underground reservation." In 1907 the tribal council hired a full-time field inspector, who quickly proved his worth by uncovering an oil scam by the Uncle Sam Oil Company. In 1911 several tribal members formed the Osage Protective Association, saying the Osages' "meek submission to departmental action is a thing of the past"—tough rhetoric that foreshadowed the energy tribes' statements of the 1970s. The Osage association members effectively backed up their rhetoric. They succeeded in blocking the renewal of the blanket lease in 1912 and in getting some protection for surface owners.[50]

In the 1920s Navajo leaders also were involved in and knowledgeable about oil and gas decisions—to a limited degree compared with today, but more than commonly thought. Historians Lawrence Kelly and Peter Iverson do not believe that the tribal council created in 1922 was composed of "yes men." The tribal council that approved the first lease on August 13, 1921, was an already existing body representing one geographic area of the treaty reservation. These early Navajo councilmen had their own ideas. Midwest Refining Company, a division of Standard Oil Company of Indiana and the holder of the first Navajo lease, wanted the Navajos to grant adjoining leases only to Midwest's friends. The Navajos and their sympathetic BIA superintendent, Evan W. Estep, did not want to grant any more leases to anyone until Midwest determined whether there was oil; they realized the benefits of competition and figured they could get higher bonuses in a proven area. On September 23, 1922, the council granted a lease to a company not associated with Midwest. The day after the decision, when the company knew it had lost the battle of wills, the company announced that it had struck oil.[51]

Although royalties were distributed to Navajo Indians in other areas of the reservation, the local council made all the leasing decisions. Then in January 1923 the Department of Interior drew up papers for a new tribal council, representing all the Navajo areas, to make decisions on oil contracts. At Interior's insistence the department's representatives had to attend all of the council's meetings. The department coerced tribal members to consent to the federal government's oil leasing plan by promising to help get new lands. Although the council delegated to the superintendent authority to sign oil and gas leases, the Navajos stayed involved in their mineral decisions. Kelly said records show the new council members had definite opinions, which they expressed clearly and forcefully. Even though they were rarely listened to by BIA, they were not quietly obedient.[52]

MINERAL REFORMS FOLLOW IRA

When Congress passed the Indian Reorganization Act (IRA) in 1934, it repudiated the policies of the past and avowed its commitment to tribal control of resources. Section 16 said constitutions of organized tribes would give them the power to "prevent the sale, disposition, lease, or encumbrance of tribal lands, interests in lands, or *other tribal assets* without the consent of the tribe" (emphasis added). This section was designed to give tribes the right to control timber and mineral leasing as well as the right to control their own purse strings. Nevertheless, the historic struggles continued between tribes and the federal government over who should decide how to spend the tribes' trust funds. Federal proponents insisted that Congress must retain control over appropriations despite the fact that tribal trust funds were not public funds.[53]

Soon after passing the IRA, Congress passed the Omnibus Tribal Leasing Act (1938), which dictated until 1982 how Indian minerals and oil and gas would be leased. This act was a critical step toward greater tribal control. A sense of outrage at past exploitation had erupted, both in Congress and in the administration. Sells's assertion in 1914 that tribes should have no power of consent was no longer accepted as official federal policy. When the Navajo BIA superintendent proposed taking coal from tribal land without payment and without tribal consent for use by BIA, a high Interior official said, "Taking minerals from tribal land without the consent of the Indians is the sort of thing for which past administrations have been severely and I think justly criticized."[54] On some reservations the department had no law to back up this philosophical position to prevent mineral development without tribal consent. The IRA had provided organized tribes with the power of consent, but tribes that had not organized under the act, which included several energy tribes, needed specific authority for controlling mineral development.[55]

Secretary of Interior Charles West believed the omnibus tribal leasing bill then pending before Congress would resolve this dilemma. In a June 27, 1937, letter to Congress, he argued for the bill's passage, saying that under the 1919 act, neither the secretary of Interior nor the tribe could prevent a person from getting a lease if he or she had complied with the law: "In several instances, it has been necessary to grant the lease notwithstanding the fact that the Indians of the reservation were opposed to leasing the lands." He said he supported the omnibus tribal leasing bill because it would bring all mineral leasing matters into harmony with the IRA. At the same time developers would benefit from having one comprehensive law for different minerals and different types of reservations. The bill would also provide mineral access to reservations that were limited to only oil and gas exploitation.[56]

The final terms of the Omnibus Tribal Leasing Act required that tribal coal be leased rather than located (public coal had been leased since 1920). It required surety bonds, in amounts satisfactory to the secretary of Interior, to guarantee compliance with the lease terms. For oil and gas leases the law required competitive bidding. Although the law excluded allotted lands, the department imposed the same protection for those lands in its regulations. The law excludes lands in Alaska. In a 1985 interpretation of the law, the U.S. Supreme Court ruled that the Omnibus Tribal Leasing Act had, by implication, disallowed state taxation of the tribes' royalty interests. Thus Congress in 1938 had stopped the states' drain on tribes' funds that Collier and others had instituted with the 1924 and 1927 Indian oil and gas laws.[57]

In later years Indian leaders and their advocates criticized the 1938 act, saying that the terms were too long ("not to exceed 10 years and as long thereafter as minerals are produced in paying quantities") and that the royalty levels set by regulations under the act were too low (minimum of 10 percent, or 10 cents a ton for coal). In the 1980s the Council of Energy Resource Tribes (CERT), a consortium of tribes, vigorously criticized the act—and especially the administration's interpretation of the competitive bidding requirement—for relegating tribes to the position of "passive beneficiary" rather than positions of authority and control. Nevertheless, the significance of the law and especially its consent provision cannot be overemphasized. Because of that provision the 1938 act must be ranked just under the IRA in its tangible and symbolic importance for energy tribes, a significant indicator of the supportive mood of the nation at that time.[58]

TERMINATION ERA REDUCES
MINERAL OWNERSHIP

During the termination era of the 1950s, when national sentiment once again favored assimilating Indians into the social mainstream, the federal government attacked tribes' communal mineral rights as well as their land rights. In 1953 Congress extinguished the Wind River Tribes' land and mineral rights in a portion of the 1905 ceded area and restored them to the public domain for a reclamation project, paying $1 million and promising 90 percent of the royalties. After signing the agreement the Shoshone Tribe hired a new attorney, Marvin J. Sonosky, who said the Indians had made a mistake. Also known for championing tribes' fight for return of their submarginal lands, Sonosky ultimately won this battle, too. Sonosky says that because BIA did not want to bother with administering mineral rights, the agency had not told the tribes of the disadvantages of having public land leases. Not only did the tribes lose their right to help set royalties and other lease terms, they also sacrificed 10 percent of their royalties for administrative costs. At first the courts would not listen to the complaint because the tribes had already agreed to the legislation. But Sonosky convinced Congress that a mistake had been made, and mineral ownership was returned in 1958.[59]

Another government attack on tribal mineral rights involved allotted lands. As under the first termination policy of "forced competency" in the 1910s, BIA in 1954 adopted a regulation saying that any allottee who removed trust protection from any of his or her land must remove it from all of the property. In keeping with this policy Congress took away tribal mineral rights under many allotments on the Fort Peck Reservation that it had granted in 1927, awarding them to the allottees, usually without trust protection.[60] In 1955 the Department of Justice instituted a successful suit on behalf of allottees on the Fort Belknap Reservation to get the minerals under certain allotments despite the 1921 allotment act, which reserved the minerals to the Gros Ventre and Assiniboine Tribes. The Interior solicitor said the suit was necessary to clear the title because oil and gas had already been leased to Phillips Petroleum Company. Consistent with opinions on other reservations the solicitor said the tribes had no rights to minerals unless the lands had been previously classified as mineral lands.

Anthropologist Loretta Fowler attributed the federal government's suit to the termination era, however, pointing out that between 1947 and 1954 the local superintendent dissolved the four tribal enterprises, leaving only oil for the tribal government to rely upon. In 1958 a U.S. district court ruling denied the tribes even that option. Fowler said this mineral rights controversy effectively destroyed the political authority of the tribal council and the ability of councilmen and other Indian leaders to generate consensus. As desired by the termination advocates the Indian people began to consider tribal income and property as assets in which they should have a per capita share, instead of as resources for future generations or a buffer against further exploitation, as they previously had believed.[61] Fortunately for the Osage, Northern Cheyenne, and Crow Tribes, Congress did not have to act during the termination era upon their tribal mineral rights because none came up for extension. Years later, when the mood of Congress had switched toward tribal self-determination, Congress perpetuated those communal mineral rights under allotments.[62]

LEASING POLICIES EXAMINED

After the 1950s the tribes no longer had to absorb themselves in constant battles for ownership and consent authority. Finally the energy tribes could begin to focus upon terms of development, such as royalties, employment of tribal members, water rights, and environmental and cultural protection. As they took the first tentative steps toward asserting more control, however, they realized that consent authority was not enough. They needed information and expertise. They also found that they were bound by the terms of contracts that had not kept pace with changing economic conditions and environmental awareness.

In the 1960s and early 1970s the public became increasingly concerned with the exploitation of public resources and with the protection of the environment. No longer would the American people tolerate the mineral policies that allowed minerals to be developed with no regard for land, air, and water quality. The National Environmental Policy Act (NEPA) of 1969 was a hallmark of this growing sentiment, with its requirements for public involvement in decision making and for environmental impact statements prior to development. Congress passed the Clean Air Act in 1963, the Clean Water Act in 1972, the Safe Drinking Water Act in 1974, and a hazardous waste law in 1976. The public's concern also resulted in studies of mineral development on public and Indian lands.

Two independent studies damned federal leasing programs for Indian and federal minerals, saying they were mismanaged and resulted in prices far below fair market value, in vast expanses of scarred lands, and in thousands of acres of leases being held for speculative purposes. In 1974 the Council of Economic Priorities (CEP) issued a study, "Leased and Lost," by James Cannon, that scrutinized federal and Indian coal leasing. The Federal Trade Commission (FTC) looked at coal as well as uranium and oil and gas and published in 1975 a report on Indian lands, separate from the national study, entitled *Staff Report on Mineral Leasing on Indian Lands.*[63] When the General Mineral Lands Leasing Act and the Omnibus Tribal Leasing Act were written in 1920 and 1938, respectively, no one realized the need for land and resource use planning. Consequently neither the laws themselves nor the regulations required a broad, long-term overview before leasing decisions. Thus many of the problems discovered by the two studies were common to both Indian and public minerals leases.

However, standard Indian lease procedures did differ from federal lease procedures in four significant ways that were detrimental to the tribes. First, federal leases were subject to renegotiation every twenty years, but most Indian leases could be held with the same terms as long as the land was producing minerals in paying quantities. This prevented the tribes from changing rents and royalties to reflect current market conditions and from adding environmental stipulations to reflect new standards. On the federal leases, CEP found that the Department of Interior was "extremely lax" about modernizing lease terms; in many cases it simply renewed the old, archaic royalty rate for another twenty years.[64]

In the Indian section of the CEP analysis, the authors said they were mystified about why a coal lease held since 1941 by the Redcreek Corporation on the Uintah-Ouray Reservation had not been readjusted or canceled by the tribe. The situation exemplifies the confusion caused by the ownership history of Indian mineral tracts. The Redcreek coal was actually developed under a federal lease before the coal ownership was restored to the tribe. Enforcement and reclamation on that lease continued to suffer from jurisdictional questions in the 1980s because the federal government leased the coal, the tribe owned it, and the state of Utah owned the surface over the coal. The lease produced only 2,637 tons of coal in thirty-three years.[65]

Second, the Indian coal regulations encouraged companies to place power plants or synthetic fuels plants—which have much more serious social and environmental impacts than mines—on reservations. If coal was used on the reservation, the company could pay lower royalties in some cases. Furthermore, companies could get leases larger than 2,560 acres only if they planned coal-fired power plants or other industrial facilities on or near the reservation. The notices of sale said nothing about the veto power of the tribe if it decided such plants would not be in the tribe's best interest. Thus in addition to encouraging more intensive development, this loophole resulted in tying up much larger acreages under one company. The average Indian lease in 1974 was eight times larger than 2,500 acres and fifteen times larger than the average public land lease.[66]

Third, because of understaffing, the U.S. Geological Survey rarely explored Indian lands prior to oil and gas lease offerings; when they did, the studies were generally cursory and often incomplete. In contrast, USGS routinely explored onshore federal oil and gas tracts to be offered for lease and publicized its findings. The FTC staff suggested that this situation resulted in two problems: less competition at the Indian sales (and thus lower bonus bids) and limited capacity for evaluating bids at Indian sales. After bids were submitted for Indian oil and gas lands, USGS was to determine whether they were high enough. FTC characterized USGS bid evaluations as "essentially guesswork by trained geologists who often have not even seen the tracts in question." USGS rarely rejected bids on Indian leases, but the Southern Ute and Ute Mountain Ute tribes—which FTC considered "quite sophisticated" in oil and gas leasing—had rejected several inadequate bids.[67]

Fourth, industry, rather than the tribes, usually decided when the Indian lands should be opened to development by nominating tracts for lease sales. Tribes often could assert their will only afterward—if tribes did not want nominated lands to be leased, they could reject all nominations. This right was exercised by the Santa Domingo, San Felipe, and Taos pueblos. When a tribe wanted an oil and gas lease sale, neither the tribe nor the federal agencies usually had the resources to explore for potentially valuable tracts to lease. The Navajo Tribe nominated its own lands for development, based upon its own geologic information, but received fewer bids in some cases because industry did not have access to the information. Inadequate advertising also contributed to the lack of competition at sales initiated by industry.[68] Also peculiar to Indian leases was the question of employment—federal leases had no employment requirements. Although standard Indian leases required Indian preference, BIA was not required to monitor compliance. FTC said the companies implemented employment clauses in oil and gas leases poorly; they were "fairly successful" in coal leases, however.[69]

The two studies found that the federal government was equally derelict in other ways when it came to fulfilling its responsibilities concerning public and Indian minerals.

Royalties. The FTC staff found that the federal government had actually collected slightly higher royalties for coal and oil and gas on Indian lands per unit produced than it had collected on public lands. Nevertheless, both federal and Indian coal royalty rates were much too low, primarily because they were set as flat per ton rates rather than percentages. When the average price of coal increased by 237 percent over twenty years, the average royalties increased by only 35 percent. Standard Indian uranium leases required a minimum 10 percent royalty, and FTC found only slight differences in royalties collected per pound of federal uranium than per pound of Indian uranium (higher one year and lower another). The lack of a readjustment clause, however, was destined to create greater disparity between federal and Indian royalty rates in later years.[70]

Environment. The standard Indian lease's environmental regulations were similar to federal requirements; for coal and uranium they were identical. When the public-leasing environmental regulations were slightly improved in 1969, the Indian regulations were also slightly improved for coal and uranium. A 1972 study by the General Accounting Office (GAO), however, said the federal government did not abide by the new regulations; subsequent studies confirmed this. CEP said the Department of Interior had totally disregarded until 1973 the National Environmental Policy Act for federal coal leasing; it had not prepared impact statements for any of the twenty leases issued between 1970 and 1973, insisting that NEPA did not apply to Indian lands. BIA, declared CEP, also had an "abysmal record in enforcing the environmental guidelines." The bureau did not perform the technical examinations necessary to control mining's impacts and did not require adequate bonds to assure compliance. Indian coal-lease reclamation clauses generally mimicked the wording of public coal leases, setting no guidelines nor specific requirements. "In most cases . . . , reclamation clauses have remained vague enough to be circumvented," stated CEP. Even a general requirement for reseeding (inserted by the Navajos in one lease) represented an improvement at that time, although it did not require the company to successfully reclaim the arid land. FTC pointed out that the Navajo Advisory Council routinely added environmental stipulations to all mineral leases prior to signing them, whether negotiated or competitive. Smaller tribes, however, lacked the expertise or aggressiveness to do this.[71]

Diligence. Standard Indian leases had stronger terms than public leases to force companies to produce rather than holding leases to sell later for higher prices. The strong terms were not enforced, however, according to the FTC staff, which said, "Rigid enforcement of diligence requirements could make leased Indian land far more productive than it is now."[72]

Unlike public land leases, Indian leases could be negotiated under certain circumstances. The 1938 Omnibus Tribal Leasing Act required competitive sales for oil and gas leases, but the secretary could reject all bids and either readvertise or (if the tribal council agreed) negotiate a lease. For mineral leases, tribes could negotiate leases with the companies without a competitive sale. The FTC and CEP studies both said the authority to negotiate gave the tribes a distinct advantage over the public—they theoretically could have contracts with more safeguards and better financial returns. In practice, however, the tribes usually did not utilize this potential for better terms in the early contracts. Occasionally the tribes negotiated some terms that were weaker than standard terms. For example, the Navajos agreed in several instances to forego their right to tax mineral production, which was not addressed at all in most other mineral contracts.

In some cases the Indians did negotiate good concessions in these early contracts, although they were not always enforced. For example, Consolidation Coal, a subsidiary of Continental Oil, agreed to build a $1.5 million health center for the Northern Cheyenne Tribe. Even though BIA's standard lease form only required a lessee to "give priority" to hiring members of the lessor tribe, through negotiations the Navajo and Hopi Tribes convinced Peabody Coal Company to agree to hire a 75 percent Indian work force at Black Mesa. Peabody also agreed to pay an average royalty of 25 cents a ton (compared with 17.5 cents per ton for most standard leases) and to dig new wells for the tribes' water supply if the mining lowered the water table below the reach of existing wells.[73]

Instead of the standard flat-fee royalties, the Southern Ute Tribe negotiated a coal contract with a percentage royalty with Peabody Coal in 1968. If coal had been mined under the lease, the 5 percent royalty would have resulted in a per ton royalty of 75 cents; the standard per ton royalty was only 15 to 17 cents a ton. Peabody also agreed to pay a higher than normal rental. Both the Spokane Tribe and the Laguna Pueblo negotiated complicated percentage royalty rates for their uranium leases. Anaconda Minerals Company agreed to adjust its royalty payment to the Laguna Pueblo as market values changed, without waiting for ten-year intervals. FTC found that Anaconda's royalty was the only one of the old coal and uranium leases that had been adjusted.[74]

Without the information and expertise to put them on equal footing at the bargaining table, the tribes had to rely upon the advice and integrity of industry and DOI advisers when setting contract terms. The power to negotiate was useless without the proper tools. FTC said, "The tribe, the BIA, and USGS generally have less business experience and acumen than the representatives of the private companies with whom they negotiate." CEP mentioned the pitfalls of bad advice and false promises, as well as the possibility that the tribes would make decisions out of "economic desperation," which history had already proven true.[75]

TRIBAL DECISION-MAKING PROCESS FLAWED

During the 1950s and 1960s when many of the early mineral contracts were negotiated, the tribes had few alternatives to relying upon BIA. They had only rudimentary tribal governmental structures and facilities, if any. When the Anaconda agreements were first signed, the Laguna Pueblo had no tribal government office; all the contract papers were kept in a trunk. And most of the tribal council members were illiterate. Although the Laguna Pueblo held a vote on the proposed mining, few of the tribal members could read the referendum question, much less comprehend the ultimate impact on the land from uranium development. Even literate non-Indian westerners had little concept of the impacts from uranium production at that time.[76]

Largely because of the mineral revenues that the Navajo Tribe had received since the 1920s, the Navajo tribal government was more fully developed than those of other tribes. Nevertheless, the tribal council members suffered from poor advice and from ignorance about the true value of their resources. Both the Navajo and Hopi governments lacked a process for accommodating different points of view in their decision-making process. Thus they relied upon secrecy—just as the federal government and industry did—to mask decisions that directly affected the lives of their members. The Black Mesa coal leases resulted from this flawed process.[77] Despite the broad indictment of federal policies contained in the FTC and CEP studies, federal policies toward Indian energy development might not have changed significantly if it had not been for Black Mesa. Through the critics' efforts Black Mesa became the most notorious mine in the country within a few years after mining began there, a symbol of the exploitation of Indian tribes for energy resources.[78]

In 1966 Peabody Coal Company (then a subsidiary of Kennecott Copper Corporation) signed two negotiated leases with the Navajo and Hopi Tribes for 40,000 acres on Black Mesa in the Joint Use Area, an area shared by the two tribes. Traditional Navajo and Hopi peoples consider Black Mesa a sacred area. The Navajos were promised over $2 million per year in royalties for thirty-five years at a time when their oil and gas resources—which had paid 70 percent of the expenses of the Navajo government—were being depleted. Later, interviews with the largely illiterate tribal council members revealed that they did not know the value of their coal, the potential impacts of mining, or the alternatives to coal development. They believed the ludicrous assertion that nuclear power would soon make coal obsolete. The Interior Department, under the direction of Stewart Udall, worked with industry and the tribal attorney to convince the council to act immediately, without deliberation.[79]

Outside of the Hopi Tribal Council, the Hopi people knew almost nothing about the contracts until a non-Indian anthropologist, Richard O. Clemmer, and a Pawnee attorney, John Echohawk, provided them with information in 1970. Echohawk was later to become executive director of the Native American Rights Fund (NARF). There were no open hearings, no community discussions, and no administrative disclosures by BIA or the Hopi Tribe, according to Clemmer. Even more serious, the attorney who represented the Hopi Tribal Council, John S. Boyden, reported in the Martindale-Hubbell directory that he also represented Peabody Coal during the same period, a clear conflict of interest. The tribal council approved the lease, based upon an IRA constitution adopted in 1936. The constitution required that the council members be certified by Kikmongwi (Hopi religious leaders who made tribal decisions prior to 1936), but opponents of the Black Mesa leases said some were not.[80]

Frustrated by federal and tribal decision makers' unresponsiveness, the Navajo and Hopi dissidents sought the help of outsiders, including the American Indian Movement, several urban support groups, environmental groups, the federal courts, and Congress. In 1971 the Black Mesa Defense Fund convinced the Senate to investigate Southwest energy plans, and the House Committee on Government Operations asked the General Accounting Office to investigate federal coal mining administration. Also in 1971 traditional Hopi tribal members sued the secretary of Interior, challenging his approval of the Peabody contract. The suit was dismissed on a technicality. Black Mesa became a cause célèbre for the emerging ranks of environmentalists across the country through articles in the *New York Times, Washington Post, Audubon,* and *Wassaja* and shows on CBS and ABC. Although the publicity did not stop the mining of Black Mesa, it had repercussions among both Indians and non-Indians that are still felt today.[81]

To prevent further exploitation along the lines of Black Mesa, tribal members became more involved and tribal officials more informed about the serious implications of energy decisions. They forced changes in political processes and federal and tribal regulations and dramatically improved the terms of the Black Mesa and other tribal contracts. (The renegotiations are discussed in Chapters 3 and 8.) In 1972 the Committee to Save Black Mesa invited a delegation of Northern Cheyennes to come from Montana to the Southwest to see the strip mine. That invitation radically changed the fate of the Northern Plains coal tribes.

Because of the Black Mesa exposé, many non-Indians still assume that Indian energy development necessarily involves exploitation of an ignorant and helpless people. The wide exposure of the tribal and federal political abuses regarding Black Mesa contributes today to a widespread misunderstanding of the role of tribal councils and how it differs from reservation to reservation. The publicity given the Black Mesa case—crucial at the time—has never been matched by subsequent media efforts to show how Indian energy development has changed.

CONCLUSION

As their knowledge and skepticism increased in the coming years, the tribes could not shake the shackles that remained from the rubber-stamp era, limiting both their ownership and jurisdiction. The boundary changes, fractionated heirships, and split estates clouded the tribes' and the allottees' mineral titles and the tribal jurisdiction. These questions slowed development, sometimes despite the tribes' wishes, and hampered their decision making.

The nation's desire to open lands for mineral development resulted in terms that were inappropriate for public lands and even more destructive for Indian lands. As the CEP study said, "The Department has acted as if it is better to have leased and lost than never to have leased at all." After such studies and the Black Mesa exposé some Indians and non-Indians came to the conclusion that energy development on Indian lands necessarily resulted in victimization and exploitation. The FTC staff, however, concluded that the problems could be solved by more tribal control. For the first time, a federal agency publicly stated that what was good enough for the public was not good enough for the Indians. Reiterating a historic complaint, FTC said that to allow the tribes freer reign with their resources, the Department of Interior should not require BIA approval of expenditures of mineral revenue. FTC recommended that the tribes consider forming an organization to share information, gain political clout, and employ experts. The month before the FTC report was publicly released, the tribes held the first meeting of the Council of Energy Resource Tribes.

The FTC staff castigated BIA's hostility toward Indian self-development and foretold a time when tribes would control their own development:

> As tribes develop the organization and expertise to handle mineral leasing on their own, . . . the role of the BIA should become increasingly passive . . . active only with regard to allotted land leasing and the technical examinations and environmental impact statements required by environmental regulations. . . . Given the flexibility of the current mineral leasing regulations, *the eventual usurpation by the tribes* of nearly all of the functions now performed by the BIA could probably be accomplished with no major changes in the regulations; the only necessary change would be the elimination of the need for BIA approval of certain tribal decisions [emphasis added].[82]

FTC's predictions came true to some extent over the coming years. With the help of their new organization and other advocates, the tribes usurped many of the functions performed inadequately by federal agencies. Despite the fact that the Interior Department retained its approval authority, the tribes began to make dramatic changes in the way they approached energy development after Black Mesa.

CHAPTER THREE

Early Horse Trading: Tribes Begin Setting the Terms

What is rightfully ours, we must protect; what is rightfully due us, we must claim. What we depend on from others, we must replace with the labor of our own hands and the skills of our own people. What we do not have, we must bring into being. We must create for ourselves.

—Peter MacDonald[1]

The early 1970s became a turning point for American Indian energy tribes. With the rest of the West they began to realize the social and economic costs of energy development. They suffered the same impacts as other rural communities, in addition to the cultural impacts as Indian people. But unlike other westerners they had no place else to go if mining destroyed their limited land base. On the other hand, the tribes had an advantage. Although other westerners had to depend exclusively upon political pressure to affect development on federal lands, the tribes owned desired minerals themselves. Thus as mineral owners they could participate in the "horse trading." A few years after the Navajos and Hopis signed the Black Mesa coal leases, the tribes began testing just how much control they as owners could exert over the energy demands of the nation and the multinational corporations at their doors. The tribes that were not bound by preexisting contracts could make choices: To gain control, many of them called for a hiatus in the midst of the uranium, coal, oil, and gas boom.

The era was also a turning point for industry and government officials, who found they could no longer treat Indian lands as federal lands. Although many companies fought the change with rhetoric and litigation, others soon took their places and continued to seek Indian minerals under the new rules.

CHEYENNE COAL LEASES

In the beginning Indian coal sold deceptively quietly. With only one or two companies bidding for tracts on some of the reservations in the 1960s and early 1970s, coal prospects seemed unimportant to both industry and the Indian people.[2] Nothing hinted that Lame Deer, Montana, the capital of the Northern Cheyenne Reservation, would become the site of a revolution in Indian energy policies. A small town, Lame Deer lies in south-central Montana among gentle hills covered with ponderosa pines, cattle, and occasional outcroppings of coal. Peabody Coal Company representatives certainly did not suspect anything dramatic when they submitted the only bids for Northern Cheyenne coal in 1966 and 1969, although they no doubt felt smug about their foresight. Peabody (then a subsidiary of Kennecott Copper Corporation and soon to be the largest federal and Indian leaseholder in the country) knew that before long, others would recognize the value of the 5 billion to 10 billion tons of strippable coal under the reservation. Someday the company's small investment in Cheyenne coal would pay off.[3]

The Bureau of Indian Affairs regarded Peabody's 1966 bid of 12 cents an acre as "very good" and accepted the bid again in 1969. Naively BIA considered Cheyenne coal a "white elephant" and wanted to make it as attractive as possible to industry. Only later did BIA realize that similar coal already had received bids of $16 to $100 an acre—100 to 1,000 times higher than Peabody's bid. The federal government had consistently undervalued Cheyenne coal. Although federal investigators recognized in 1928 that thick veins of coal lay under the entire reservation, the U.S. Bureau of Mines believed the field was not only too small but also impractical to mine without railroad access.[4]

For many years demand for western coal had been weak because of its low heat value and its distance from major markets. By the mid-1960s, however, big oil firms began to buy up western coal properties in anticipation of a coal boom caused by several expected political and economic changes. Cheyenne coal—and much of the rest of the coal in the West—became a potential bonanza long before government policies for either federal or Indian coal reflected its true value. Western coal's value rose as railroad access and the technology of the giant dragline shovels improved; strip mining cost about half as much as underground mining. The Clean Air Act of 1970 made the low sulfur content of western coal even more attractive because it required reduced sulfur dioxide emissions from power plants; most midwestern and eastern coal, in contrast, contained relatively high levels of sulfur. The companies also expected demand for electricity to continue escalating. American consumers, already the world's biggest per capita users of energy, increased their energy demands in the early 1970s twice as fast as the population increased. Analysts' predictions that demand would triple by the end of the century fueled the ambitious dreams of utilities and mining companies, with coal the leading actor in the drama. Coal was estimated to contain nearly 90 percent of the energy value of all fossil fuels in the United States. Discarded several decades earlier in favor of cleaner fuels for railroads and power plants, coal had obtained a competitive cost advantage when oil and natural gas prices rose, important both for coal-fired power plants and for conversion to synthetic fuels.[5]

By the third Cheyenne coal sale in 1971 industrial giants and speculators recognized coal's potential even though the Department of Interior and

Northern Cheyenne President Allen Rowland takes time out to enjoy a birthday party in Lame Deer, Montana, in 1981. (Photo by Sara Hunter-Wiles.)

the tribe still did not. BIA offered the remaining reservation acreage, and twelve firms submitted bids. Four received exploration permits, including AMAX Coal Company (under the bucolic name of its subsidiary, Meadowlark Farms), Consolidation Coal (a subsidiary of Continental Oil Company), and some speculators.[6] By now coal companies had won rights to mine 56 percent of the Cheyenne Reservation, with bids averaging only $9 an acre. The tribal council happily distributed most of the $2.25 million bonus monies as per capita payments to tribal members. Tribal President Allen Rowland and several council members began asking questions about why their coal was worth only 17.5 cents a ton in royalties when their gravel was selling for 18 cents. Yet the tribe hesitated to look a much-needed gift horse in the mouth. In fact, the BIA area director later said the tribal council had insisted upon having the lease sales when they saw nearby state lands being leased. More than half the people were unemployed, and the average per capita income of $1,152 a year was less than half the Montana average.[7]

In 1972 the Indians' eyes were opened when Consolidation Coal (Consol) returned to the reservation with an astounding offer: a bonus of $35 an acre (four times the previous average), a royalty of 25 cents a ton, and—important for the isolated, rural reservation—a $1.5 million community health center. Consol wanted to build four plants to gasify Cheyenne coal on the tiny, 400,000-acre reservation. Saying it could not afford the delay of a competitive sale, Consol sent an urgent letter giving the tribe fifteen days to decide. "If Consol cannot conclude negotiations with the Northern Cheyenne Tribe at an early date, Consol will be forced to take this project elsewhere. If it becomes necessary to do this, this project will be lost to the Northern Cheyenne, and it may be a long time before a project of this magnitude comes again, if ever." In the letter the company said the project would solve the unemployment problem for the tribe and raise the standard of living; however, it offered no compensation for its use of Cheyenne water or for the land on which the giant gasification plants would be built.[8]

This arm twisting did not work. Consol's extraordinary offer and its urgency tipped off tribal officials about the true value of their coal in the marketplace. As the tribal council negotiated with Consol for several months, worries about possible social and environmental impacts grew. The Northern Cheyenne Research Project, a federally funded group of primarily non-Indian scientists, distributed information about possible coal impacts. Cheyenne allottees—people who owned reservation land individually—became upset about exploratory drilling on their lands and on Indian burial grounds. They formed a landowners' association to fight the coal development and to explore economic alternatives. With only six hundred families on the reservation, it was obvious that outsiders would fill most of the gasification plant construction jobs, and the Cheyennes did not want to become a minority on their own reservation. Consol itself said a new town of about thirty thousand might be necessary on the reservation for its incoming workers, and Peabody also planned gasification plants for the Northern Cheyennes' coal. Despite the possibility that each family might receive $500,000 from Consol, tribal member Ted Risingsun spoke for many others when he said, "I would rather be poor in my own country, with my own people, with our own way of life than be rich in a torn-up land where I am outnumbered ten to one by strangers."[9]

Tribal Councilman Edwin Dahl took the coal contracts to a young Osage lawyer, George Crossland, who advised the tribe to challenge them. On March 5, 1973, the Northern Cheyenne Tribal Council voted 11 to 0 to seek cancellation of all of the permits and leases. The new tribal attorneys, Alvin J. Ziontz and Steven H. Chestnut, prepared a petition asserting that the permits and leases violated thirty-six federal regulations. The Cheyenne petition surprised and alarmed federal and industry officials across the country who knew that many of the same coal companies and speculators had acquired thousands of other Indian coal acres elsewhere under similar terms. The government had set the terms years earlier—before the renewed interest in coal—to encourage federal and Indian coal development.[10]

On the neighboring Crow Reservation in Montana, AMAX, Peabody, Gulf Mineral Resources, Shell Oil Company, and Westmoreland Resources held coal permits and leases for 234,787 acres. Consol held 34,231 acres on the Fort Berthold Reservation in North Dakota. Peabody Coal was exploring 19,000 acres of the Southern Ute Reservation in Colorado. The only producing Indian coal leases were in the Southwest: Peabody's 1964 Kayenta lease with the Navajos; Peabody's 1966 Black Mesa lease with the Navajos and Hopis; Utah International's 1957 lease with the Navajos; and Pittsburg & Midway Coal Mining Company's 1964 lease with the Navajos. Westmoreland's 1972 lease with the Crows authorized construction of a mine on the Crow ceded strip, a 1-million-acre parcel of land outside the reservation boundaries where the tribe retained coal ownership.[11]

History proved that concern for these other Indian coal leases was justified. Aside from the mining that had already begun, none of the Indian coal contracts existing at that time ever produced a shovelful of coal, principally because of tribal opposition. In one case, a 1967 Peabody lease on the Southern Ute Reservation, economics dictated the decision—the site was too far from railroads and population centers.[12] These leases violated the same federal regulations and trust responsibilities listed in the Cheyenne petition: no site inspections, inadequate bonds to assure compliance, no environmental impact statements, vague reclamation clauses, and inadequate royalties. They did not comply with the new federal environmental regulations that had been adopted in 1969. The leases were all large, averaging 23,523 acres (eight times larger than the 2,560-acre limit in the regulations). The Utah International mine on the Navajo Reservation was the largest strip mine in the United States.[13]

Instead of percentage royalty rates, eight of the eleven Indian coal leases had flat royalty rates, usually 17.5 cents a ton, which deprived the tribes of thousands of dollars in potential royalties when the price of coal escalated. The average price of coal at the mine mouth rose from $4.40 in the 1950s to $15.00 in 1974; thus although the average royalty per ton increased only 35 percent, the average value of coal increased 237 percent. The companies obviously could afford to pay more; some companies paid speculators, such as Norsworthy and Reger, 5 cents a ton overriding royalty on top of the tribal royalty.[14]

BIA had not attempted to share information on good lease provisions, and the companies, of course, did not volunteer the information to the tribes. For example, both the Southern Utes and the Navajos successfully negotiated for percentage royalties prior to the Cheyenne and Crow 17.5 cents a ton coal sales. An unnamed officer of the Billings BIA area office responsible for the Crow and Cheyenne coal sales said that he was not aware of the percentage royalty concept, even though it had been adopted for federal coal in 1971. When federal auditors asked officials of the central BIA office in Washington why such information was not shared, they heard a strange, patronizing explanation. If information about mineral deals was shared, BIA officials said, a tribe might believe that what was obtained on one reservation could be obtained on its reservation, without considering different circumstances.[15]

Despite the reluctance of BIA to help share lease information among tribes, the tribes learned from each other through the "moccasin telegraph." Activists on the Northern Cheyenne, Crow, and Fort Berthold reservations talked to others in the Black Mesa area in the Southwest and visited strip mines to find out about potential problems. Scientists with the Northern Cheyenne Research Project and the Crow Research Project shared information. The Cheyennes invited Alvin M. Josephy, Jr., who had been instrumental in exposing the Black Mesa lease problems, to speak in Montana. Council members at Fort Berthold contacted tribal member Thomas Fredericks, who directed the Native American Rights Fund in Boulder, Colorado, to critique the coal leases. NARF had helped both the Northern Cheyennes and the Crows to analyze their coal contracts. Anson Baker, another tribal member and also the BIA superintendent at Fort Berthold, had learned about coal lease violations when he worked for BIA on the Crow Reservation.[16]

Once the tribes learned from one another about the lease problems, they responded somewhat differently. In the end the Northern Cheyennes, Crows, and Fort Berthold's Three Affiliated Tribes succeeded in stopping coal development on their reservations, with the sole exception of the Westmoreland lease on the Crow ceded strip. This battle against inequitable leases tied up tribal coal rights for almost a decade.

THE CONTEXT OF THE TIMES

The tribes' resistance coincided with protests from other westerners. When the Northern Cheyennes filed their petition in March 1973, federal coal leasing was already in serious trouble. Interior Secretary Rogers C. B. Morton had imposed a moratorium in February 1973 as the result of pressure from members of Congress and environmentalists. In June 1973 the Sierra Club and several local rancher-conservationist groups sued Morton, saying the federal government should study the regional implications of massive coal development in the Northern Plains—including the impacts on water— instead of relying upon isolated, site-specific environmental impact statements. The critics said the department was leasing coal far ahead of market demand at bargain-basement prices and without adequate environmental precautions.[17]

The coal boom threatened the way of life of not only Indians but also many farmers and ranchers in the intermountain West. The isolated, secretive project planning posed a cumulative impact to the entire region; residents began to realize this when they discovered the North Central Power Study, an innocuous-looking and little-publicized federal-industrial document that summed up the plans for the Northern Plains. Extolling the advantages of increasing the nation's power supply the 1971 study predicted forty-two power plants in the Northern Plains, with a combined generating capacity of 200,000 megawatts—greater than that of any country in the world except the United States and the Soviet Union. Thirteen of those plants would produce 10,000 megawatts each—five times the size of the Four Corners plant, which was already a notorious polluter. These plants' water consumption would exceed by 80 percent the municipal and industrial requirements of New York City, reducing by half the flow of the Yellowstone River in dry years. Strip mines would consume as much as 175,000 square miles of surface in the region over thirty-five years, and the population would increase by 500 percent. Another federal study predicted sixteen coal gasification plants in the region by the year 2000. In the Southwest, where most of the coal was owned by the Navajo and Hopi Tribes, a consortium of utilities and government agencies planned to add 36,000 megawatts of generating capacity by 1990.[18]

The Organization of Petroleum Exporting Countries (OPEC) embargo in 1973 turned the country's ambition into panic. In November of that year President Richard Nixon declared the nation's intention to achieve energy independence by 1980. To free the United States from dependence upon OPEC's stranglehold, Nixon said the country would rely upon its strongest suit—coal—and double coal production by 1985. As part of the strategy, he intended to abandon price controls on coal, thus improving the coal companies' incentives for patriotism.[19]

Thus the energy crisis undermined the efforts in the West to control energy development. In 1979 it looked as if Washington would use its power to override westerners' resource decisions. President Jimmy Carter proposed drastic measures to push the nation toward energy independence, including an energy mobilization board that would have the power to override state, county, and tribal governments in order to expedite energy projects. Congress narrowly defeated that proposal after western officials became alarmed at the board's proposed powers. Congress and Carter agreed, however, to subsidize synthetic fuels. The Energy Security Act of 1979 provided for putting $20 billion a year for five years and $80 billion thereafter into developing synthetic gas and oil products from coal and oil shale. The Western Governors Association gloomily predicted that forty-one synthetic fuels facilities would result from the subsidies, many of them on or near reservations.[20]

The federal government's blind boosterism of energy development must be considered when evaluating DOI's efforts to promote Indian coal leasing. For example, the Senate Interior Committee told the General Accounting Office in 1974 to review BIA's efforts to encourage development of natural resources, saying the committee wanted the tribes not only to contribute to national production but also to benefit from the employment, income, and economic development that would result.[21] Although BIA was justly criticized for failing to foresee negative impacts on the Northern Cheyenne land and people, Indian tribes and BIA were not alone in their ignorance of such problems from energy development. Nor were they alone in excluding affected people from participation in or information about energy decisions. The nation's leaders and many western communities believed western energy development would bring only needed jobs and money. Despite the passage of the National Environmental Policy Act in 1969, state and federal agencies generally had not adopted the processes that later provided for public participation in decision making.[22]

Faulty Indian coal contracts resulted from ignorance and outmoded policies that did not distinguish between Indian and federal minerals. Although the coal bidders may have been greedy and although their contracts did not abide by all federal regulations, they did conform to the unwritten philosophical orientation of the Department of Interior. It should have surprised no one that the tribes' rebellion caused so much confusion and anger among energy companies.

A FAR-REACHING DECISION

Although the Black Mesa contracts made the problems clear, the Northern Plains tribes were the first that could act to avoid them. The Cheyennes' petition affected not only their own reservation but also several other coal tribes, the federal government's Indian leasing program, and Congress. Most important, it set in motion a tribal movement to slow development of energy minerals. A year after the Northern Cheyenne petition Interior Secretary Morton astounded onlookers by issuing a statement that, in effect, nullified the contracts. Without admitting any wrongdoing by his department, Morton said the acreage of the Cheyenne coal contracts should be reduced and environmental studies prepared. He called upon the companies and the tribe to reach agreement and said he would pay legal costs if the tribe decided to sue.[23]

The Cheyennes decided not to sue, afraid that if they lost, they would end up with six coal mines on the reservation. When the companies realized the Cheyennes were not interested in renegotiating, they supported special legislation—finally passed by Congress in 1980 despite BIA's opposition—that canceled the Cheyenne leases and provided federal coal leases for companies that reached agreement with the tribe. Peabody, AMAX, Chevron, and Consol received federal coal leases and paid the tribe millions of dollars as compensation for its lost opportunities, exploration damage, and legal costs.[24]

The Cheyenne Tribe's decision to get out of the coal contracts and a later decision to preserve its clean air made it famous as the antidevelopment tribe of the Northern Plains, a reputation it never sought. This image haunted it years later when the tribe began considering other coal and oil and gas development options. The Cheyennes opposed the coal contracts primarily because they were unfair; the tribe's unanimity in its opposition to these contracts came from the lack of tribal control. Thus although the Cheyenne leaders were concerned about preserving the productivity of their land, they also wanted to mine the coal themselves. When later proposals for energy development on and near the reservation provided for greater tribal control and more monetary return, they caused great dissension within the tribe.[25]

The Crow Tribe renegotiated its contract with Westmoreland Resources. Unlike the other coal companies with Indian coal contracts in the Northern Plains, Westmoreland had already begun construction of its mine, spending nearly $40 million on a railroad spur line, a dragline, and a large mine facility. The new contract rectified many problems of the previous ones, including providing for a percentage royalty. Westmoreland claimed the new rates made it difficult for the company to compete; its royalties were the highest in the West at that time.[26]

Although the Interior Department issued a statement on Crow leases much like Morton's statement on the Cheyenne leases, the Crow Tribe chose to sue DOI and the other companies in September 1975. The tribe omitted Westmoreland from the litigation, afraid that sympathy for the company's sizable investment would jeopardize the case against the other companies. The Crows' complaint echoed that of the Cheyennes, saying the Department of Interior seemed "unduly concerned for a quick sale of coal" and failed to advise the tribe about economic and business alternatives. Unlike the Cheyennes, Crow tribal leaders—at least one faction—were interested in renegotiating the coal contracts and proceeding with mining. The tribe reached agreement with Shell in 1980 on a large, new coal contract, but it continued its suit against the other companies. In 1984 and 1986 it settled with Peabody and AMAX because the tribe and the companies could not agree upon development contracts.[27]

The Three Affiliated Tribes of the Fort Berthold Reservation in North Dakota asked BIA Superintendent Anson Baker not to allow Consol to mine coal on the reservation, and when Baker complied in 1975, Consol was irate. Consol had spent more than $1 million on exploration under the assumption that the company had a contract with BIA. Normally prospecting permits were automatically converted to leases when coal was discovered. Baker told Consol that its contract was with the Three Affiliated Tribes and the individual mineral owners, not BIA, and the Indians did not want to proceed. Baker's superiors backed up his action. Recognizing the benefits of networking with one another, the coal tribes' leaders were instrumental in founding both the Native American Natural Resources Development Federation (NANRDF) and the Council of Energy Resource Tribes (CERT) in 1974 and 1975, respectively. They began to rely more and more upon each other rather than upon the federal government for information.

As the tribes changed their attitudes toward federal protection, the Inte-

Anson Baker, a Mandan-Hidatsa, served as BIA superintendent on his home reservation at Fort Berthold, North Dakota, before becoming area director in Billings, Montana. (Photo taken in 1981 by Sara Hunter-Wiles.)

rior Department was forced to change its own, especially with the storm of litigation against the department that began in the 1970s. The Cheyenne petition gave national visibility to the Interior Department's Indian mineral practices. Interior Secretary Morton ordered his department to revise its Indian leasing regulations because of the problems the Cheyenne petition pointed out. The department published new leasing regulations in April 1977 that would minimize environmental and cultural impacts and ensure fair market value, but because of many political and logistical problems, the regulations still had not been adopted eleven years later. Despite the lack of revised regulations, Interior policies for Indian leases improved with better implementation of the old regulations, such as the requirement for environmental studies. The Interior Department also raised the minimum acceptable royalty rate for Indian coal to 12.5 percent after Congress raised the royalty rate for federal coal in 1976.[28]

Ranchers and environmentalists in the Northern Plains pushed for a federal law requiring coal reclamation and surface owner protection. The publicity surrounding the Indian coal problems added to the impetus for a federal coal strip mining law, which Congress passed in 1977 with a special section providing for Indian lands. The Indian Mineral Development Act of December 1982 resulted from the awareness, which began growing in 1973, that the tribes could get the control and the monetary return they wanted only by becoming partners in development rather than being passive beneficiaries of royalty payments.

OTHER TRIBES ALSO CALL FOR HIATUS

In the years following the Northern Cheyenne petition many other tribes imposed moratoriums on energy development, either formally by council resolutions or informally by refusing or failing to act on industry proposals. Outsiders frequently misinterpreted these moratoriums, thinking that the tribes were holding out to provide cartel-like power to their organization CERT; that they were acting out of unpatriotic defiance; that they were antidevelopment; or that the tribes simply wanted more money. Tribes that delayed development, however, had four principal reasons: They lacked the data they needed; they were sometimes torn by internal tribal dissent; they were intimidated by the nation's demand for their resources; or they did not want to repeat the mistakes tribes had made in past resource decisions and therefore they felt inhibited. Economic pressures forced other tribes to proceed with development, ready or not.

After discovering how poorly prepared BIA was to monitor environmental damages and to evaluate lease terms for the Consol coal contracts, Fort Berthold's Three Affiliated Tribes imposed a moratorium on both coal and new oil and gas development in 1973. The moratorium did not affect existing oil and gas leases, many of which had been signed in the 1950s. Tribal council members said the tribes did not want any coal development until the federal government improved strip mining regulations, the tribes could participate in a coal joint venture, and the tribes evaluated the desires of their people. The council also wanted a coal management plan, including adequate information on coal reserves, the effects of coal mining, and alternative methods of development. After a mineral engineering and planning firm (hired by BIA) conducted a survey of the Fort Berthold oil and gas potential and recommended which sections should be developed first, the Three Affiliated Tribes lifted the oil and gas moratorium in August 1975.[29]

The Navajo Tribal Council imposed a short moratorium on new energy development in May 1980 while its staff prepared to implement a tribal energy policy. As part of its policy the tribe created a new board, the Navajo Energy Development Authority, through which all new energy proposals were to pass. The council also required the tribal staff to prepare several reports, primarily concerning economics, before the moratorium was lifted. The moratorium did not affect the fifty leases and rights of way then being negotiated. The Blackfeet Tribe imposed a year-long moratorium in 1979 while the tribe prepared minimum criteria for new oil and gas development. Concerned about water pollution from previous development the Cheyenne

River Sioux Tribe stopped oil and gas development for two years until the tribe adopted its own environmental regulations.[30]

Other tribes' moratoriums, such as that of the Hopis, were designed to give the tribes time to get necessary mineral data. Many tribes felt vulnerable without such information. When Indian leaders formed CERT in 1975, they could not get figures from the federal government on collective Indian mineral ownership. Although BIA hired grazing and timber experts and prepared plans for systematic management of grass and trees, minerals were a low priority despite their greater potential value. Some within BIA recognized the need for information. Congressional investigators discovered a June 1974 internal memo that said, "Data on the availability of energy resources on Indian lands is urgently needed . . . in order that the tribes, for economic development reasons, and ourselves, for trust and planning reasons, can intelligently determine priorities, plans, and budgets for the development of these resources."[31]

At that time industry knew more about Indian resources than either the federal government or the tribes did. Although the federal government would explore federal oil and gas lands before scheduling a sale, it did not routinely explore Indian lands before sales, which resulted in lower bids. After exploration the companies would sell geologic information to each other; sometimes they tried to sell such information back to the tribes. Occasionally they also obtained information from illegal surveys on reservations. Information about minerals such as coal and uranium was much more difficult and expensive to obtain than information about oil and thus often not known by either the government or industry.[32]

Few tribes had the money to obtain their own information. On the other hand, the Navajo Tribe, which in 1974 was collecting $15 million from mineral leasing, was hiring geologists to explore before nominating land for lease sales. Phil Reno, an economist with Navajo Community College and one of Navajo Tribal Chairman Peter MacDonald's advisers, said in 1978 that the Navajos had "a good bit of information about the quantity of their resources, a fair amount about quality, and not nearly enough about the value."[33]

To figure out their minerals' value in the marketplace, tribes needed to know the sulfur content and heat rating, production cost, market potential, transportation rates, and what royalties were being paid to other mineral owners, such as private owners and foreign countries. They also needed to understand how public utility commissions indirectly controlled prices paid for coal by controlling electricity rates. Without such information neither the tribes nor the federal government could evaluate bonus and royalty rates, causing tribes to be seriously handicapped at the bargaining table.[34]

Just getting the basic information about quantities of reserves would have been a big step for most tribes. The federal government offered broad estimates of reserves that indicated the vast extent of Indian holdings, but the estimates varied greatly from year to year and agency to agency. The Interior Department said thirty-three reservations had as much as 200 billion tons of coal, which represented as much as 30 percent of the coal west of the Mississippi. Federal estimates of Indian uranium holdings ranged from 16 to 37 percent of the nation's total. The department said forty Indian reservations held reserves of 4.2 billion barrels of oil and 17.5 trillion cubic feet of gas—3 percent of the nation's known reserves. Most of these minerals still lay underground; so even if the tribes had been politically able to operate as a cartel, they could not have influenced energy fuel prices. Nevertheless, they represented the largest mineral owners in the country outside of the federal government and the railroads.[35] Such broad estimates could not be used by either BIA or the tribes for planning. In 1974 the Interior Department began a mineral inventory to get more detailed, reservation-by-reservation information. In later years the inventory did not receive the high priority funding that it deserved and consequently was always several million dollars behind requests.[36]

In addition to not having data, tribes did not have the infrastructures necessary for handling the difficult energy decisions, which contributed to internal dissent and consequently to slowing development. Most tribal governments were established in their new form in only 1934 and thus were still getting on their feet in the 1970s, compared with state governments, which were 100 to 200 years old. Thus tribes generally lacked the administrative, legislative, and judicial structures to deal adequately with many energy problems, especially with resolving conflicting rights of surface users and mineral owners, distributing benefits, and minimizing impacts.

In the 1970s Navajo coal and uranium development frequently was disrupted and sometimes slowed by political protests and lawsuits. Other tribes became increasingly aware of such problems and wanted to avoid them. According to commentator Lynn A. Robbins, by mid-1978 all of the affected chapters (political subdivisions of the Navajo Reservation) in the Four Corners and Crownpoint, New Mexico, areas had passed antienergy development resolutions, opposing coal-hauling railroad spurs, uranium mines, and coal mines. The local Navajo people in areas slated for development had many legitimate concerns that their tribal government had neglected to address. Burial sites had been desecrated by the Black Mesa mine; the tribe had not even consulted with residents of the Shiprock, New Mexico, area to see if they would be willing to move to make way for Exxon Corporation's uranium mining. Residents near Aneth, Utah, suffered from environmental problems and cultural conflicts with oil-field workers while receiving no commensurate benefits.[37]

Plans for seven coal gasification plants drew much of the protesters' wrath because of the potential social impact and the amount of water the plants would consume. Two plants were proposed by Consolidation Coal and El Paso Natural Gas Company (CONPASO) in connection with a controversial coal mine near Burnham, New Mexico. At around the same time that it was rebuffed by the Cheyennes, Consol and El Paso planned the gasification plants near Burnham. Farther north, near Farmington, New Mexico, another four plants were planned by Western Gasification Company (WESCO, a joint venture of Pacific Lighting Corporation of Los Angeles and Transwestern Pipeline Company). The WESCO gasification plans proceeded much further than those of CONPASO, which never got beyond the preliminary planning stages.[38] The Burnham chapter fought both the mine, which was to involve more than forty thousand acres, and the gasification plants. In 1978 thirteen Navajos and the National Indian Youth Council filed a suit in federal court to halt the strip mine. When the lawsuit failed, several Navajos took over the mine site in July 1980 until they were arrested by the Navajo tribal police.[39]

The companies scrapped plans for all of the synfuels plants on the Navajo Reservation, partly because of tribal reluctance and partly because of national politics and economic forces. WESCO reported that it had spent over $32 million planning and designing its proposed plants before they were voted down by the Navajo Tribal Council in 1978 by a margin of 48 to 8. The tribe did not want to pay hundreds of millions to finance construction of and services to a new town of 40,000 for mine and gasification plant workers when most of the revenue would have gone into state and not tribal coffers through taxation. In the continuing game of musical reservations, WESCO moved its synfuels proposal to the Crow Reservation.[40]

At Burnham the protesters' efforts combined with market conditions to block mining indefinitely. Because of the efforts of local dissidents, tribal officials, and federal officials, the final lease agreement included important concessions, such as compensation for people who had to be relocated and numerous reclamation safeguards. Yet in 1989, when CONPASO still had not begun mining, many unemployed Navajos resented these achievements. The companies lost coal sale contracts during the delays, and then the coal market went sour, eliminating the possibility of new contracts.[41]

Why all the protests? Local people had genuine, serious concerns. They did not feel involved in decisions, and they suffered from the impacts of those decisions without realizing benefits. Some protesters felt deep religious and cultural objections to development that disrupted or destroyed their life styles. The question of how to share benefits among tribal members troubled many of the energy tribes. The wealth of Indian resources sharply contrasted with the poverty of the Indian people. On the Navajo Reservation in 1975, $19 million flowed into tribal coffers from mineral leasing. By tradition the Navajo Tribal Council did not distribute royalties to tribal members through per capita payments, as some tribes did. The royalties instead supported government and social services for the members, including Navajo Community College and scholarships. Thousands of Navajo miners averaged wages of $30,000 a year in 1978, but only a small portion of the

Navajo dissidents raise their fists as they leave the courtroom in 1980 after taking over the Consolidated Coal Company mine in Burnham, New Mexico. (Photo by Paul Natonabah.)

147,210 Navajo people held jobs in the energy industries and 45 percent were unemployed. A study by the Shiprock Research Center found that between 1970 and 1978—when tribal energy revenue increased—the rich got richer and the poor got poorer. Although the mean household income increased, the median household income declined to $2,520—$563 less than 1970. Although Navajo energy resources provided enough electrical power for 5 million American homes in Southwest urban areas, only 40 percent of Navajo homes had electricity. To make way for the mines, hundreds of Navajo people had to leave their grazing lands. Forty families had to move from the mine site at Burnham alone.[42]

The protests split communities, turning Navajo miners against their neighbors, slowing development and thus the inflow of revenue, and alienating industry. Yet these protests ultimately paid off. As a result of the efforts of the dissidents and the concerns of tribal officials, the Navajo Tribe developed some mechanisms for increasing public involvement in decision making and improving the distribution of benefits. Although those issues continued to cause problems in later years on the Navajo Reservation and wherever energy development occurred, significant progress was made in the late 1970s and 1980s.

Gradually Navajo officials and dissidents established precedents. As a result of the Black Mesa protests the tribe created the Black Mesa Review Board, an organization of tribal, local, and company officials to address problems. More money from energy projects was funneled to local chapters. Autonomous and partially autonomous organizations, such as the Shiprock Research Center and the DNA-People's Legal Services, began to inform the Navajo public about decisions being made and their ramifications. The Shiprock center disbanded in 1977 when its founder, Harris Arthur, went to work for the Carter administration. DNA (Dinebeiina Nahiilna be Agaditahe—"attorneys who contribute to the economic revitalization of the people") continued to help protect local interests in energy battles. By the time the tribe became a partner with industry in planning a giant coal-fired power plant in the 1980s, the Navajo tribal government was well aware of the necessity of satisfying local concerns.[43]

Regardless of the dispute resolution mechanisms in place at the time, the protests would have been fueled by fears that the Navajo tribal government was playing into the hands of multinational corporations and of an energy-craving nation. John Redhouse, a Navajo and associate director of the National Indian Youth Council, led the effort to challenge the Burnham lease. He said, "When President Carter referred to the implementation of his national energy policy as the moral equivalent of war, we knew then that the Indian wars were not yet over and that we must do battle with the bureaucratic cavalry and the corporate cavalry in order to protect our land, our way of life, and our future as Indian people."[44]

Some tribal leaders themselves wondered just how far the United States would go to get Indian energy resources if the energy shortage was considered a national emergency. Did the tribes really have veto power? Or would financial pressure from Washington be used to force development? Indian advocates were wary of the immense power of the multinational corporations active on Indian lands.[45] The CERT staff and the Department of Energy fed the fears by touting development and boasting about how much more energy would be produced by Indians for the nation.[46]

Tribes knew that a national sentiment to encourage mineral development had resulted in the poor terms of existing federal and Indian contracts. They did not wish to rush headlong into new contracts that might be equally detrimental. Despite the federal and private reports of the early 1970s damning the terms of the old contracts, most of the major coal contracts for producing mines remained in effect well into the 1980s. Unlike federal leases, Indian leases did not come up for adjustment every twenty years. The Interior Department either had removed the adjustment clause entirely from Indian leases or had rewritten it to allow discretionary royalty adjustments, according to the Federal Trade Commission staff analysis. The 17.5 cent coal royalty contracts had met the federal minimums for the time and complied with the national orientation to encourage development.[47]

When coal prices skyrocketed above $20 a ton and the federal minimum royalty became 12.5 percent, the tribes did not want to continue settling for royalties that represented less than 1 percent of the coal's price. Some members of Congress agreed that the contracts should be renegotiated, and they approved canceling the Northern Cheyenne leases that had not been developed. Others frowned upon the idea of Indians "breaking contracts." Senator Peter Domenici of New Mexico said, "If you make a bad deal, you should live with it." Most companies strongly opposed changing old contracts and criticized the Northern Cheyenne and Crow Tribes for reneging on theirs. They had good reason to oppose the changes: Once the royalties had been set and utility contracts signed, the coal companies were required to deliver coal at a certain price to utilities for twenty years or more. The Indian affairs representative for Gulf Oil Corporation, Kent Ware, however, said that tribes should not be criticized for trying to escape bad contracts when market conditions changed given that industry regularly did the same thing. After the price of uranium rose from $6 a pound in 1972 to $41 a pound in 1977, Westinghouse Corporation, formerly the largest purchaser of uranium, brought suit against Gulf and virtually every other uranium company in the world, charging them with illegally raising and fixing prices.[48]

The Navajo Tribe did not tear up contracts. Instead it eventually—twenty to thirty years after the contracts had been signed—coaxed companies to the bargaining table to renegotiate, using whatever political or legal handles it could to get higher royalties, safer reclamation protection, and better labor provisions. One company approached the tribe to renegotiate only because its lease had nearly expired in 1973 due to nonproduction.

Other companies needed tribal approval for acreage increases, plant sites, or transmission-line rights of way. The Navajo and Hopi Tribes eventually convinced Peabody to renegotiate its southern Black Mesa leases because its northern Navajo leases included renegotiation requirements. Peabody and the tribes completed those negotiations in 1987; the other companies completed their Navajo coal lease renegotiations in 1984 and 1985. In each case the Navajos' bargaining position was severely weakened by the lack of readjustment requirements, and the tribe had to make concessions for water and transmission-line rights of way that would have otherwise been unnecessary. Nevertheless, the tribes eventually succeeded in improving the terms tremendously as well as improving their monetary return through taxation.[49]

In the mid-1970s Charles Lipton, director of an independent New York consulting firm and a consultant for the United Nations, appeared on the scene to heap fuel on the tribes' burning resentment of the way they had been treated in the past. Lipton had advised foreign countries in energy negotiations prior to serving CERT and several tribes, including the Crow and Blackfeet, in the late 1970s. He was appalled that American companies were offering the Indians the kind of deals that foreign nations had been rejecting as inequitable for some time. Lipton told the tribes not to accept such "trinkets and lollipops."[50] He said their bargaining position for renegotiations depended upon publicly embarrassing the companies, and he took on that task personally. Lipton was the key speaker at gatherings sponsored by several different Indian organizations, and his advice was widely publicized in Indian journals. Lipton's aggressive, didactic style inspired awe in some Indian people and distrust in others, but it always got attention. His message was not new or unique. Several economists and tribal attorneys had already pointed out the advantages of international energy contract models. Yet Lipton's charisma and wide exposure made him a catalyst for change in the Northern Plains and Rockies. Not surprisingly, Lipton aroused antagonism among energy executives, who grumbled for years afterward that Lipton had encouraged tribes to think they somehow could have the best of both domestic and international minerals agreements.[51]

Ironically, by offering a formula for tribes to use in evaluating proposals, Lipton may have contributed to the lifting of some tribal moratoriums. He suggested eighteen points that ought to be included in any oil and gas contract. In 1979 the twenty-eight tribal chairmen on the CERT board of directors endorsed Lipton's list. Several energy tribes subsequently adapted the suggestions to their own needs and began entertaining development proposals from industry. The Blackfeet of Montana, for example, expanded Lipton's suggestions and adopted twenty-six points, and the Gros Ventre and Assiniboine Tribes of the Fort Belknap Reservation in Montana took fourteen. Lipton's eighteen points covered monetary return (profit sharing, bonuses, royalties, land payments, and water payments); tribal access to geologic and financial data; Indian employment preferences (including provisions for education and job training); and tribal involvement in management decisions. These points not only addressed problem areas the tribes had discovered in the past. They also went a giant step further—they provided tribes with the means to become partners in future development.[52]

The eighteen points represented a formula approach and so were not necessarily ideal for all tribes or all situations. They tended to scare off oil companies because those companies assumed the tribes that used them would not consider negotiating to resolve specific problems. Later, as they became more knowledgeable, tribes abandoned the formula. The eighteen points were important because they offered security—a comprehensible checklist to determine whether an offer was reasonable or not—when the tribes most needed it.

INDUSTRY REACTION

For energy companies interested in Indian minerals the tumultuous 1970s inspired reactions ranging from anger to fear to excitement. Problems between industry and the tribes were occasionally caused by greed and racism on both sides. More often they occurred because the changes had come too rapidly, communications were too difficult between the different cultures, the parties felt hampered by excess federal red tape, or the tribes lacked the information and expertise they needed to be equals at the bargaining table.[53]

With OPEC member countries beginning to take as much as 60 percent of the corporations' net profits, many multinational energy companies felt they were being assaulted from all directions and that their survival—or at least their profit margins—were at stake. Until 1973, companies could acquire Indian coal leases for 17 cents a ton as easily as they acquired federal leases. Suddenly they were faced with tribal moratoriums, followed by demands for a share in profits and management decisions. At the same time new federal environmental laws put increasing restraints on their operations on both federal and Indian lands.[54]

Reflecting the attitude of the Neanderthals in the energy business, a representative of one major company based in Denver said in 1980, "Those dumb Indians ought to be shot for not letting us dig on their lands. Everything—national parks included—ought to be exploited, or we'll retrogress back to the bears." The progressives in the industry heeded the words of Ralph E. Cox, executive vice president of Atlantic Richfield Company, who said companies must learn to deal with rapid change for their own good. When Fortune compiled its twenty-fifth Fortune 500 list, it omitted half the companies that had been on the first list, primarily because they could not deal with change, Cox said. "Rapid adaptation is not easy for an Indian tribe with its traditions, and it's not easy for a large corporation."[55]

As part of their attempt to adapt, industry and Indians tried to learn more about each other. In many cases they started from ground zero. Most American schools do not teach about tribal governments and their unique powers, about concepts such as self-determination and sovereignty, or about contemporary Indians and the fact that each tribe is different. Because the concept of tribal sovereignty was new to them, many industry officials feared it meant that the tribes could expropriate with impunity company mines, wells, and other properties on reservations.

Communication was made more difficult by tough rhetoric on both sides. The Navajo director of economic development, for example, would tell a room full of energy attorneys that he advocated tearing up contracts, not because he really planned to do so but because he believed that overstatement was a valid tool in bargaining. Tribes sometimes sought symbolic coups in contracts, even if it meant the monetary returns would not be as high. Companies also worried about the symbolic importance of contract terms, fearing their peers might think them too generous. One tribe's bluff easily became industry's stereotype of what the Indians really wanted.[56] Before they had access to adequate data and negotiating expertise within their own ranks or BIA, tribes did not always know what they wanted, causing problems for all parties at the negotiating table. Sometimes they made unrealistic demands. Other times they had second thoughts because they feared they might have negotiated too much away. When the tribes' expertise improved, so did their relations with industry.

Many company officials wanted desperately to be able to generalize and say, "This is what the Indian tribes want," or at least to be able to generalize about individual tribes and say tribes X, Y, and Z support energy development, whereas tribes A, B, and C do not. Some tribes could be categorized—they refused to participate in even preliminary mineral estimates. Few tribes, however, took a consistent, hard-line stand for or against energy development. The only generalization possible was that all of the tribes wanted control. They were struggling to determine how much development their culture, their religion, their people, and their land could tolerate. When they hesitated, their uncertainty added to the confusion, especially when, as in some cases, they simply did not bother to respond to industry proposals. When they acted to set limits, their actions were sometimes misinterpreted by people who did not share their values. One skeptical oil man said, "Indians think that sacred land leases for more money per acre."[57]

At other times months of negotiations increased cultural understanding. On the Southern Ute Reservation in Colorado, for example, the tribal council rejected in 1982 an offer with an up-front bonus of $5 million from Amoco Production Company. Amoco regional land manager Claude Neely had negotiated with tribal staff and council members for over a year and a half on an oil and gas contract for a pristine area, and by all accounts it was a good, generous contract. After polling tribal members on the proposal the council rejected the deal, partially because many tribal members wanted to protect that portion of the reservation. Some council members were also reportedly concerned that, because they had not yet planned how to use the money, political pressures would force them to distribute the bonus monies in per capita payments to the members. They remembered the 1950s when many tribal members spent their land claim monies imprudently, alcohol-related problems erupted, and people abandoned their agricultural subsistence way of life.[58]

Neely said that he was disappointed in the 1982 council decision but that he understood it. "They did what I say tribes ought to do—they made the toughest deal they could, they got us down to our bottom line, and then they sought to see if their people would approve it," he said. Amoco and the tribe emerged with positive feelings about each other, and they successfully consummated other deals in 1988.[59]

Many energy company spokesmen found the federal government presented more of a roadblock to implementing energy plans than the tribal governments. Spokesmen for WESCO, Exxon, and Consolidation Coal said the Navajo Tribal Council approved their projects several years before the federal government did.[60] Federal law requires the Interior Department's approval for any actions that result in the sale of "trust resources," that is, land, minerals, or water that are held by the federal trustee for Indian tribes or individuals. Stung by lawsuits and administrative appeals, such as those of the Northern Cheyennes, that discredited their fulfillment of their trust responsibilities, Interior officials naturally wanted to be sure that subsequent deals were fair to the tribes. At times government officials did seem to overuse this authority, delaying approvals because they were reluctant to trust the tribes' abilities or were jealous of tribal initiatives. At other times federal intervention clearly did save the tribes from imprudent decisions.

In 1976, for example, Interior Secretary Cecil Andrus refused to approve the Burnham lease until it had been renegotiated by the Navajo Tribal Council. After the lease was finally approved, the federal Office of Surface Mining (OSM) took an additional fifteen months to approve the mine plan. Marcus Wiley, Consolidation mine superintendent, later said, "If they [the Navajos] had been sovereign, we would have been in business three years sooner." Yet both federal delays seemed well justified. During the lengthy negotiations leading up to the proposed 1976 lease, the federal government had raised federal coal royalty rates, and Andrus thought the Indians should get at least as much as the federal government for their coal. Later, Tribal Chairman Peter MacDonald praised the Interior Department for its role in improving the lease terms. The OSM delay stemmed from serious doubts that the Burnham site, which received fewer than six inches of rain a year, could be reclaimed. When OSM finally issued the permit, it included several unusually tough reclamation stipulations.[61]

Although industry spokesmen often said they would just go to private or federal lands if Indians demanded too many concessions, most did not follow through on such threats. Reluctantly at times, they recognized the extent of the reserves that lay under Indian lands. The companies also knew that, unlike most private owners, the Indians often had the rights to the surface lands and water necessary for the development. As one representative of a major oil company put it, "We wouldn't choose the North Slope of Alaska either, but we go where the oil is."[62]

Deregulation of oil and gas prices also helped. Congress enacted a schedule in 1978 for gradually lifting price controls on natural gas. President Reagan lifted oil price controls on January 28, 1981. The Indians' game rules had changed, but with the rise in oil prices, so had the stakes. Eager companies pushed aside those reluctant to adapt to the tribes' new demands and inundated tribal leaders with development proposals. Industry negotiators attended CERT meetings, hoping to make contact with tribal officials and to learn tips on how to conduct business with tribes. One commentator said half of the five hundred people at the 1981 meeting represented business corporations. Crow tribal officials reported that they received fifteen to twenty proposals in a six-month period in 1980, ranging from oil leases to synthetic fuels plants. Hundreds of companies bid against one another in some reservation lease sales in 1980.[63] Whenever advice was offered on how to get along better with Indian energy tribes, people listened. Bar associations, insurance companies, and industrial and tribal groups sponsored frequent advisory forums in the 1970s and early 1980s.

In 1980, soon after Shell had successfully concluded five years of negotiation for a large coal contract with the Crow Tribe in Montana, company spokesman Kirk Blackard gave pointers at a forum.[64] The fact that Shell and the tribe had succeeded in their negotiations amazed many onlookers who had watched from afar as both parties broke all the rules. Shell was one of the companies included in a lawsuit that the Crows had filed in 1975 charging that its exploration permit was "inequitable and illegal." In its attempt to renegotiate, the company insulted the tribe by sending letters just prior to the Crow Fair promising a bonus of $200 to each tribal member if a new contract was signed. The Crow Tribe has a tradition of factionalism and instability, exacerbated by its system of government. Unlike other tribes, the Crows make their business decisions at quarterly meetings of the general council, at which all seven thousand adult tribal members are eligible to vote. To fulfill his agenda for economic development, a tribal chairman must be sure his or her faction dominates each quarterly meeting. Thus, in effect, he or she is up for reelection every four months, as one commentator pointed out. During a two-year term the chairman exercises the authority of a monarch between the council meetings. At the time of the coal negotiations, two factions of the tribe battled over who was authorized to represent the tribe, ousting the tribal chairman twice in one year and earning a reprimand from the assistant secretary of Interior.[65]

Blackard referred only obliquely to this turmoil in his talk at the Federal

Bar Association conference in April 1980 in Phoenix, saying problems were caused by cultural differences and both parties' initial lack of credibility in the other's eyes. After talking around the table for many months, the Crows and Shell began to understand each other. He said negotiations progressed more smoothly when the tribe gained expertise. Before that time, the tribe's unrealistic expectations caused both parties to take extreme positions.

The stakes were high, and both the tribe and Shell knew the potential benefits made it worth their while to reach agreement. The Crows owned so much coal that the *Wall Street Journal* said they could qualify as the world's ninth largest coal-owning country. Recognizing the potential of Crow coal the Department of Energy awarded the tribe $750,000 in 1980 to study the feasibility of a coal-fired power plant and $2.7 million in 1981 for a synthetic fuels plant study; either or both plants could have used coal produced by Shell. The plants were to be constructed on the reservation. DOE, CERT, and tribal officials touted the ambitious, high-technology Crow plans as a showpiece of industry, government, and tribes working together. The combined Crow plans required more money and resources than any of the other Indian energy plans discussed at that time or since, and the Crow Tribe planned to be a major participant, not just a passive beneficiary of coal royalties.[66]

The synfuels plant was a grandiose plan with proportions typical of the era. It would cost $3.65 billion to build and would consume 7,500 acre-feet of water and 5 million tons of coal per year to produce 125 million cubic feet of synthetic gas per day. Occupying 1,000 acres of land, the plant would provide 500 permanent jobs, with an annual payroll of over $10 million, and thousands more jobs during construction. If the plant was later expanded to its full potential size, most of these figures would be doubled. Aside from tribal ownership, it was basically WESCO's plan moved north, with the same operator (Pacific Lighting Corporation), the same designers (Fluor Engineers and Constructors), and the same main customer (Southern California Gas Company).[67]

The synfuels study concluded that the Crow synthetic gas would be too expensive, beyond the reach of federal price and loan guarantees. The power plant study said the plant was feasible, but the tribe had to find a utility interested in participating in a joint venture. Plans for both plants were hampered by the extra cost of meeting strict air pollution standards. The neighboring Northern Cheyenne Tribe had changed the classification of air over its reservation so that hardly any degradation would be allowed under federal law. Consequently, builders of the coal-fired power plant would have had to spend as much as $282 million for pollution control equipment—24 percent of the cost of the total facility.[68]

To address investors' and industry's concerns, the power plant study recommended forming a relatively autonomous tribal corporation to insulate the development plans from the turbulent Crow tribal politics. A prestigious group of corporate executives and former government officials—including Cecil Andrus, former secretary of Interior, and Frank Zarb, former federal energy administrator—agreed to serve on the board of the Crow Development Corporation, which was formed in August 1982. The bylaws for the new corporation seemed to give these men (none of them tribal members) almost unlimited powers in all aspects of exploration, development, and conversion of coal, oil, natural gas, and minerals and their by-products. The corporation also was given authority to deal in water and water rights and to mortgage property. The executive committee of the tribe was to act only as an advisory board. Many tribal members did not believe in granting such broad powers to the new corporation, and the council dissolved it at the January 1983 quarterly meeting. Although dissidents opposed the synthetic fuels plant, many thought the smaller-scaled and less risky power plant might be acceptable.[69]

Without the expected energy demand and with the insolvency of the Washington Public Power Supply System, plans for power plants throughout the region, including those of the Crows, were put on hold. All the planning money that went to CERT, energy-related firms, and coal consultants through the years inspired Crow Tribal Chairman Donald Stewart to comment, "The only ones who seem to be making money off of our coal are the consultants." CERT and industry had capitalized on the national interest in both Indians and synfuels. Because tribal members were not deeply involved in the studies, they did not gain expertise on the technical aspects. On the other hand, a tribal coal expert, Ellis Knows Gun, said the tribe itself could not have conducted the studies that it needed for planning purposes. The only other alternative to using consultants would have been using the Department of Interior to conduct the studies, and Knows Gun thought the consultants did a better job than DOI would have done.[70]

The shelving of the Crow industrial plans, combined with the downturn of the coal market, scuttled Shell's mining plans. To avoid paying millions of dollars to the Crows for coal it could not sell, Shell tried to renegotiate the contract. When company and tribal officials could not agree, Shell surrendered its rights in December 1985. In effect Shell tore up the contract despite the time it had invested and the $7 million already paid to the tribe in bonuses. Both Shell and the tribe left the door open for negotiating again when economic conditions changed.[71]

ALTERNATIVES TO STANDARD LEASES

Despite the ultimate fate of the Shell-Crow contract it represented a new trend in Indian mineral contracts for tribes and industry. Tribes had found that standard leases forced them to be passive beneficiaries of royalty payments rather than active partners with the opportunity to participate in decision making and in profit making. The only other mineral property owners with holdings as large as the Indians were the railroads and the U.S. government. The railroads, which obtained minerals as part of their land grants during western expansion, made fortunes by developing their own minerals or going into business with other companies. The federal government limited itself to leasing and so, for administrative convenience, Congress limited the tribes to leasing, too.[72]

Although many of the tribes had been negotiating individual variations of the standard lease forms for their minerals for years, until the mid-1970s they had stayed within the general constraints of the leases used for federal minerals.[73] When they heard about Third World energy contracts, more and more of the tribes decided to try them. The Navajos negotiated the first contract that provided them with an ownership option in 1972 with Exxon Corporation for uranium. In 1975 the Blackfeet negotiated an oil and gas contract with Damson Oil Corporation of New York, which provided for joint management and profit sharing. In 1976 the Jicarilla Apache Tribe negotiated an oil and gas contract with Palmer Oil, which later resulted in the tribe taking over ownership of several wells.[74]

The Interior Department vacillated about whether to approve such alternative contracts and finally decided it could not legally do so without congressional·authorization. Beginning in 1949 BIA had refused several tribes' requests to negotiate oil and gas contracts, based primarily upon its interpretation of the 1938 Indian Mineral Leasing Act and individual tribal governance documents. The Omnibus Indian Mineral Leasing Act of 1938 said that oil and gas could only be offered at competitive sales and that other minerals must also be offered competitively—unless the commissioner granted prior, written permission to the tribe to negotiate.[75]

For a few years in the 1970s, however, BIA approved many of the negotiated contracts. In fact, in 1977 and 1980 the department published proposed leasing regulations that recognized the tribes' authority to negotiate contracts.[76] Depending upon local field office attitudes, some tribes had to fight long battles for approval. When the Navajos negotiated their uranium agreement with Exxon in 1972, BIA refused to grant its approval until January 1977, partly because the lease was not obtained by bidding. The tribal minerals department believed that the officials balked because they had been left out of the negotiations and said the delay in payment of the bonus bid cost the tribe $1.6 million in interest.[77]

On other reservations the department tried to impose its preference for oral auctions over negotiated oil and gas contracts. The chief of the bureau's Energy and Minerals Division, David Baldwin, had succeeded in getting outstanding prices for the Osage Tribe in Oklahoma when he served there as superintendent by requiring oral auctions, using written bids as floor prices for leases. In fact, during the oil boom of the late 1970s oral auctions did bring record returns for many Indians, sometimes ten times the bonus bids they previously received. BIA suggested that tribes combine the high bonuses with variable royalties in some cases—16.7 percent escalating to 25 percent, based on the amount of production. If they had been offered several years earlier, the tribes probably would have welcomed the BIA initiatives eagerly.[78] By the late 1970s, however, many of the more experienced tribes were ready for something radically new. They knew that even large bonus payments amounted to only a small proportion of the potential revenue they could earn from production if they shared in the profits. Just as important, alternative contracts could provide greater tribal control over management decisions than leases.

Until 1980 the Department of Interior allowed alternative contracts despite the doubts of some officials. Interior's ambivalence came to a head in September of that year when department attorneys reviewed an oil and gas deal struck between the Northern Cheyenne Tribe and Atlantic Richfield Company (ARCO). Interior Solicitor Clyde O. Martz concluded that the Interior Department did not have the authority to approve the contract. With the decision, Martz called into question the legality of most of the alternative contracts approved earlier. Interior's change in stance shocked tribal leaders. CERT Board Chairman Peter MacDonald said it represented the "final betrayal" by making the tribes' primary goal—developing their own energy resources—illegal. The issue was not a minor quibble. The whole orientation of the major energy tribes had become self-development, toward which they felt they had been making gradual but steady gains.[79]

Interior officials admitted the department's reversal was humiliating. The decision came just one month after the department published the leasing regulations authorizing such agreements and after countless, fervent statements by Interior officials that they agreed with the concept behind them. Interior officials said the ARCO agreement and similar agreements could not be approved because of legal technicalities that could subject the department to litigation. Thus the department's hands were tied by the statutes, and Congress would have to act.[80]

The confusion about whether or not federal law allowed tribes to negotiate angered industry as well as the tribes. Some industry officials, such as Barrie Damson of Damson Oil, had embraced alternative contracts enthusiastically, excited about being pioneers and about helping tribes to become self-sufficient. Federal requirements for competition after Martz's decision forced tribes to go through complicated machinations, laundering agreements through bogus lease sales and manipulating the language to make them look like leases, sometimes adding pages of stipulations to the three-page standard form. Interior's indecisiveness penalized the companies that were willing to work with tribes and break away from the archaic restrictions of the standard leases. Some companies dropped their negotiations in frustration, waiting for Interior to explain its policy.[81]

The oil and gas agreement negotiated between the Ute Mountain Ute

Tribe and Tricentrol Resources presents the most dramatic example of the type of machinations required by Interior at that time. Tricentrol, a British independent oil company, negotiated a 94,000-acre lease with the southern Colorado tribe in 1982. The company and tribal officials agreed upon terms that gave the tribe an outstanding advantage over a standard lease, including an escalating royalty, a work commitment, and a bonus of several million dollars. Despite the fact that the tribe and the company had established rapport through the negotiations and were prepared to work closely together, BIA required a competitive auction, using the negotiated terms as the terms of the sale. Tricentrol, concerned that another company would outbid it, offered even more than it had negotiated—$7.7 million. To the company's surprise there was only one other token bid. Other companies interested in gaining access to Paradox Basin did not bid because they assumed it was a setup and that the tribe would go through the formalities and then reject all bidders. The agreement was finalized just a few months before Congress authorized such negotiations without going through competitive sales.[82]

Interior officials said that only a legal technicality stood in the way of alternative contracts and insisted that they believed strongly in tribal self-development. Behind the scenes, however, they were not nearly as enthusiastic as they appeared. David Harrison, an Osage tribal member and director of trust responsibility in the Interior Department, was one of the main proponents of alternative contracts, along with Tim Vollman in the solicitor's office, who drafted the regulations and later the legislation allowing tribes to negotiate. Harrison said many upper echelon Interior officials opposed allowing tribes to negotiate their own agreements for three main reasons: The U.S. Geological Survey did not feel competent to account for income that went beyond bonus bids and percentage royalties; BIA felt uneasy about its staff's ability to evaluate such agreements—bonus bids were easier to compare; and many people within the agency sincerely believed that competitive sales resulted in the highest possible returns. The first two concerns were legitimate problems—the department indeed lacked the expertise and the staffing necessary.[83]

To solve the impasse, Interior officials met with the Northern Cheyennes and ARCO. When ARCO agreed to assume the risk that the contract would be later ruled invalid, Interior approved the agreement, allowing ARCO to pay the bonus to the tribe and proceed with exploration. Then Interior drafted general legislation to authorize tribes' negotiations. Although many tribal advocates thought the department was wrong and that negotiated agreements of many types were authorized by existing law, most decided to support a new law.[84]

The proposed legislation suffered numerous delays caused by the change from the Carter to the Reagan administration and by substantive questions about the legislation. One of the biggest controversies was the fate of eight alternative agreements previously approved by Interior: Blackfeet-

Damson, Jicarilla-Palmer, Jicarilla-Odessa, Navajo-Exxon, Northern Cheyenne–ARCO, Jicarilla-ARCO, Wind River–Wesseley Energy Corporation, and Navajo–Petroleum Energy. During the period of legal limbo between 1980 and December 1982 several others were not approved by Interior, including the Crow-Shell, three agreements with the Navajos, one with the Blackfeet, and one with the Northern Ute. All involved oil and gas except the Exxon uranium and the Shell coal agreements. In December 1982, more than two years after the Martz opinion, Congress finally passed the Indian Mineral Development Act.[85]

Under normal conditions a two-year delay in Indian energy development would have been less significant. However, conditions were far from normal. The boom of the 1970s had begun to bust. In 1981 it became clear that Americans were actually reducing their energy consumption, partly because of economic restructuring away from energy-intensive uses such as heavy industry. By 1985 Americans were using 15 percent less than what the Ford Foundation Energy Project predicted would have occurred under the most conservative, "no-growth" path possible. Oil and gas prices ended up as small percentages of what had been predicted, with oil peaking out in 1981 at over $40 a barrel before starting a precipitous fall. Uranium brokers' predictions that the uranium market would recover within a year and a half after the Three Mile Island nuclear power plant accident of March 1979 proved untrue; utilities stopped ordering nuclear power plants, and uranium prices fell from $43 a pound to less than $15.[86]

Congress had amended the Clean Air Act to require pollution control equipment, whether or not low sulfur coal was used. Coal also suffered because of the price drop in oil and gas and because Congress in 1986 dismantled the Synfuels Corporation. Only one of the many synfuels plants predicted for the Southwest and Northern Plains regions—the Great Plains Gasification Plant beside the Fort Berthold Reservation in North Dakota—was built to commercial scale. Congress in 1984 created a "clean coal" subsidy program that benefited some gasification projects, but it offered much smaller subsidies than the Synfuels Corporation had.[87]

Before the bust, the West had increased its energy production tremendously but not nearly as much as predicted. The six western coal states' production more than quadrupled between 1972 and 1982, growing to one quarter of the nation's total, and their coal-fired generating capacity increased from 9,000 megawatts to 16,000. Although no parallel figures are available for each Indian mineral, the total revenue from these minerals increased by ten times between 1970 and 1982.[88]

By the time Congress acted on the Indian Mineral Development Act in December 1982, the companies were no longer lined up at the tribes' doors in such numbers. Ironically, many of the companies found themselves in the position of holding contracts with terms that no longer fit the market conditions. Abrogating contracts, which caused such alarm a few years earlier when tribes talked about it, became commonplace. Pipeline companies refused to pay the agreed-upon high prices to producers despite "take-or-pay" contracts; utilities refused to pay contract prices for coal. Companies such as Wintershall (which had purchased Tricentrol) and Shell asked tribes to change the terms of their contracts. The Crows refused to change the terms for Shell, which had not started mining. The Ute Mountain Ute Tribe agreed to allow relief from work commitments for Wintershall Corporation, which had already conducted extensive oil and gas exploration.[89]

Despite the financial problems caused by the drop of interest in new contracts, some tribes said they appreciated the 1980s respite. They still needed time to get ready, and so did the Department of Interior. The department had not adopted regulations implementing the Indian Mineral Development Act of 1982, and it still was experiencing major problems accounting for royalties.[90]

CONCLUSION

The improvements made in tribal control during the 1970s catapulted the tribes far beyond the rubber-stamp era. Because of changing national policies toward self-determination and because of the expertise they found within their own ranks and from non-Indians, many of the tribes succeeded in gaining more power over their resources than was possible before. With their moratoriums Indians proved that at least they had the power of consent, the most basic authority of all. Increases in control sometimes facilitated the tribes' response to constituents; when the federal government had determined the tribe's goals and objectives and provided all tribal funds, it was almost impossible for the tribe to be responsive to its people.

The Indians' expertise as horse traders continued to grow as they inked agreements providing for much better monetary return. They forged new relations with industry and prepared the way for branching into new areas of control, including royalties, taxation, and tribal regulation. Critical to this progress was their new organization, the Council of Energy Resource Tribes (CERT).

CHAPTER FOUR

Indian OPEC? The Council of Energy Resource Tribes

An iron bond has linked Indian people to exploitation, paternalism, and dependency. Development of Indian resources has the potential for breaking that pattern, for balancing social, economic, and political development.

—A. David Lester[1]

"What have we done?" Charles Lohah asked LaDonna Harris as they walked across a parking lot in Washington at 1 o'clock one cold September morning in 1975. About twenty tribes had just agreed to a charter for a new organization, the Council of Energy Resource Tribes (CERT). Although Lohah and Harris remember that their hearts soared like eagles with dreams for the group, they did have misgivings. Lohah (Osage) asked Harris (Comanche), "Is this turkey gonna fly?"[2]

Others had even more serious misgivings. Students of Indian history knew that, although Pontiac and Tecumseh had put together short-lived intertribal alliances, such attempts usually had failed. As individuals, Indians traditionally did not speak for one another. As tribes, their differences often outweighed the desirability of bonding together against a mutual enemy. Although the energy tribes shared some common threats, many remembered battling over hunting grounds in the past or, more recently, over water. The National Congress of American Indians (NCAI) had survived since 1944, but many other intertribal groups had disintegrated. The energy tribes had different goals in 1975. Some tribal chairmen felt a mandate to encourage development and bring jobs and money to their reservations in the near future. Others felt compelled to stop inappropriate development.

As it turned out, these fears were realized in the coming years. Some early CERT staff members and board members tried to speak for the other tribes without authorization. They forgot the importance of focusing on political and social development as well as economic development. They created the public image of a cartel of rich, greedy tribes and later of a broker of Indian resources promoting energy development. The organization's dependence upon federal funds contributed to the image problems and undermined the tribes' efforts to set their own agenda. Nevertheless, the tribes had created CERT themselves, and when they became unhappy with its direction in 1982, they changed it. CERT survived because the energy tribes needed to share information, obtain scientific expertise, and gain political clout, and the organization enabled them to do so.

THE BIRTH OF CERT

For years critics portrayed CERT as a creation of the federal energy department. Yet its history shows clearly that the tribes formed CERT of their own volition. Leaders of energy tribes felt an urgent need to work together. With the energy crisis exacerbated by the 1973 embargo of the Organization of Petroleum Exporting Countries (OPEC), the nation had focused upon western resources—especially coal—to reduce its dependence upon foreign oil. Because they had vast quantities of energy fuels under their lands, Indians faced pressure to develop. The coal tribes had come to realize the serious economic, environmental, and social problems that energy development could bring if it was not properly controlled. They had won crucial battles against exploitation with the help of their attorneys, but they needed the political clout of collective action. The tribes also wanted to share technical experts, whom they could not afford alone. Equally important, as the coal tribes had recently learned, they could learn from one another how best to protect their interests.

In 1974 twenty-six tribes of the Northern Plains made the first attempt to form an organization, the Native American Natural Resources Development Federation (NANRDF). Many of the same tribal leaders were later to form CERT and abandon NANRDF. Their NANRDF founding document—the "Declaration of Indian Rights to the Natural Resources in the Northern Great Plains States"—emphasized water and agriculture as well as minerals. It also documented the threat posed to the tribes' water and other rights by energy development and stressed the tribes' resistance. The two organizations' reasons for being and their goals were similar: to obtain data, to share expertise, and to help the tribes protect their rights.[3]

With their ambitious agenda the Northern Plains tribes soon recognized the desirability of a national Indian energy organization, especially given that several of the nation's top energy producers at that time—the Navajo Tribe, the Hopi Tribe, and the Laguna Pueblo—were located in the Southwest. In the Northern Plains, Indian energy development had been limited to oil and gas. Coal development, with the exception of one mine, had been stalled. The Northern Plains and southwestern tribal leaders who attended the early meetings of the new national organization remained prominent in the Indian energy field for years to come, including Northern Cheyenne President Allen Rowland, Navajo Tribal Chairman Peter MacDonald, Blackfeet Chief Earl Old Person, Laguna Pueblo Governor Roland Johnson, Fort Peck Tribal Chairman Norman Hollow, and Jicarilla Apache Vice President Dale Vigil. The other charter members of CERT were from the Hopi Reservation in Arizona; Nez Perce and Fort Hall reservations in Idaho; Crow, Fort Belknap, and Rocky Boys' reservations in Montana; Acoma Pueblo, Jemez Pueblo, Santa Ana Pueblo, and Zia Pueblo in New Mexico; Uintah and Ouray Reservation in Utah; Cheyenne River Sioux Reservation in South Dakota; Colville, Spokane, and Yakima reservations in Washington; Fort Berthold Reservation in North Dakota; Wind River Reservation in Wyoming; and Southern Ute and Ute Mountain Ute reservations in Colorado. After the organization had been named, Vigil remembers MacDonald passing around a box of Certs, the breath mint, saying, "This is what we will be called."[4]

The tribes agreed upon objectives and sent them to Federal Energy Agency (FEA) Administrator Frank Zarb on September 18, 1975. That mission paper included no threat of a tribal cartel. Nor did the tribes express a wish to rush headlong into development to get rich from their resources. Instead the tribes focused on their right to ownership and control of the coal, uranium, oil, and gas beneath their lands: "If our energy resources are to be developed at all, they are to be developed with our informed consent and participation." They agreed with the nation's goal of making the United States independent of foreign energy sources, but they said emphatically that they did not want to be victimized by it: "Historical and economic forces have combined to create these problems for the people of America and the world. We have combined to see that these same forces, which we have dealt with before under many forms other than energy, do not cause the United States to react in its historical pattern of wasteful and unlawful exploitation of Native American resources for immediate needs."[5]

This original document showed remarkable foresight. Although energy markets, tribal capacities, and federal funding priorities changed drastically during the intervening years, the recommendations submitted to Zarb named the same goals and objectives that still drove CERT in the late 1980s. The tribes said in 1975 that the resources should be administered in ways to improve the economic, environmental, and social conditions on reservations. They asked for funding for mineral inventories so they would not have to conjure up figures on what they owned. They told Zarb that they had formed CERT because they wanted resource-related education; a clearinghouse for exchanging information; help arranging financing for development; expertise on alternative contractual arrangements; studies on using resources on the reservations rather than continuing the tribes' role as colonies exporting energy; impartial environmental studies; and a means for advising the federal government about Indian energy development.[6]

LaDonna Harris was instrumental in getting CERT off the ground by contacting FEA, a connection that proved crucial to the organization's existence for the next decade. Harris founded Americans for Indian Opportunity (AIO) in 1970 to focus on tribal self-government. She became concerned in the early 1970s when AIO studied the tribes' poor resource agreements and discovered that "in the whole world, there were no contracts as bad as the tribes' leases." As a member of the FEA Energy Consumer Advisory Council, Harris had come to know Zarb and several of his staff, including Ed Gabriel, who had decided Indians should be represented on the FEA advisory council and had arranged for Harris's appointment.[7]

To find out how much oil, gas, and coal the tribes owned, AIO had hired three interns from Dartmouth University to go through federal records prior to the first CERT meeting. From the scanty information they found, Harris "made up" some percentages and marched into Zarb's office. "You can't have an energy policy without Indians; collectively, they're the biggest private owners of energy in the country," she told Zarb. She claimed the tribes owned one-third of the nation's low sulfur coal, one-half of the uranium, and large amounts of oil and gas (CERT used these figures for many years). The federal government's vague estimates through the years sometimes agreed with Harris's, with the exception of uranium. The federal Department of Energy (DOE—a successor of FEA) estimated in 1979 that Indians held 37 percent, not 50 percent, of the nation's uranium, and CERT revised its promotional material accordingly a few years later.[8]

Apparently impressed with Harris's figures, Zarb agreed to bring the tribes to Washington in September 1975 to form an organization. After the first day of the initial CERT meeting, Harris took a small group of volunteers to her office to mold the ideas of the tribal leaders into a charter. She remembers Lohah (then with the Native American Rights Fund) lying on the floor with his green tennis shoes propped on a desk, dictating, while a "little old attorney" for the Jicarilla Apache Tribe, Robert Nordhaus, typed. She did not know then that their volunteer typist was an already renowned attorney who was later to win a landmark Indian tax case in the U.S. Supreme Court. Lohah later served as the acting director for CERT until it received federal funding in 1977. The organization received this funding because of its connections with FEA. Although the ties with FEA and other federal energy agencies aroused considerable criticism of CERT, these agencies provided CERT with much financial help and moral support for many years.[9]

NANRDF lost its federal funding after CERT was formed. Northern Cheyenne President Allen Rowland—a founder of both organizations—said NANRDF died because the tribes did not support it. The principal NANRDF founder, Thomas Fredericks (Mandan-Hidatsa), believed that CERT had swallowed up his organization, and he resented it. He continued to believe a regional consulting group could more effectively provide technical expertise needed in local areas. The rift between CERT and Fredericks seriously threatened the organization in 1980, when Fredericks became the assistant secretary of Interior in charge of Indian affairs.[10]

From the beginning the Bureau of Indian Affairs resented CERT and attacked the use of federal funds for the tribal organization. At nearly every

CERT annual meeting for several years, a BIA official would tell the organization that BIA planned to stop providing funds to CERT, and an official of the Department of Energy would commend CERT for the fine job it was doing and promise continued funding. Many BIA officials felt threatened by CERT. Its very existence challenged the adequacy of BIA's services. Tribal leaders often said that if BIA had been doing its job, the tribes would not have needed CERT.[11]

In response to the tribes' requests FEA initially agreed to fund a $250,000 study. Instead of cooperating with FEA's plan to infuse new money into Indian resource data, BIA demanded that the General Accounting Office block any FEA appropriations for an inventory or even for a review of the literature on tribal resources, saying the inventory was not one of FEA's mandated responsibilities and that BIA was conducting its own review of available literature. Prior to 1974 BIA had not attempted in more than a general way to determine the extent of tribal mineral ownership, but just prior to CERT's founding the Interior Department had initiated an interagency mineral resource inventory. The mineral inventory was an absolutely essential first step for the tribes to begin to control development on their reservations. As CERT consultant Charles Lipton frequently told the tribes, "You have to know whether you have a Volkswagen or a Rolls Royce before you know what to sell it for." Although BIA continued the inventory work in later years, it was perpetually underfunded.[12]

Under protest BIA provided a portion of the interagency grants and contracts that supported CERT each year, starting in 1977. Frequently BIA denied funds to tribes, saying the money had gone to CERT. One BIA official who used that argument was Ken Fredericks, brother of Thomas Fredericks. Ken Fredericks (Mandan-Hidatsa) admitted that BIA had failed to fulfill its responsibilities in the past, but, he said, giving so much money to CERT necessarily limited the bureau's ability to do so in the future. He believed BIA should not transfer its duty to serve all tribes onto the shoulders of a private organization that served only its members. Fredericks argued that federal funding created a conflict of interest within CERT: "Anytime you start getting money from the system you're criticizing, you're beholden to them."[13]

No matter what might have motivated the general BIA resentment of CERT, Fredericks's criticism was warranted. Problems caused by dependency upon federal funds plagued CERT throughout its history. Although CERT's membership expanded beyond the original twenty-five founding reservations to forty-three in 1988, it never encompassed all tribes with energy resources. Thus despite BIA's inadequacies, many tribes and especially allottees still had to rely upon the bureau. In an ideal world, funds directed to CERT would not have affected BIA. In the world of continuing federal budget cuts, however, they were directly related.

CERT'S OPEC ERA

When CERT was organizing and fighting for funding, the public all but ignored the fledgling tribal organization. Then in March 1977 in Phoenix, Peter MacDonald compared CERT with OPEC, and headlines across the country shouted the news. The tribal leaders who made up the CERT board had elected MacDonald, a powerful speaker and master at his own aggressive, quotable style, as their chairman in 1976. In his historic speech, this leader of the largest Indian tribe in the nation slammed both industry and government:

> The DOI and the BIA signed perpetual leases which gave companies the right to extract all of our coal at 15 cents per ton for as long as it lasted. Why? Because we lacked the expertise, the knowledge, the information, and the sophistication necessary to protect ourselves. We also lacked the legal ability to do it because we were wards of the federal government, presumed incompetent to act on our own behalf. That was until recently. Now, as some of you may know, a dozen Indian nations have formed a domestic OPEC. We call it CERT.[14]

MacDonald pointed out that the tribes had suffered from the same poverty and resource exploitation that OPEC countries had. The media focused not upon the similarity in backgrounds but instead upon the implications of a domestic energy cartel because MacDonald had threatened to withhold tribal resources if the tribes did not receive the federal assistance they needed: "We ask now quietly and constructively. We will not ask much longer. We will withhold future growth at any sacrifice if that is necessary to survival."[15]

In retrospect MacDonald's rhetoric was not any more threatening—or any more articulate—than that in the NANRDF declaration of rights. Because he was MacDonald, however, his words were heeded. The public began to notice CERT. Seeing how voraciously the press reacted to the OPEC analogy, MacDonald repeated the implied threat, saying CERT was meeting with OPEC representatives for information, technology, and financial assistance because the U.S. government had been so unresponsive. The tactic seemed to get results. In November 1977 the federal government provided funds to hire CERT's first executive director, Ed Gabriel. At FEA, Gabriel had directed the office of impact analysis and served as acting director of the technical assistance office. Having played a key role in CERT's proposal at the agency, he was a logical choice. In December CERT opened its Washington office and the following year opened a second office in Denver.

In 1979 CERT hired Ahmed Kooros as its chief economist, which furthered the OPEC image because Kooros had served as Iran's deputy minister of economics and oil under both Shah Mohammad Reza Pahlavi and

Lucille Echohawk, a Pawnee, confers with Ahmed Kooros, formerly of Iran, at the CERT meeting in 1986. On the right is Glenn Ford of the Spokane Tribe. (Photo by David Hardesty.)

Ayatollah Ruhollah Khomeini. Kooros had been instrumental in arranging the 1971 Tehcran agreement in which the Persian Gulf oil-exporting countries raised the price of oil. The decision to hire Kooros was not a publicity stunt, however, and the years have proven that it was one of the organization's wisest decisions. Educated in the United States before serving in the Iran government, Kooros came to CERT eager to tackle a challenge even bigger than what he had faced in the Middle East. He believed that American Indians in the 1970s could be compared to the OPEC countries in 1968 in two ways: They had a similar level of underdevelopment, and they had the resources to do something about it. Kooros wanted to help the tribes get benefits commensurate with the real price of their energy resources. Still with CERT in 1989 he had worked for the organization longer than any other employee, turning down many more lucrative offers, some from OPEC members. During those years his negotiating skills had helped tribes gain millions of dollars more than they were originally offered for their resources.[16]

Soon after MacDonald called CERT a "domestic OPEC," some tribes and outsiders accused CERT of being unpatriotic. Six members of Mac-

Donald's own tribal council wrote to Interior Secretary Cecil Andrus expressing concern about talks with OPEC. MacDonald said the tribes were seeking a self-sustaining economy—the essence of capitalism and hardly something that could be termed unpatriotic. MacDonald also frequently pointed out his own claim to patriotism. After enlisting in the marines at the age of fifteen, he became the youngest of the famed Navajo Code Talkers, whose code was the only one in the annals of World War II that was not broken by the Japanese. Kooros also defended CERT's patriotism, saying he did not believe that helping the tribes get fair prices would hurt the United States. He pointed out that low prices only encouraged consumption and thus the rapid depletion of oil. Kooros said that, unlike OPEC, the tribes wanted to help the United States meet its energy needs: "The OPEC countries wanted to get even with the West, but the American Indian doesn't want to do that. I think they want to prove to the American people that, even though they were pushed into a corner and abused, they now can come to the rescue of the American people. And they take pride in that. They really do."[17]

CERT needed all the public relations it could muster to counter the OPEC image. When Kooros was hired, the *Denver Post* wrote a particularly vitriolic editorial questioning how the tribes could demand $1 billion in federal funds to help the Indian tribes "export" their oil to other Americans at the OPEC price:

> Supposedly we are to pony up cheerfully so the noose of escalating energy prices can be tightened around our necks. . . . *The people who manipulate Indian policies* are indulging in much nonsense. . . . Admittedly, justice has not always been dispensed equitably. But it is the sufferance of our national government—dedicated to tribal advancement—that gives the tribes leeway to act with more independence than other Americans.
>
> But limits there are. Imagine what would happen if *some adviser* persuaded a tribal group to sign a treaty with Libya under which Colonel Quaddafy was to ship Russian missiles to the reservation to guarantee the tribe's integrity? (Emphasis added.)

The editorial went on to suggest that tribal groups pursue development without seeking "disproportionate benefits." The tone of the editorial confirmed the truth of its final message: "The energy crisis is too important for confrontational politics, which, if pursued, likely will boomerang and hurt the Indian cause rather than help it."[18]

By the end of 1979 the OPEC image was clearly an albatross for the organization of tribes, especially in their pursuit of federal funds. In November an Iranian mob seized fifty-two employees at the U.S. Embassy in Teheran after President Carter allowed the deposed shah to enter the United

States. Americans' outrage brought bursts of patriotic fever, as in war times, eventually contributing to the electoral defeat of President Carter. Although the comparison between the past exploitation of OPEC countries and of the tribes was accurate and helpful, the cartel threat was foolish. Later, CERT discovered another unfortunate residue from the Indian OPEC–image era. After reading articles about "fuel-rich tribes," the public and some of the nation's leaders perceived tribes as affluent—or at least potentially affluent—and imagined tribal sheiks riding in limousines paid for by American consumers. President Reagan told an audience in Moscow that some Indians had become "very wealthy" pumping oil. Federal officials later justified massive budget cuts to energy tribes by reminding them of their vast mineral wealth.[19]

The impression of affluence, which sometimes encompassed all tribes and not just CERT members, could not have been more untrue. Tribes had only different levels of poverty. The 1980 census found that 40 percent of reservation Indians lived at below the poverty level, compared with 12 percent of the country as a whole, and the Northern Plains and Navajo Indians exceeded even the 40 percent level. A BIA task force concluded that a small percentage of Indians—less than 15 percent—lived on reservations where natural resources could ever make a significant contribution to their economic development. Because of the ratio between the large population of some energy reservations and their energy income, only two reservations—the Osage and the Jicarilla Apache—had incomes that represented more than $10,000 per capita per year.[20]

CERT AS A BROKER OF INDIAN ENERGY

In 1979 MacDonald abandoned the OPEC image and the threat of withholding resources to drive prices up and instead took the opposite tack. In a July 20, 1979, letter to President Jimmy Carter, he promised that in exchange for a few million dollars the tribes could increase their production to relieve the nation of its dependence upon foreign oil. President Carter's historic energy message a few days earlier asked the nation to commit $140 billion as an investment in national energy security. The money would be used to reduce the country's dependence upon foreign oil by 2.5 million barrels of oil per day by 1990. Carter had proposed major new initiatives to encourage domestic energy production, including overriding regulatory obstructions with an "energy mobilization board." A new synthetic fuels corporation would subsidize development of synthetic fuels—such as gas from coal—and oil shale.[21]

Conveniently forgetting his own use of the OPEC image MacDonald told the president, "I must ask why—at the very time you speak of unity and of forging a national energy consensus—you have appeared to look upon

the Native Americans of these United States as foreigners." MacDonald
went on to say that Indian resources could contribute the equivalent of an
additional 2 million barrels of oil per day by 1990 if Carter would put one-
half of 1 percent of his proposed national energy security budget—$700
million—into Indian energy development programs over the next ten years.
Explaining his proposal to the National Press Club, MacDonald offered a
promise and a threat: "Today I offer my support, *and that of the 24 other
CERT energy-producing tribes*, to the President and his Administration,
and will await his direction. But we will not wait forever" (emphasis
added).[22]

MacDonald's rash promise was based upon very rough estimates of
tribal reserves, which no one could either challenge or substantiate with
available data. Even if they were accurate, MacDonald implied that CERT
could speak for all its member tribes' elected leaders, that their constituents
would support such development, and that the tribes could speak for individual Indian mineral owners, the allottees. None of these three assumptions
proved to be true. Nevertheless, the letter got results. Newly confirmed Energy Secretary Charles Duncan met with CERT executive committee members within a month after MacDonald's letter. DOE took an interest in
CERT after the MacDonald letter because it was a well-thought-out offer to
help and because it made the agency aware that it would be difficult to develop energy in the West without involving the tribes or tribal water, according to Duncan's assistant at DOE, Richard Stone.[23]

At the annual meeting in Phoenix in December 1979 Stone promised
$24 million in federal energy aid to CERT members for the following year.
In a speech remarkably free of threats, MacDonald reiterated his promise
that the member tribes were "posed to make a massive contribution toward
the national effort." CERT unveiled six new energy projects on member reservations, including two large, coal-fired power plants, a refinery, a hydroelectric project, and a coal gasification plant. It distributed a chart showing
the member tribes' energy reserves, which critics of CERT referred to sarcas-
tically as the "Sears and Roebuck catalog of reservation resources."[24]

REASSESSING THE MISSION

Judging from the media blitz, it appeared that CERT had reached new
heights of success. Yet behind the scenes there were rumblings in the tribal
ranks and growing distrust by outsiders. A group of people from the Big
Mountain area of Arizona had staged a protest at the Phoenix meeting that
the CERT staff tried to ignore. The group, which included three Indians and
one non-Indian, had claimed to represent thousands of Navajo and Hopi
shepherds whom they said were being driven out of their traditional lands to

Ed Gabriel (left), CERT's former executive director, listens with Don Ami, a Hopi, at the 1986 CERT meeting. (Photo by David Hardesty.)

make way for a strip mine. They said CERT "was a front being utilized by the multinational corporations and the U.S. government . . . to justify the land rip-off and to disguise the corruption and the political manipulation of the resources of the indigenous people."[25]

Several articles in 1980 in *Akwesasne Notes*, a national Indian newspaper, created an uproar in Indian Country when they accused CERT of being a broker of Indian resources. These critics and others believed that CERT represented a centralized negotiating arm for the tribes, an image that Mac-Donald had fostered with his all-encompassing offer to President Carter. At one CERT meeting soon after Ahmed Kooros was hired, an *Akwesasne Notes* editor accused Kooros of acting in a "neocolonial role," a charge that deeply offended Kooros.[26]

Three years later Gabriel said the Phoenix meeting was a low point in the organization's history. "We got what we asked for, but it took us more than a year to recover," Gabriel said. He admitted DOE had asked Mac-Donald to cool his rhetoric in exchange for the agency's help even though the $24 million that DOE announced was largely not new money; from one-third to one-half of it would have gone to Indian projects anyway. CERT came out of the meeting looking like an elitist broker of Indian resources, ignoring the interests of grassroots tribal people and prostituting its members' land and people in exchange for energy agency dollars. This image haunted the organization in 1980 when LaDonna Harris ran for vice president of the United States on the Citizens' party ticket, and again in 1982 when Peterson Zah, a man with environmentalist leanings, was elected chairman of the Navajo Tribe. Once again CERT had gone overboard.[27]

The Phoenix meeting's aftermath generated soul searching among the members as they tried to answer three questions and determine how much of the problem was merely image and how much was substance: First, did CERT promote energy development and was it because funding by DOE dictated its agenda? Next, was such a bias reflected in its technical assistance, or did CERT's staff offer objective, quality technical assistance? Finally, who was in charge—the board as a whole or MacDonald and Gabriel?

Tribal members began asking their leaders whether CERT controlled their resources. Some feared they would be locked into a CERT development plan. A staff member for the Cheyenne River Sioux Tribe said some members feared CERT would somehow force their tribe to develop the reservation's uranium. In fact, he said, the reservation's uranium was not worth developing, but the fear overrode his assurances to the contrary, and the tribal council voted to withdraw from CERT. As a result of the broker image and of conflicts between MacDonald and other chairmen, several tribes withdrew from CERT. By December 1980, membership had dropped to twenty-five reservations, down from twenty-eight in January. Meanwhile CERT retained its largest staff ever—sixty-five people.[28]

Some tribes recognized that the organization suffered from serious problems but decided to tackle the important ones from the inside. Caleb Shields of the Assiniboine and Sioux Tribal Council on the Fort Peck Reservation in Montana said his council appreciated the technical services CERT offered. Although he thought the board should involve tribal councils more in decision making, he disputed the charges that the organization served as a broker for tribal resources and that the board lacked control. "It would mean 25 tribal chairmen were being led around by the nose. I know many of them are pretty smart cookies," he said.[29]

In retrospect it is easier to evaluate the criticism. As suggested by Shields, some of the criticism stemmed from the critics' inclination to perceive tribal leaders as naive victims of the dominant society rather than as "smart cookies" making conscious choices. The *Denver Post* editorial referred to the "people who manipulate Indian policies." *New Age* went even further in October 1980 in an effort to discredit CERT and LaDonna Harris's campaign for vice president. Showing their own ignorance of tribal governments and of differences among reservations, the *New Age* editors said:

LaDonna Harris was instrumental in setting up and organizing the CERT *and tribal council system.* . . . The tribal councils were set up by the U.S. government to preempt the traditional governments of the native people. Harris and the other supporters of CERT and the tribal councils represent a puppet government imposed upon American Indians. To say that these bodies were organized to "wrest back control" of Indians' lands and resources is a complete fabrication; these bodies were set up to take the control of tribal lands out of the hands of the traditional leaders and place that power in the hands of a few "elected" Indian representatives. . . .American Indians throughout the continent are calling out for support in combatting the imposed puppet rule of the tribal council, and LaDonna Harris stands as a symbol of those imposed regimes.[30] (Emphasis added.)

Although some traditional people on certain reservations objected to the tribal council system, most reservation Indians saw it as the only way for tribes to control their resources. Harris was indeed instrumental in setting up CERT, but she was only three years old in 1934 when the Indian Reorganization Act provided for the modern tribal council system.

Beyond such patronizing approaches lay the substantial question of CERT's role in energy development. Did CERT promote energy development? Did its attempts to gain federal funding compromise the integrity of its services to the tribes? Both MacDonald's rash promises during the Phoenix meeting and the organization's annual reports to its funding agencies gave credence to such charges. The fiscal year 1980 report, for example, said CERT assisted the Crow Tribe in developing a coal mineral inventory proposal that was "instrumental in the Crow Tribe's decision to pursue the development of a synfuels plant."[31] At the time, DOE clearly was a favored and well-funded agency under President Carter, and Ed Gabriel was known for his ability to write attractive grant proposals. The tribal leaders on the board themselves had years of practice at opportunism. Each time the federal agenda changed, they tended to change their own tribal agendas in order to attract the federal funding necessary to keep their tribal governments functioning.

Asked about the accusation that CERT and DOE were conniving to exploit the tribes' resources, Richard Stone of DOE said that the critics "see shadows under every bed. . . . They need perspective on the way the real world works." At that time, however, many westerners felt justifiable alarm at the nation's hysterical push for development. Carter had proposed giving the Energy Mobilization Board immense powers to override western states' and tribes' authority, and Congress only narrowly defeated the proposal in late 1980. To its credit CERT fought alongside western states to protect state and tribal regulation authority, rather than worrying about offending the president or DOE.[32] On the other hand, CERT and Congress supported

Carter's federal grant program for synthetic fuels, and CERT succeeded in its efforts to get tribes included as potential benefactors. The CERT organization profited from that effort when DOE awarded a $2.7 million feasibility study grant to the Crow Tribe, which hired CERT to manage the study.[33]

For many years DOE carried the largest share of CERT's budget—close to $1 million in 1979 and 1980. Asked about the interrelationship between the tribes' goals and DOE's, Stone said he believed both benefited from the ties. "From their perspective, energy offers one of the last potential ways to develop their economies that doesn't depend upon a handout, which they don't want. From our perspective, we're accomplishing some social good while also advancing the development of energy." Nevertheless, he admitted that DOE funding persuaded some tribes to focus on innovative technology such as coal gasification, which required huge amounts of money and Indian resources, instead of upon more practical applications of conventional technology, such as coal mines or hydroelectric plants. Federal subsidies generally were not available for conventional technology projects.[34]

In 1978 the federal Office of Surface Mining (OSM) awarded CERT a large contract to study issues related to tribes' taking over regulation of coal surface mining on their lands. The study was conducted by the coal tribes and Doug Richardson of the CERT staff. It represented one of the organization's few comprehensive attempts during that period to focus upon tribal governmental powers instead of just economic development.[35]

Whether or not individual feasibility studies were affected by CERT's quest to impress DOE is not easy to judge. Of ten tribes contacted in 1980–1981, six expressed unequivocal satisfaction with the technical assistance they received. The scientists of the CERT staff who wrote the studies clearly did not always agree with management's priorities. Although CERT's management touted the potential for a coal-fired power plant on the Southern Ute Reservation, the staff prepared a "prefeasibility analysis" of coal options in 1981 that did not attempt to obscure the negative environmental impacts of such a plant. The study said, "Generating plants tend to be dirty environmentally, releasing large amounts of pollutants into the environment. These releases can be properly controlled but at a high cost." The study found economic conditions at that time did not favor any of the options, which included a mine, different-sized power plants, and a synthetic natural gas plant. A former Southern Ute staff person who favored a methanol-to-fertilizer plant said, however, that he felt CERT had skewed the results of a study on that proposal. Another former Southern Ute staff member said young CERT staff members had tried to obscure their ignorance on some aspects of a study.[36]

In contrast to the complaints of other critics, some ex–CERT staff members and oil and gas company officials believed that CERT was too environmentalist in its leanings. They objected to the development moratoriums imposed by several member tribes as they worked with CERT to decide upon acceptable development terms. CERT encouraged terms that provided the tribes with far more monetary return and control, to the consternation of many industry people. Through the years, CERT consultants and staff members frequently advised member tribes, "If you can't sell it for what you want, sit on it." After being fired, five disgruntled ex-staff members for CERT said in 1980 that they thought CERT had a bias against development and should have done more to expedite it. Initially excited about the prospect of being involved in developing the biggest unexploited reserves in the country, these five people later became upset with CERT's "anti-corporate attitude," which insinuated that every company in the United States was "trying to rip them off." They said that when their technical reports were reviewed by nonscientists on the Washington staff, they were changed to "justify certain preconceived notions" before being given to the tribes.[37]

If CERT had attempted to influence tribes to drastically increase their energy production, it failed in this mission. In the first place, CERT created false illusions of potential increases in energy production in MacDonald's letter to President Carter and in the Phoenix meeting boasting. CERT implied that it could speak not only for all member tribes but also for allottees. Carol Connor, an allottee on the Fort Peck Reservation and then an attorney with the American Indian Law Center, persistently pointed out that allottees owned as much as 40 percent of the minerals CERT claimed to control.[38]

Nor could CERT speak for the tribes, especially before completing any feasibility studies. By 1988 none of the six projects heralded at the Phoenix meeting had been built nor were they likely to be built—partly because of falling energy prices in the 1980s and partly because of tribal resistance. The Nez Perce Tribal Council, for example, chose not to develop a 60-megawatt hydroelectric plant on its reservation in Idaho, one of the projects listed at Phoenix. Nez Perce Tribal Chairman Wilfred Scott said a CERT study concluded the dam would harm the salmon and steelhead fishery. Southern Ute Tribal Chairman Chris Baker said his tribe decided to forego its coal-fired power plant—another of the listed projects—because it was not economically viable nor environmentally acceptable at the time.[39]

MacDonald had promised President Carter additional production equivalent to 2 million barrels of oil a day by 1990. By 1980 CERT claimed the tribes were producing the equivalent of almost 1 million barrels, an increase of 100 percent since CERT was formed. By 1985 CERT admitted their members' production was down by two-thirds to 356,000 barrels a day because of falling prices. Federal royalty figures confirmed that Indian energy production increased until their royalties represented 5 percent of the total Indian and federal oil royalties, 2 percent of the natural gas, and 17 percent of the coal in 1981. In its effort to enhance its own image, however, CERT neglected to mention that this continued a trend that started before CERT was born. In the ten years prior to the formation of CERT, Indian coal production increased by 709 percent and Indian uranium production by

30 percent, much more than public land production increased during the same period.[40]

CERT's efforts to enhance its production record while neglecting the facts indicated a more general problem. Although any organization dependent upon outside funding must devote considerable time and expense to public relations, some tribes and staff members felt that CERT focused on self-perpetuation, public relations, and lobbying at the expense of its technical services. CERT did not devote enough effort to its relations with member tribes, where serious problems developed. Tribes began to chaff under the autocratic leadership of Board Chairman MacDonald and the insubordination of Executive Director Ed Gabriel, who seemed to serve MacDonald, not them. They had formed CERT because they did not want to rely upon the decisions and judgments of BIA, and they did not intend to hand such authority over to an employee or a board chairman. The two men ran meetings with an iron hand. For example, at the Phoenix meeting, Gabriel quickly changed the subject when two board members tried to introduce the sensitive subject of allottees. Carol Connor had sought CERT board support for her proposal for BIA funds to study allotted resources and related legal questions. Gabriel opposed it, she said, apparently believing she was competing for federal money then being allocated to CERT.[41]

In 1980 Gabriel and MacDonald secretly fought the confirmation of Thomas Fredericks as assistant secretary of Interior despite the fact that the CERT board took no position on the confirmation and that five member tribes endorsed Fredericks at the hearing. Fredericks had made no secret of his opposition to CERT and especially to giving CERT federal grants. Although Gabriel swore that neither he nor MacDonald had interfered with the confirmation, several congressional and federal agency staff members said Gabriel and MacDonald convinced Senator Dennis DeConcini of Arizona to block Fredericks's confirmation for four months until the presidential election, when it became moot because of the pending change in administrations. Although Fredericks then served as assistant secretary under a recess appointment and continued to oppose CERT funding, Undersecretary of Interior James Joseph kept his commitment to provide CERT with $1 million in grants and contracts.[42]

In 1981 CERT's budget reached its highest ever—$3.9 million, with the federal government providing 74 percent of the total and the tribes providing 26 percent. Half went to salaries (CERT employed twenty-three people on its Washington staff and forty-two in Denver). Although auditors found that salaries compared with other consultants with similar responsibilities, some tribal leaders and even staff members complained about the lavish offices that CERT rented for $181,000 a year and the extravagance of some staff members who, they said, "spent money like a bunch of sheiks."[43] CERT had raised enough suspicion that Congress decided to look into CERT finances. The Joint Congressional Budget Committee in 1980 called for a General Accounting Office (GAO) investigation because committee members were concerned that the organization did not represent all the tribes with significant energy resources, yet it received most of the federal money designed to help develop these resources. The committee asked GAO to investigate how CERT spent its money and other sources of income and how effectively the organization benefited all tribal energy resource interests.[44] After the investigation was completed a GAO investigator said that GAO would not release the results of the investigation because the problems were not serious, and DOI and DOE assured GAO that the agencies would solve them. As a result, the agencies tightened up their financial controls and reporting requirements and improved safeguards to avoid duplication among agencies. Although CERT passed the government auditors' tests, its own auditors found serious problems that threw the organization into turmoil just two years later. The January 11, 1980, report to the board of directors from Touche Ross & Co. said, "The Council had a substantial excess of expenses over revenue."[45]

At the November meeting in 1982 CERT announced it had hired a new executive director, A. David Lester. Lester, a member of the Creek Tribe from Oklahoma, had been involved with CERT funding since 1978 as commissioner of the Administration for Native Americans. When the CERT board decided it wanted an Indian director, MacDonald asked Lester if he would consider the job. Coincidentally, MacDonald presented his swan song as board chairman at the same November 1982 meeting at which Lester was first introduced as the new director. MacDonald had been defeated in a tribal election earlier that month and thus could no longer serve on the CERT board, which included only elected tribal officials. The board subsequently elected Wilfred Scott, chairman of the Nez Perce Tribe, to chair the CERT board.

In a typically provocative speech, MacDonald reviewed CERT's history, including its critics. Given that he had been defeated by Peterson Zah, who was believed to be more of an environmentalist, his remarks seemed also to be addressed to his own critics within the Navajo Nation. MacDonald said he believed that CERT's critics had oversimplified the question of CERT's purpose. Although MacDonald himself had inspired the oversimplifications, his analysis of CERT's problems had much truth to it.

> We were charged with capitalism in the first degree. . . . CERT is not in the business of advocating rampant exploitation of the land . . . but it is more comfortable to think in terms of dichotomies. Us versus Them. Spiritual versus Material. Cultural Preservation versus Subsistence. Tribal Sovereignty versus Corporate America. They believe commercial exploitation of mineral resources is somehow inherently evil. CERT has challenged all of these assumptions; we had no choice. Why? Because

poverty was no longer fashionable, and hunger was no longer fashionable.[46]

In retrospect MacDonald and Gabriel both played crucial roles in CERT's early years. The tribes initially perceived MacDonald's flamboyance as dynamic leadership and elected him chairman. As CERT outgrew his type of leadership, it began to see him as arrogant, self-centered, and autocratic. Because the public closely associated MacDonald and CERT, allegations of fraud and corruption against MacDonald tainted CERT's image, too. In 1989, long after he served as spokesman for CERT, the Navajo Tribal Council removed him from the tribe's highest office because of corruption charges. At the same time, the council fired Ed Gabriel, whose public relations firm had been receiving $1 million a year from the tribe.[47]

In the 1970s, however, MacDonald's involvement was critical to getting CERT on its feet. He succeeded in getting more news coverage of not only CERT but also related Indian issues, such as unemployment and income levels on reservations, health problems, and the need for tribes to become economically stable and provide their own government services. Despite his image as a no-holds-barred energy promoter, from the beginning MacDonald emphasized tribal control and environmental and cultural protection in his talks. By constantly reminding the federal government to remember Indians when formulating national energy policy, he succeeded in getting himself appointed to several energy task forces.

MacDonald's attention-grabbing public appearances combined with Gabriel's fundraising abilities to bring large infusions of federal money into CERT. Gabriel helped select a staff from among Washington area agencies and consultant groups that could produce "good paper"—they knew what the agencies wanted to hear in funding proposals. Gabriel had the political savvy to make Indian energy a priority in Congress and to convince several new agencies, including DOE, the Community Services Administration, and the Department of Agriculture, to provide Indian energy funding. He said CERT had "leveraged" $17 million in loans and grants from federal agencies and private industry that would not have otherwise been available.[48]

Although Gabriel's staff members were good at raising money, they were also good at spending it. The board did not know the severity of the financial problems until 1983, when the realization forced it to abruptly close its Washington office in August and lay off its staff there with just two days' notice. Many outsiders thought the coalition of energy tribes had breathed its last breath. Lester later said CERT had a cumulative deficit of $1.2 million in overhead and indirect costs. In March 1982 CERT had staged an extravagant, black-tie fundraising event at the Kennedy Center in Washington called the Night of the First Americans. CERT took in $300,000 and did not realize until the books were straightened out later that the organization lost

$250,000 in that one night, according to Lester. He said they should have named it the Night of the Living Dead.[49]

Ross Swimmer, principal chief of the Cherokee Nation and a banker by trade, had just joined the CERT board, and he was appalled by CERT's financial condition. He said CERT's staff, under Gabriel's direction, was drawing down on contracts in advance before the contract was earned. Although legal, this practice made it impossible to know what the organization's cash position was. Gabriel defended his management, saying the $1.2 million was not a debt; CERT could renegotiate its overhead costs with the U.S. government and prove the $1.2 million was necessary to fulfill the contracts.[50]

A CHANGE IN DIRECTION

The financial crisis forced CERT's board to evaluate its priorities and direction. The years 1982 and 1983 marked the end of an era for CERT. During its early years federal monies were relatively abundant and energy markets were thriving. By August 1983, however, both sources of revenue looked less promising. Prior to the election in 1980 candidate Ronald Reagan sent Senator Peter Domenici of New Mexico to the annual CERT meeting to represent him. Tribal representatives recognized Domenici's message for what it was— Reagan's dressed-up threat to cut federal financial assistance for Indian energy development. Domenici said, "Under the present administration, CERT and other Indian tribes are being wooed with trinkets. Million dollar grants here and there are being touted as proof of the Carter Administration's desire to help." Domenici promised that instead of grants and contracts, Reagan would help the tribes by improving energy markets.[51] After Reagan was elected he fulfilled his promises to restrict regulation of energy development and prices and to drastically cut funding for tribes and Indian organizations.

The markets did not respond as promised, however. By November 1982 the price of oil was $34 a barrel and falling. As the prices plummeted and there seemed to be a glut of oil, the nation lost its sense of urgency to become independent of the OPEC stranglehold. Reagan had come into office bent upon eliminating DOE, Carter's favored agency. He abandoned that mission and instead shifted DOE's emphasis even more toward weapons. Reagan and congressional fiscal conservatives did succeed in killing the Synfuels Corporation and many DOE grant programs. The time had obviously passed for CERT to invoke cartel threats and to depend so much upon the federal largess.

Nevertheless, the changed conditions did not require a major rethinking of CERT's reasons for being. Although a child of the energy crisis of the early 1970s, CERT's mission was broad when the founders wrote the objectives in 1975: building tribal infrastructures, finding financing for projects, protecting the reservation environment, sharing technical expertise, and lobbying. CERT's new leaders embraced those goals. With its 1983 budget crisis CERT had to close either the Washington or the Denver office. Gabriel would have favored lobbying, his own area of expertise. In 1983, however, the tribal leaders chose technical assistance over lobbying and policy analysis and closed the Washington office. Their first priority was reservation economic development, and the federal agencies did not provide funds for policy analysis.[52]

By laying off fourteen people in Washington and nine in Denver and eliminating all services not directly related to providing technical assistance, CERT cut its labor overhead and other indirect costs by 50 percent. Among the casualties were several well-qualified staff members, whom Lester praised, and *CERT Report*, an excellent newsletter edited by Richard La-Course, a Yakima Indian.[53]

The board recognized that CERT, along with other Indian organizations and advocates, had scored several major legislative victories through the years and especially in 1982, including Indian language in the Federal Oil and Gas Royalty Management Act, the Nuclear Waste Policy Act, the tribal tax status act, and, most important for CERT, the Indian Mineral Development Act. The IMDA allowed tribes to escape the restraints of old leasing laws and to negotiate their own agreements. Nevertheless, the board wanted CERT's technical help most. It could leave most of the lobbying to other, experienced groups in Washington, such as the National Congress of American Indians, the National Tribal Chairmen's Association, the Native American Rights Fund, Americans for Indian Opportunity, and individual tribes' lobbyists.[54]

As federal money became harder to get, good relations with tribes became increasingly important to CERT. To meet the challenges of the new era, CERT had new leadership. Both Wilfred Scott, the new board chairman, and David Lester were more cautious toward energy development as well as being more amicable, more conciliatory, and less aggressive than MacDonald and Gabriel. From the time he first joined the board, Scott had pushed for more board involvement. Although not the articulate, public relations man that MacDonald was, Scott did not impose his vision on other board members, resulting in much smoother relations. Scott's successor as board chairman from 1986 through 1989, Judy Knight, chairman of the Ute Mountain Ute Tribe, continued with the tradition of soft-spoken peacemaker. When MacDonald was reelected chairman of the Navajo Tribe in 1986, he became vice chairman of the CERT board but played a low-key role with the organization. Staff-board relations also improved when Lester became director in 1982. Although perceived as an urban Indian, Lester was more attuned to the wishes of the reservation Indians on the CERT board than Gabriel, a non-Indian, had been. During this period tribal leaders and tribal attorneys began giving most of the presentations at annual meetings, with CERT staff members assuming much lower profiles than before. At tribes' request the staff would provide assistance to allottees with royalties and right-of-way negotiations.[55]

By 1988, the membership had grown to forty-three reservations, including several that had only solar energy potential and no fossil or nuclear fuels and several east of the Mississippi. The member reservations were Ute Mountain Ute, Navajo, Nez Perce, Yakima, Acoma Pueblo, Cherokee, Jicarilla Apache, Oglala Sioux, Standing Rock Sioux, Blackfeet, Chemeheuvi, Cheyenne Arapahoe, Cheyenne River Sioux, Chippewa Cree, Coeur d'Alene, Crow, Flathead, Fort Belknap, Fort Berthold, Fort Peck, Hopi, Hualapai, Jemez Pueblo, Kalispel, Laguna Pueblo, Muckleshoot, Northern Cheyenne, Pawnee, Penobscot, Ponca, Rosebud Sioux, Santa Ana Pueblo, Saginaw Chippewa, Seminole of Florida, Shoshone-Bannock, Southern Ute, Spokane, Tule River, Turtle Mountain Chippewa, Umatilla, Uintah and Ouray, Walker River, and Zia Pueblo.[56]

Partly because of the change in leadership and partly because of the changed economic conditions, CERT made a noticeable shift during this period from its previous public position of energy advocate. As chairman of the Nez Perce Tribe, Scott had been part of the decision to forego economic return from selling hydroelectricity if it meant sacrificing the reservation fishery. Lester was especially outspoken about the dangers of energy development. Before joining the CERT staff he challenged CERT's adamant advocacy of energy development in a speech he presented at CERT's 1980 annual meeting in Washington. Setting tribes up for boom and bust without important safeguards is "unconscionable," he said. In some ways Lester was giving in 1980 the same message that the traditionalists from Big Mountain had tried to present in 1979 when they were rebuffed. Lester, however, said he believed that economic development—under Indian control—was desirable. "When a community is not producing more wealth than it consumes, the deficit is made up through either a lower standard of living or grants," he said.[57]

Without economic development, tribes are held down by an "iron bond linking exploitation, paternalism, and dependency," Lester continued. To escape, they first needed to improve their tribal governmental infrastructures. As commissioner of the Administration for Native Americans (ANA) for four years, he had tried to increase energy tribes' ability to achieve economic self-sufficiency through a grants program that also focused on improving social conditions. When Lester joined CERT, he continued that effort.[58]

After coming into office Lester and Scott faced two major challenges to prove that CERT had changed—one from the Navajo Nation and the other from the Interior Department. The new Navajo tribal chairman, Peterson Zah, threatened to withdraw his tribe from CERT because he believed that MacDonald had neglected to involve the Navajo Tribal Council in CERT decisions; in fact, Zah could not find a resolution authorizing the tribe to even join. Zah questioned whether the tribe needed CERT's technical ability and challenged the organization to prove that tribal leaders rather than staff determined CERT's direction. He sought assurance that CERT would not try to push his tribe into undesirable development.[59]

Although Scott said with bravado that CERT could live without the Navajos, the loss would have substantially weakened CERT both financially and politically; the Navajos were the largest American Indian tribe and owned more minerals than any other. During personal visits to the reservation Scott and Lester convinced Zah the board would be in control. Zah's staff, in turn, convinced him that, although the Navajos were more sophisticated than most tribes on energy matters, they still needed CERT's technical assistance. They told him CERT had helped to set up a Navajo Energy Development Authority, to get the reservation designated as an energy impact area to qualify for federal assistance, to evaluate a consultant's proposal for a new coal processing technology, to teach computer analysis, and to analyze the marketability of Black Mesa coal. The organization had provided invaluable support to the coal tribes—especially the Navajos, Hopis, and Crows—by preparing a study of tribal authority for coal regulating and a strategy for getting the legislation through Congress. CERT's Ahmed Kooros had assisted the Navajos with pipeline, coal, and uranium negotiations. For example, the Navajos with Kooros's assistance had negotiated a right-of-way agreement for the Four Corners pipeline that would provide $92 million to the tribe over the life of the easement, instead of the $20,000 that the company had paid the tribe for the first twenty-year easement.[60]

Ultimately, Zah and the Navajo Tribal Council decided to remain in CERT. Speaking at CERT's November 1983 annual meeting, Zah explained why he had changed his mind. He had come to see CERT as a "quasi-independent agency" that could provide him with information quickly. "Lester's job is not to create policy for the Navajo Nation," he said. "His job is to give information that we need." Zah agreed with Scott and Lester that energy resources were not the answer to all problems and compared dependency upon energy income with dependency upon the federal government. He said, "It's nice, and it's good when the dollars are there. But when they are not, then we hurt."[61]

As soon as that battle had been won, CERT faced another attack from the Interior Department. Assistant Interior Secretary Ken Smith told CERT at the 1983 annual meeting that with the lack of an energy market, BIA should withdraw its support from CERT, leaving it to the tribes to support if they believed they needed it. CERT Board Vice Chairman Ross Swimmer (Cherokee) objected, with words that would later haunt him after he took Ken Smith's job. He said that CERT would one day become self-supporting but that as long as Congress continued appropriating funds to help tribes get a better deal, some of the money should go to CERT. In 1985, with energy prices still falling, Swimmer said that CERT was needed to help tribal governments take advantage of the market interlude to strengthen their tribal governments so they would be ready when development pressure increased again.[62]

The day following the 1985 meeting, President Reagan announced Swimmer's appointment as assistant secretary, and the CERT board members expressed elation that one of their own would be at the helm of BIA. When Swimmer attended the next board meeting in 1986, however, his speech contained much the same message as Smith's. He suggested that because of the poor energy market, CERT should either dissolve itself or broaden its mission. After substantially cutting funds to other Indian organizations, he apparently did not want to show favoritism. Despite Swimmer and despite general federal belt tightening, Congress believed that CERT was still needed and instructed BIA to provide CERT with at least as much money in the 1987/88 fiscal year as it had the previous year.[63]

RETURN TO BROADER OBJECTIVES

CERT at that time did not need admonitions to return to its original, broad mission. However, this federal pressure did encourage CERT to be more public about its broader functions. In a few cases prior to 1982, CERT had helped tribes set up civil service personnel systems and improve management techniques behind the scenes. Although that work had contributed greatly to tribes' satisfaction with CERT, it was not publicized. During those early years CERT sought acclaim with flamboyant, pie-in-the-sky development proposals. After the energy market downturn, when more funds were available for environmental programs than for energy subsidies, CERT spokesmen publicly acknowledged the hazards of uncontrolled development and touted the organization's efforts to help tribes gain control, especially through environmental regulation.

In October 1983 CERT completed a pivotal study confirming many of Lester and Zah's fears that members of energy tribes had not benefited from past development on their lands. The study measured the self-sufficiency of six energy reservations with differing levels of development. The authors of the study, Ahmed Kooros and Anne C. Seip, found that because the reservations were not under tribal management, energy revenues and federal grants generally did not result in increased economic viability. Without tribal control, there was no transfer of knowledge, and the money "leaked" out to non-Indian border towns or financial centers. St. Louis (Peabody's corporate headquarters) benefited more from Peabody Coal's mine on the Navajo Reservation than the Navajo people did. Explaining the significance of the leakage study, David Lester said, "Imagine a garden hose full of leaks that results in only 22 percent of the water ever reaching the other end. If you want to increase the volume of water coming out, do you turn the faucet up and increase the water pressure? Or do you begin repairing the leaks?" The study results received much attention in Washington and helped to change CERT's image.[64]

Lester and the CERT board used the results of the study when setting the organization's direction. Instead of emphasizing royalties ("increasing the water pressure"), the CERT staff focused more of its efforts on increasing tribal control, improving tribal infrastructures, and encouraging tribes to invest their income from energy development. The Laguna Pueblo, for example, later tripled its income from a right-of-way settlement by investing it, according to Kooros. Other tribes sought CERT's assistance in improving their governmental efficiency, developing their regulatory mechanisms, improving their relations with financial institutions, and monitoring income from existing development.[65]

CERT provided management assistance to Fort Peck's Assiniboine and Sioux Tribes. Diana Smith of the CERT staff designed and installed a computerized financial management system that included accounting, purchasing, property management, budgeting, and payroll. Smith, a non-Indian with a master's degree in public administration, spent much of one year on the reservation. The Fort Peck model was later adopted by other CERT members. With the new system Tribal Chairman Norman Hollow said he could get information on income, expenditures, and account balances within minutes—information that previously took BIA months to provide. Within half an hour he could get oil and gas information, such as the number of wells in a certain field, the daily production, and the daily income. Speaking at the 1985 annual meeting, Hollow—one of CERT's founders—said such a system helped the tribes get on an equal footing with industry and helped them toward self-sufficiency.[66] For several years CERT worked with the Blackfeet Tribe to help it successfully open its own bank in 1987, the first tribally owned bank in the country. The only local bank had been shut down in 1983 by federal regulators.[67]

CERT's announcement in 1985 of plans for a giant solar plant on the Chemehuevi Reservation seemed to be a throwback to earlier times. CERT had received a $125,000 federal, interagency contract to look at the feasibility of a photovoltaic system for the Chemehuevi Reservation on the Arizona-California border. Despite an oversupply of electricity in that area, CERT tried to find investors to spend $50 million to $80 million to build a 20- or 30-megawatt plant—the world's largest photovoltaic system, according to the CERT staff. Over 200 acres of the 33,000-acre reservation would have been covered with rotating rows of modules containing photovoltaic cells and reflecting lenses. Capital- rather than labor-intensive, it would have employed just ten to fifteen people, although tribal officials hoped it would attract other industry to the reservation.

Unlike the Crow synfuels project, the Chemehuevi feasibility funds did not represent "new money" because they were part of CERT's interagency contract and would have been spent on some tribal project anyway. Nevertheless, the organization seemed once again to be devoting inordinate amounts of time and energy to a showy, high-technology project. Because of the lack of electricity demand and grandiose financial dimensions of the project, it was unlikely to ever come to fruition. Lester said the tribe, CERT, and the funding agencies knew from the beginning that the project was a long shot. He believed, however, that a "reasonable man" under the same circumstances would attempt it because the site was ideal, the tribe had no other resources, and the unemployment rate there was close to 50 percent. Lester said the staff felt a responsibility to plow new ground for the tribes.[68]

During the 1980s CERT was able to get new money from the Environmental Protection Agency (EPA) rather than depending entirely upon the original participants in the interagency contracts. Ironically, considering President Reagan's known antipathy toward environmental programs, such funds became more available to CERT under Reagan. In 1983 Congress redirected much of DOE's civilian budget toward an environmental function—waste management. Some of this money went into CERT's coffers when the Confederated Tribes of the Umatilla Reservation and the Nez Perce Tribe hired CERT to help study the proposed Hanford nuclear waste site. In 1984 EPA adopted an Indian policy to address the regulatory vacuum on many reservations. With that policy initiative and some statutory changes in environmental laws came new federal funding for CERT and member tribes.[69]

EPA became CERT's new "glamour girl," in the words of Marie Monsen, local and Indian affairs chief for DOE. Monsen, who had coordinated CERT's interagency contracts since 1981, said the original federal funding agencies had always intended to reduce their involvement in CERT as the tribes assumed a greater share of CERT's operating costs. In 1986 DOE still provided $500,000 to CERT through the interagency grant. The new participant, EPA, provided half that or $250,000.[70]

Since 1981 CERT had devoted large portions of its annual meetings to environmental issues, especially air quality, because several tribes had air quality monitoring projects funded by EPA. By 1984 the number of tribal requests for CERT's environmental technical assistance had mushroomed; six staff members now worked in CERT's environmental branch. In that year EPA awarded CERT $125,000 to study hazardous wastes on twenty-five pilot reservations. Later, EPA provided $90,000 to establish an environmental information base and provided other, relatively small contracts for regional information meetings. CERT competed for such EPA contracts with other Indian organizations, such as the National Congress of American Indians and Americans for Indian Opportunity.[71]

All of EPA's funding to CERT was earmarked. With the federal deficit, CERT and the tribes could not expect any federal agency to continue supporting CERT's general operating budget at previous levels. The reliance on federal dollars had always been frustrating because of BIA's threats, the public's criticism, the agencies' delays in delivering promised funds, and the Office of Management and Budget's (OMB) cumbersome requirements. That some of CERT's funding had shifted from DOE to EPA made little difference in that regard. "Every time there's a major scandal at the Pentagon, OMB slaps more requirements on CERT's contracts," Monsen said. Lengthy negotiations delayed funding for the 1985 contract by seven months, and CERT nearly had to lay off some staff members as a result.[72]

By 1986 Lester said tribes were providing nearly half of the budget through dues; contributed time, travel, and expenses; and contracts for CERT services. To give the organization more independence from federal funding, the CERT board that year created two subsidiary corporations. The CERT Education Fund was to continue CERT's fundraising and assistance for students, such as its college preparatory summer institute and its scholarship program. To increase tribal support the CERT board created the CERT Technical Services Corporation, which was designed to market technical assistance. This new subsidiary bid competitively on contracts to serve any tribes, whether or not they were members. CERT hoped tribes would give its technical services subsidiary priority over other firms because CERT was directed by an all-Indian board and because half of CERT's staff members were Indians, including most of the senior management staff. The parent organization, CERT, was still to protect energy tribes' interests by monitoring federal agencies and Congress and attempting to influence their decisions.[73]

CONCLUSION

Over its first decade, CERT grew from a boisterous, young organization that threatened to withhold its resources if it did not get funded to a more reserved, mature one that promised full-scale development. As its members' awareness of the pitfalls of energy development grew, CERT returned to its founding document's broad mission statement. With their growing sophistication the tribes took over some of the functions for which they formerly depended upon CERT, such as negotiations, and took a more active role in directing their organization.

As it approached 1990 CERT faced the possibility of increasing competition from other consultants for providing technical assistance to tribes. Whether CERT survived another decade would depend more upon its ability to sell its usefulness to tribes than to the federal government. Despite their growing expertise the tribes continued to need a pool of technical experts to draw upon because they could not afford to each hire their own and because the federal government continued to be unable to provide them.

Federal budget cuts, combined with plunging energy income, made some tribes even more dependent upon outside experts at a time when they were less able to pay for either CERT's experts or others. In order to better shepherd their existing leases, the tribes, with CERT's backing, were putting more emphasis upon royalty accounting both in the accounting books and in the field.

CHAPTER FIVE

Who's Minding the Store?
Indian Royalty Management

If I owned a butcher shop or a gas station and asked people to come in and take what they wanted and to pay their bills a month later, how long do you think I could stay in business?

—Chuck Thomas, USGS inspector[1]

In the oil fields of the Wind River Indian Reservation on June 13, 1980, the sage- and sulfur-scented winds carried the rhythmic thumping sounds of the oil pumps and occasionally the yip of a coyote. The bobbing pumps lay far from the bustle of the reservation capital, Fort Washakie, and, until that day, far from the scrutiny of prying eyes. Then federal oil-field inspector Chuck Thomas, acting on a hunch, pulled an oil tanker over to the side of the road to check its credentials, shattering the calm of the high plains. Thomas, a Cherokee, worked for the U.S. Geological Survey (USGS), which was responsible for preventing thefts of federal and Indian oil. He figured the pipelines could carry nearly all of the oil produced on the reservation, meaning few if any trucks should be hauling oil from wells there. Just as he suspected, the driver did not have a "run ticket," the document authorizing him to remove oil from a federal or Indian well.[2]

Thomas sparked a full-scale investigation that spread quickly from the oil fields of Wind River to encompass the whole federal inspection and accounting system for Indian and federal lands. He also shifted the attention of Indian tribes across the West, which, with a few exceptions, had been focused primarily upon getting better terms in new mineral contracts. The tribes and individual Indian mineral owners (allottees) came to realize that they were not getting even the relatively small shares of revenue promised by their oil and gas contracts. Thomas's employer, the USGS, had not been doing its job. Inspectors had not assured that the oil fields were secured against thefts. The agency also had neglected to reconcile royalty accounts to be sure the companies were paying properly. Instead it had relied upon a seriously flawed honor system, and that system had failed.[3]

To states and tribes, oil and gas revenues were of vital importance. The states received 50 percent of the royalties from federal leases within their boundaries, except for Alaska, which received 90 percent. The Indians received all of the royalties from Indian leases. In the peak year 1982 the Indians earned $198 million in royalties from oil and gas production. With the drastic drop in federal grants and contracts under the administration of President Ronald Reagan, tribes began looking in the 1980s for ways to streamline their operations, limit their services, and make more efficient use of limited funds. Tribes with energy income became even more dependent upon that income, which also was dropping; by the spring of 1986 energy tribes had suffered a 30 to 60 percent decline in oil and gas income. Making sure that they were actually receiving the revenue to which they were entitled became ever more important.[4]

The energy tribes, like the states, needed the full value of this royalty income to help provide basic government services such as police and fire protection, natural resources management, road maintenance, health care, and education. On the Jicarilla Apache Reservation, for example, the tribe received about $25 million in oil and gas revenue in 1986, with which the tribe paid for police and fire protection, schools, water and sanitation, and other social programs. Oil and gas royalties and taxes provided 85 percent of the tribe's total income. The Blackfeet Tribe derived 90 percent of its total income in 1985 from oil and gas royalties and taxes.[5]

To protect their ownership interests the tribes turned to their governmental powers. The sudden realization of the federal government's failings in royalty accounting forced the tribes to act. Often building new governmental institutions from the ground up, the tribes approached royalty management on two fronts. They hired energy police to provide site security in the oil fields, and they hired accountants to check the books for paper losses. As part of their accounting investigation they discovered serious inadequacies in the way that the government determined the value of the resources and in the government's lack of enforcement of diligence. In each of these efforts they relied upon political pressure and media exposure to force the government to recognize the differences in its responsibilities for the Indians' and for the public's minerals. The public royalty management controversy forced many painful reforms upon the Interior Department, including providing a bigger role for tribal governments. Yet the original problem—the lack of field security—was nearly forgotten.

SITE SECURITY INVESTIGATION

Although Chuck Thomas pulled the tanker over in June 1980, the story did not leak out until more than four months later when the *Rocky Mountain Journal* reported on the grand jury investigation. When his superiors would not act on his suspicions, Thomas had blown the whistle, taking his briefcase full of incriminating photographs to the Federal Bureau of Investigation (FBI). The following spring, as part of an investigation for the Senate Select Committee on Indian Affairs, Montana Senator John Melcher grilled USGS officials about their cover-up of Thomas's discovery. Although the unauthorized truck could have carried seven thousand dollars' worth of oil stolen from the Indians, USGS officials explained lamely that Thomas's report was ignored through an "oversight."[6]

USGS officials were not the only ones who hesitated to rock the boat and dive into the morass that such accusations of theft suggested. Wind River's Shoshone and Arapahoe Joint Business Council had known earlier about the possibility of oil thefts. In 1979, a year before Thomas pulled over the tanker, John King wrote to the tribes. King, a Denver oil man who had served time for stock fraud, said he thought 30 percent of their oil was being stolen, based upon differences between reported production and spot checks of wells. He thought the tribes should be suspicious because when oil prices increased by 500 percent in four years (from $8 to $48), the tribes' annual royalties climbed only 20 percent. The tribes ignored his letter, possibly because they did not want to be associated with King and his criminal record and possibly because they figured it was just another consultant trying to sell his services by "helping" the Indians. Tribal council members were skeptical that problems were widespread, and when the proof surfaced, they reacted defensively.[7]

Industry representatives also reacted defensively as the accusations of widespread thefts grew. Interior Secretary James Watt appointed a commission headed by David Linowes in 1981 to investigate oil-field management and accounting procedures for the Interior Department. In addition to congressional and departmental hearings in 1981, the FBI's energy fraud unit conducted an investigation in Colorado and Wyoming, and the USGS stepped up its inspection program. None of the industry spokesmen appearing before the Linowes Commission believed that oil theft was widespread or significant. Industry representatives felt satisfied with their own arrangements for security against theft despite the fact that they could have been bilked by other companies or individuals out of a much larger amount of money than mere royalties. Although federal government leases yielded one-eighth royalties and Indian leases generally one-eighth or one-sixth, the leaseholders—the companies that shared the lease ownership—were supposed to receive seven-eighths or five-sixths of the production sales. Nevertheless, the petroleum industry persisted in opposing proposed reforms.[8]

Congress should have recognized the grave oil-field problems years earlier and should have provided sufficient funding and a mandate to USGS to resolve the situation. The General Accounting Office had reported to Congress many times, as far back as 1959, about serious accounting problems for federal and Indian oil. In 1974 and 1976 USGS told congressional auditors that it lacked the staff to prevent oil from being stolen; with the money it had, USGS could only perform 10 to 20 percent of the required inspections. On the Jicarilla Apache Reservation in New Mexico, GAO auditors reported in 1976 that they had discovered a well that had been flaring gas for over six months without approval. Because they had not been out in the field, USGS inspectors were unaware of the flaring. USGS had been forced to rely upon industry to inform it if oil or gas wells were producing. In one case on the Uintah and Ouray Reservation in Utah, the USGS district office was unaware that a well had been in production for over nine months, meaning the company could have avoided paying the tribes their royalties for that long.[9]

Meanwhile the price of oil climbed dramatically, from less than $3 a barrel in 1971 to more than $40 in 1981. Oil had become black gold—an increasing temptation for private contractors and operators in the oil fields, who knew that nobody was minding the store. At the same time, royalties from federal and Indian oil and gas rose from less than $500 million in 1971 to $5 billion in 1982, greatly increasing the workload for the already overburdened USGS record keepers.[10]

It should have been obvious long before Thomas pulled over the tanker that the situation would not improve when temptation increased, without substantially increasing management funding. Although USGS's budget grew between 1970 and 1982, the money did not pay for nearly enough inspectors or accountants. Wes Martel of the Shoshone Oil and Gas Commission told Melcher in 1981 that the regional USGS office in Wyoming had two technicians who were responsible for inspecting nearly five hundred federal and Indian leases, including ninety-four on the Wind River Reservation. When that office requested an accountant, USGS headquarters told them they could get by with a clerk instead.[11] The Bureau of Indian Affairs shared some of the surface inspection responsibilities for Indian leases, but BIA also suffered from a lack of funding and staff. For example, although royalty income to the tribes and allottees on the Wind River Reservation totaled $17 million in 1980, the Wind River BIA Agency had only one mineral staff member; the same was true of other agencies responsible for millions of dollars in minerals.[12]

Except on the Osage Reservation, tribes generally had not taken on inspection duties. While investigating theft problems in the Indian oil fields in 1981, Senator Melcher asked the tribes with the largest production why they had not. With the potential benefits, tribes certainly could justify funding an inspection program, he suggested. The tribes and states explained that before losing their innocence in 1980, they had relied upon the federal government. A spokesman for the state of Utah said that only four or five states previously had done anything more than wait for the checks to come in and carry them to the bank.[13]

Some tribes had expressed concern many years before but had met with no response and no cooperation from the Interior Department. USGS and BIA refused to let tribes see their own lease files, saying the information was proprietary. As early as 1951 the Navajo Tribal Council asked the Navajo chairman to hire a consultant to investigate whether the tribe and allottees were being paid the proper royalties for their uranium. In 1977 a Navajo tax document said auditing lease payments should be the top priority of the Navajos and other tribes. Between 1974 and 1977, after several years of unsatisfactory responses to tribal complaints about the department's handling of royalties, the Jicarilla Apaches sued a dozen major petroleum companies and the U.S. government, challenging accounting methods. Even before the Wind River scandal, the Northern Ute Tribe on the Uintah and Ouray Reservation in Utah had established a data system to monitor oil and gas development and production.[14]

In Osage County, Oklahoma, the tribe and BIA—not USGS— were responsible for protecting Osage oil and gas, and they had been doing a much better job than USGS. They had many more years' experience, and the responsibilities were not split among several federal agencies. Ever since the Osage allotment act of 1906, separate legislation had governed Osage oil and gas. Before the 1930s, when USGS first began to regulate Indian mineral lands elsewhere, BIA's Osage Agency already had an active minerals branch. Unlike USGS, which apparently had no idea what to do with Thomas's allegations, the Osage Tribe had instituted a clear theft process, according to ex-BIA Osage Agency Superintendent, David Baldwin. An Osage tribal member himself, Baldwin said that if BIA learned of any possible theft, the lease was immediately sealed off by the superintendent, who notified the FBI and Osage County sheriff. Over the ten years prior to 1981, Baldwin said, BIA received many reports of theft, all but one of which resulted from "poor communications." One thief was prosecuted. The Osage Tribe partially supported the BIA agency with the tribe's share of royalties. In 1987 the BIA Osage Agency minerals staff included five petroleum engineers and two lawyers—more professionals familiar with oil and gas than in any other BIA office in the United States. Even the Osage staff was not big enough, however. Six people had to inspect 11,000 wells, which the Linowes Commission said was not sufficient.[15]

After the Wind River exposé, BIA and the tribes looked to the Osages for help. The Osage Agency minerals staff trained nine individuals from other tribes and agencies between December 1980 and March 1981 on field inspection, compliance, and surveillance techniques. When BIA instituted its national lease compliance training program in the fall of 1981, it included the Osage Agency chief oil inspector as one of the instructors. BIA and USGS provided the other instructors, who conducted the classes in reservation oil fields.[16]

Several tribes tackled the problems themselves. When Shoshone and Arapahoe tribal leaders became convinced of the potential for oil thefts, they contacted Thomas, who had been transferred to Alaska, and asked if he would work for them. Although Thomas and his wife believed they had suffered attempts on their lives while in Wyoming because of his oil-field work, they agreed to return in November 1980.[17]

Before going to work for USGS in 1976, Thomas had worked in the oil fields for fifteen years as a roustabout and roughneck. He told a reporter, "I'm not a man of long words and big politics. . . . I have a worm's eye view of it (oil thefts) because I was the man in the field." When he began training inspectors Thomas advised his Indian interns, "Be suspicious and trust nothing or nobody." Thomas had the opportunity to preach his cynicism widely because soon after his arrival at Wind River, both the Blackfeet Tribe of Montana and the Northern Ute Tribe of Utah flew staff members into Wyoming to get trained by Thomas. Within five days after the training in Wyoming the Northern Utes were inspecting lease sites on their Uintah and Ouray Indian Reservation. The Blackfeet Tribe hired a four-man team.[18]

Thomas trained not only tribal members but also a U.S. senator about methods of theft. In May 1981 Thomas personally led Senator Melcher and a *Denver Post* reporter on a surprise tour of Blackfeet oil fields, where they found many of the same problems as at Wind River. The tour occurred a year after Thomas had pulled over the tanker at Wind River and months after USGS assured Melcher that problems were not so bad elsewhere as at Wind River.[19]

Because of the extent of the theft problems, tribes reacted quickly. By the fall of 1981, just as BIA was preparing to launch its training program for tribal inspectors, the Fort Peck Minerals Department in Montana had a five-member team under federal contract. Using federal funds and user fees, the Three Affiliated Tribes of the Fort Berthold Reservation in North Dakota hired six uniformed rangers to monitor oil and gas and environmental conditions on the reservation. Although their authority to do so was questioned, several tribes adopted resolutions authorizing tribal staff members to pull over tankers and demand to see their run tickets.[20]

Because most tribes did not have the already developed infrastructures that the Osages had, they sometimes found it difficult to respond to the complex oil-field questions in the crisis atmosphere of the early 1980s. On the Wind River Reservation, for example, Shoshones voted at a general meeting of the membership to disband the new investigatory body, the Shoshone Oil and Gas Commission. Some tribal members were concerned about how much the investigation was costing, how the money was being spent, and whether the commission was acting too independently of the business council, which was made up of older, more traditional tribal members. At meetings held across the reservation in the following weeks, the young, assertive tribal members who worked on the commission explained that the commission had collected over $1 million from oil companies in less than a year as a result of the investigation. Such arguments apparently convinced the other tribal members, who voted by referendum to reinstate the oil and gas commission.[21]

This upheaval eventually cost the jobs of most of the older joint business council members, who were later voted out of office. As a sign of the Indian people's growing recognition of the importance of proper man-

John Murray (above) and Gary Comes at Night, both Blackfeet tribal members, served as tribal inspectors in the Blackfeet oil fields in 1981. (Photo by Marjane Ambler.)

agement of oil and gas royalties, the Shoshone Tribe in the next few years elected three members of the Shoshone Oil and Gas Commission—John Washakie, Wes Martel, and Orville St. Clair—to the tribe's business council. The Arapahoes elected a young Arapahoe geologist, Gary Collins, as chairman of their business council.[22]

The Wind River storm continued to swirl around Chuck Thomas and his salary of $36,000 a year. When Thomas moved to the Blackfeet Reservation in the spring of 1981, controversy followed him. To lure Thomas to their reservation, the Blackfeet Tribal Council offered him the same salary that he received at Wind River, plus a percentage of the money he recovered for the tribe. Blackfeet Natural Resources Department staff members earned a fraction of that, averaging only $8,800 a year. The staff—all tribal members—felt unhappy that the council had chosen to listen to an outsider's instead of the staff's warnings of oil-field problems, and it saw Thomas as a publicity hound. By the end of the year, the Blackfeet Tribal Council had decided not to continue his contract. Regardless of his personality Thomas would have disrupted the status quo by blowing the whistle on Indian oil-field security problems, where millions of dollars were at stake.[23]

Soon after Thomas left Wind River, the Wind River tribes hired top national legal and accounting advisers. Members of the Shoshone Oil and Gas Commission learned how to check the records, comparing four monthly forms against one another for each lease—a minimum of five thousand documents for each year. From the commission's investigation, the tribes discovered especially serious problems with Amoco Production Company leases. If the Wind River tribes had had their way, Amoco's leases would have been canceled. After discovering a hidden, unmetered pipeline and hundreds of thousands of dollars in accounting errors and after not getting the response they wanted from Amoco, the Shoshone and Arapahoe Tribes took their petition for cancellation straight to Washington in 1982.[24]

Federal regulations then provided only two alternatives—minor fines or cancellation—to penalize companies for rule infractions. Interior had been reluctant to take such a drastic step as canceling a lease. The department had never canceled a lease on federal lands, no matter how serious the transgression, and had canceled only a few Osage leases in Oklahoma.[25] Amoco argued that its buried pipeline at Wind River was for waste oil. After the tribes discovered accounting problems, Amoco paid the tribes $763,606 to make up for the errors, which Amoco said resulted primarily from not keeping separate records for tribal and allotted oil production in the same field. Even though Amoco later discovered that it had overpaid the tribes by $210,000, the company did not seek reimbursement, hoping to mend fences with the tribes.[26]

In Amoco's defense it could be argued that some of the wells had been operating since 1916 when oil was selling for a dollar a barrel and expensive security measures could hardly be cost effective. At least theoretically the setup criticized in the 1980s had been inspected and approved by USGS for several decades. Nevertheless, the Amoco site was not up to code, and the buried pipeline looked like a deliberate attempt to circumvent meters. (Wind River allottees filed their own lawsuit against Amoco, which is discussed in Chapter 6.) It seemed certain that the tribes' petition would succeed. After visiting the Amoco oil fields at Wind River in the fall of 1981, David Linowes, the head of the commission investigating oil thefts, said Amoco security was "very weak." Department officials indicated they would be tough on offenders such as Amoco.[27]

Interior, however, sabotaged its own attempt to improve its tarnished image. Just as in the Thomas run-ticket incident, it turned out that the department lacked a formal process for dealing with allegations of oil-field problems. Before giving Amoco a chance for rebuttal, Deputy Assistant Secretary Roy H. Sampsel told the Linowes Commission and the press that the department would be granting the tribes' request to cancel the leases. In fact, Amoco said it first learned of the tribes' petition by reading about it in the *New York Times*. Although Sampsel later that spring signed a statement swearing he was impartial and removing himself from the decision-making process, it was too late. Amoco attorneys quickly charged "foul" and filed a lawsuit, saying they had no faith in the impartiality of Interior officials who were to rule on the tribes' petition. They questioned the department's authority to cancel leases at all.[28]

As Sampsel admitted to the Linowes Commission, "some form of due process needs to be spelled out." The cancellation petition was set aside as the court action dragged on for more than three years and as the company and the tribes tried to settle the dispute. Then, after Ross Swimmer (Cherokee) became assistant Interior secretary for Indian affairs in December 1985, he dropped the cancellation petition a month later, making the lawsuit moot. The tribes were outraged and won the support of the National Congress of American Indians in protesting Swimmer's action, which they felt sent a message to industry that Interior once again would tolerate even flagrant abuses of Indian lease terms. The Wind River tribes argued that Amoco would never have fought the cancellation so hard if it did not consider the case an important precedent. The Arapahoe Tribe's attorney said that Swimmer's office had not even consulted with them before canceling the petition; he believed Interior did not want to embarrass itself by trying to defend Sampsel's actions. Swimmer, who said he assumed the tribes had been consulted, argued that the litigation could have dragged on for ten years or more before the department would be allowed to even rule on the cancellation petition itself.[29] In the end Amoco settled with both the allottees and the tribes. Although on paper thousands of barrels of oil were unaccounted for, the tribes never had an opportunity to prove it in court. Ironically, through the settlement process the tribes and Amoco gained respect for one another and improved their working relationship.[30]

INVESTIGATION RESULTS

The idea of thieves filching truckloads of oil, especially from Indians, captured national headlines for several months in 1980 and 1981. When the

Oil cannot be stolen without detection when a numbered seal such as this is used properly. (Photo by Sara Hunter-Wiles.)

dust settled and the initial federal investigations ended in 1982, most of the provable losses turned out to be the work of a sharp pencil, not a pipe wrench. No leases had been canceled and only a few people convicted. The Linowes Commission's investigation confirmed that USGS needed more inspectors, but it also found the field inspectors needed better training and supervisors with more backbone when diligent inspectors, such as Thomas, reported violations.[31]

Investigators traced the federal safeguards that had been designed to prevent oil from being stolen but that had not been enforced. Then and now, oil is generally measured not at the wellhead but at the point of sale. It is pumped from several wells into a battery of treatment facilities and storage tanks where it is held until sold. Most large-production leases use a lease automatic custody transfer (LACT) meter, which automatically measures the sale. On smaller-production leases, measurements are done manually by gauging the depth of the oil in the tank before and after a sale, with adjustments for temperature, gravity, and impurities. LACT meters are nearly tamper-proof. Automatic, sequentially numbered printouts from these meters record sales volume—but only if the piping passes through the meter. Since 1978 the federal government has required numbered seals, similar to boxcar seals used by trucking firms and railroads, which have to be broken before valves can be opened. Thus it is clear when a valve has been used— but only if it has been sealed. Federal regulations prohibit good oil from being stored in spill ponds because it can be taken without passing through meters.[32]

Oil-field service companies—such as fresh-water haulers, hot oilers, reclaimers, and junk-oil dealers—had taken advantage of the loose controls. Junk-oil dealers, who supposedly haul waste oil out, actually in some cases pumped good oil into the junk-oil pits. By checking refiners' purchase records, the investigators found unauthorized individuals and companies had mysteriously acquired and sold large quantities of crude oil. Thefts ranged from fifty barrels to over a thousand. To avoid paying high royalties, companies sometimes transferred oil from a lease with a high royalty to one with a lower royalty. Purchasers sometimes exaggerated the amount of impurities (sediment and water) in a truckload of oil, lowering the value of the oil and thus the royalties. Dishonest operators sometimes would not report completion of a well and thus skip paying royalties on the first days of production, when flow rates are often the highest.[33]

From all indications solid energy minerals, such as uranium and coal, did not present ready opportunities for theft because of their bulk (coal was worth less than $30 a ton) and the relative security of coal and uranium production facilities. Investigators did not study the solid minerals as much because the total oil and gas royalties were so much higher. In 1980 Indian and federal coal royalties totaled only $40 million and uranium only $9 million, whereas oil and gas produced royalties of $2.6 billion. The Linowes Commission said, on the other hand, that paper theft of coal and uranium royalties was likely. The commission's preliminary review of solid minerals found that general problems of verifying production, determining fair market value, and designing effective audit programs were common for all minerals.[34]

At first it looked as if the site security investigations would result in many convictions. When USGS suddenly stepped up its inspection effort in 1980, the agency issued over two thousand citations to companies for site security violations that could have resulted in theft or mishandling of oil. The Linowes Commission heard testimony about interstate theft rings, kickbacks, and bribery. Private oil-field security investigators estimated to the *Denver Post* that 2 to 6 percent of all crude oil produced in the United States was being stolen at that time (1980–1981).[35]

When it came down to proving that oil had been physically stolen and how much, however, investigators had a more difficult task for several reasons: poor records, an undeveloped process for penalizing violators, and investigators' emphasis on paper audits over site security. Chuck Thomas described the USGS system as an unminded store, but in fact the situation was worse. Because well production was not checked regularly by USGS, the storeowner had no baseline inventory to know what was missing. Neither the Indians nor the public would ever know how many thousands of dollars they lost to thieves in the oil fields. Only four men were convicted in Wyoming—

the president of an oil reclaiming company, a contract pumper, a pumper operator, and a trucker. A handful of oil thieves were also convicted in Oklahoma, New Mexico, and California. The oil companies themselves—including Amoco—emerged unscathed, even though the Linowes Commission and most other experts considered the companies responsible for not securing the oil fields against theft.[36]

It would be impossible to assess how many more convictions there might have been if in 1981 the Interior Department had not shifted its focus in the middle of the investigation from physical thefts to paper audits. Linowes objected to the decision. Many experts—including Linowes and former USGS district engineer in Wyoming George Kinsel—felt convinced there were many physical thefts that did not leave a "paper trail." Nevertheless, James M. Yohe, assistant inspector general for investigations, decided to concentrate on audits instead. "Undercover investigating is not the only technique. It has a lot of romance and pizzazz, but I don't know if it's necessary," he said. Unfortunately, after the investigation's focus switched to auditing and accounting, federal agencies—and some tribes—forgot the importance of improving field security and inspections, which continued to be neglected for years.[37]

The paper investigations clearly offered bigger, quicker payoffs for the royalty owners, who recovered millions of dollars. For example, soon after the Wind River tribes hired accountants to check their royalty records, Conoco, Amoco, and Gulf paid the tribes more than $1.3 million in additional royalties. As with the field thefts, however, the Linowes Commission concluded that the total underpayments would never be known, much less recovered. The commission said Interior officials estimated one hundred million to several hundred million dollars a year could have been lost.[38]

The investigations proved that the Interior Department had been derelict in fulfilling its responsibilities in the office as well as in the fields. USGS had been keeping records of, not accounting for, royalty payments. The records were often inaccurate because the system relied upon manual instead of computerized entries of information. Many USGS and BIA employees, who had been handling millions of dollars in mineral accounts for years, did not understand basic accounting practices. The federal royalty program was basically an honor system given that USGS did not verify data, did not know which companies had paid, rarely conducted audits, and did not impose penalties for nonpayment or underpayments.[39]

The Osage Tribe and local BIA had discovered years earlier that penalties combined with effective monitoring succeed. A federal auditor found that only 0.3 percent of the payments to the Osages were late, compared with 70 percent for the Uintah and Ouray Reservation in Utah and 46.7 percent for the Jicarilla Apache Reservation in New Mexico. The Osage Tribe had computerized information for its 3,900 leases and required a late charge of 1.5 percent for each month the payment was late. The Linowes Commission found USGS almost never imposed penalties even for "gross, repeated underpayment of royalties," which prompted the commission to say, "It is remarkable that USGS royalty collection functions at all, considering that there are virtually no teeth to the system."[40] The Linowes Commission recommended throwing out the honor system but not because the members believed petroleum companies were without honor: "Underpayment often results from a defensible interpretation of a complex set of rules. . . . Oil and gas companies may take the same approach that most people do with their taxes: Where there is a doubt, they interpret the rules to their own advantage, guarding against overpayment."[41]

The Linowes Commission had some kind words for USGS, calling it an "esteemed scientific organization" dedicated to the pursuit of the earth sciences. USGS accomplished well its primary mission to explore and map the geologic resources of the country. Unfortunately, federal and Indian leases were lowest on the agency's priority list. Thus the commission recommended creating a royalty management agency, separate from USGS just as the Bureau of Reclamation and Bureau of Mines had been separated from USGS in the past.[42]

The Linowes Commission emphasized the importance of valuation and fieldwork, but the Interior Department gave both a lower priority in later years. The commission said the federal government should not just rely upon company-generated data. It should inspect each well each year and also periodically obtain well test data, run tickets, and LACT meter readings on a sample basis. To verify production the federal government would have to hire more inspectors so that it would be able to trace petroleum from the point of production to the point of sale, making sure that oil did not get "lost" before the royalties were paid. Most underpayments of natural gas royalties resulted from undervaluation, according to the commission. Because royalties were computed as percentages of the value of minerals sold, valuation was critical.[43]

FEDERAL ROYALTY REFORMATION

In time the Interior Department adopted most of the Linowes Commission's recommendations. A dinosaur was dragged kicking and screaming into the twentieth century despite its deeply ingrained distrust of adaptation. The changes were painful not just for the institution but also for the Indian royalty recipients as the bureaucrats tried to figure out where the tribes fit into the new system. The reforms made important changes in the ways that the federal government accounted for mineral proceeds and provided for state and tribal participation. The reformation suffered from four serious flaws, however: moving too fast without planning, treating Indian tribal governments in the same manner as states, neglecting fieldwork, and giving too much weight to industry comments and to efficiency when adopting valuation guidelines.

From the moment that Chuck Thomas blew the whistle, USGS and BIA had been under siege, resulting in an attempt to completely overhaul the royalty system in the space of a few years. With a flurry of activity, Interior showed its determination to do something—anything—quickly. Interior Secretary James Watt created a new Minerals Management Service (MMS) at Interior the same day (January 21, 1982) that the Linowes Commission issued its 267-page report. By the end of the year Congress had passed the Federal Oil and Gas Royalty Management Act (FOGRMA) to implement other Linowes Commission recommendations. Responsibility for field inspections had been shifted to the Bureau of Land Management (BLM), another agency of Interior.[44]

Unfortunately, Interior was not quite sure where it was going before it took off. The department purchased a new computer to handle royalty accounting before MMS had been created or the management team hired to participate in the decision. The system could not handle the massive load of information, and a new one was delayed for years. In an internal memo in 1984 a top MMS official, Robert E. Boldt, called the computer system a "disaster . . . a very rudimentary system, developed quickly by an inexperienced contractor." Nevertheless, Boldt said MMS tried to keep up a good front between September 1981 and February 1983 by asking companies to pay the Lakewood Accounting Center, *as if the new system were in place* (emphasis added). As a result, some royalty owners did not receive any money for several months. Later Boldt said MMS realized its second big mistake—the computer hardware that it had purchased could not handle the volume of information. Boldt said in 1984 the error-ridden files still had not been completely unraveled or reconciled.[45]

Moreover, because USGS had been completely understaffed, hundreds of new people had to be hired and trained. The agency had to start with USGS's totally inadequate data base; the old records were so far off when MMS started trying to reconcile them that MMS wrote off any balance under $100,000. Fortunately, that policy was not applied to Indian leases. For the first several years, MMS was preoccupied with training people and cleaning up data, where possible. In addition, the transfer of inspection responsibilities to BLM did not go smoothly.[46] For example, rather than admitting its problems, MMS officials turned defensive during the transitional period. Congressional investigators found that the MMS management chose not to work cooperatively with the states and Indian tribes but rather chose to "circle the wagons," repeatedly saying that "all was well within the program without stepping back to analyze whether the program was being led in the right direction."[47]

The Indians' transitional problems were aggravated by the fact that the new royalty management plan required three of Interior's "sister" agencies—MMS, BLM, and BIA—to work with each other. As one unnamed BIA employee said, "We couldn't get the three agencies to sing from the same song sheet." As the Indians launched verbal attacks against the circled wagons, each agency pointed fingers at the other to assign blame. Of the three agencies, only BIA was familiar with the federal government's special responsibilities to Indians. Possibly as a result of this ignorance, MMS initially audited only offshore federal oil leases, not Indian leases, because more dollars were at stake. Requests for audits of Indian leases became lost in the bureaucratic maze while Indian mineral owners lost their homes and cars as a result of unpaid royalties. These problems began to be resolved only after Congressman Mike Synar of Oklahoma publicly chastised MMS. He pointed out that Indians obviously suffered more than the federal treasury when royalties were not paid on time.[48]

The FOGRMA set the stage for better royalty management by providing for civil penalties of $500 to $25,000 a day and criminal penalties of jail and $50,000 for serious infractions; clarifying inspectors' authority; requiring regular audits and inspections; imposing additional site security and record-keeping responsibilities upon industry; and providing roles for states and Indian tribes.[49]

By mid-1989 neither MMS nor BLM was fulfilling all the requirements of FOGRMA, and another investigation revealed that many of the same problems of field thefts and accounting errors persisted. Nevertheless, each agency had made some improvements as a result of reforms. Interior collected millions of dollars in civil penalties for violations. Just by training companies how to fill out forms the department reduced the error rate from over 40 percent in January 1983 to 4 percent in 1987. After six years of auditing, the federal government had collected $12 million in additional royalties and $3 million in late payment charges for Indian leases—an average of $2.5 million a year that the Indians likely would never have received if Thomas had not pulled over the tanker in Wyoming.[50]

By the end of 1987 the department had more than doubled its inspection staff, but it still did not have enough inspectors to check more than 92 percent of the high priority leases, not 100 percent, as required by FOGRMA. Moreover, the department had delayed final valuation regulations because of controversy over proposed changes. Thus two of the most important reasons for royalty problems—inspections and valuation—had not been resolved by 1988. Although far from perfect, the improvements that did occur—collection of civil penalties, error rate reduction, collection of millions of dollars in royalties and late payment charges, more regular audits and inspections, and involvement by states and tribes—inspired praise from William Proxmire, the senator from Wisconsin, who said, "Instead of a lashing, they deserve a pat on the back for beginning to do a difficult job well." The Interior Department's inspector general also said in April 1988 that the department had made considerable progress.[51]

FITTING INDIANS INTO THE SYSTEM

Although MMS and BLM had serious problems fitting Indians into the federal system during the transition, the Indians started with an advantage. FOGRMA was one of the first laws that provided for tribes' having the same authority as states. In the past the federal government frequently had ignored energy tribes when planning federal policy, which had brought criticism from CERT and others. The Linowes Commission and the Congress heeded such criticism and the Indians' insistence at royalty investigations that they would not be ignored after the Wind River exposé. As a result the debates centered upon "how" rather than "if" Indian governments would play a role in deciding upon and implementing royalty management policies.

To involve tribes and states Interior adopted Linowes Commission recommendations to appoint an advisory committee and to provide access to lease and royalty information. The royalty management advisory committee included members from industry, states, tribes, allottees, and the general public, who advised the department as it implemented its royalty management plan. Interior appointed representatives of seven tribes and an allottee association. Although MMS did not always abide by the committee's recommendations, the often heated discussions forced agency personnel to face the actual impacts of their policy decisions.[52]

The State and Tribal Support System (STATSS) program gave states, tribes, and federal officials computer access to MMS financial data on leases with which they were concerned. MMS offered to train tribal royalty auditors at its offices in Lakewood, Colorado, through the Intergovernmental Personnel Act (IPA), and the two tribes that took advantage of the offer—the Navajo and the Northern Ute—gained tremendously from the training. Interior Secretary Watt invited states and Indian tribes to enter into cooperative agreements to increase oil-site inspection work and auditing.[53]

These initiatives, which treated tribes as states, signified critical progress for tribal governments given that they had previously been ignored by Congress and federal regulatory agencies. It sometimes was difficult for tribes to take advantage of these opportunities, however, because the tribes were not just like states. Congress and MMS, which had some understanding of state governments and their capabilities, designed programs with them in mind, ignoring the special constraints faced by tribes, such as the lack of a tax base and of developed infrastructures.

When Senator Melcher chided some tribes in 1981 for not taking more responsibility for site security and royalty accounting, many energy tribes, including the Shoshone, Northern Ute, Navajo, and Jicarilla Apache, testified that they would be glad to—if they had the funding, the cooperation, and the authority they needed. The tribes recognized as well as Melcher the monetary benefits of being involved, but they also recognized tribal and federal obstacles, which were different from those faced by states.[54]

For example, during the self-analysis sparked by the Wind River thefts, some states realized that they needed to increase communication among state agencies, reallocate state monies, and change administrative priorities. Although such changes were not necessarily easy for the states to make, state agencies generally had trained staffs, funding, and acceptance by the public. The tribes, on the other hand, had very limited funds and often had to develop governmental structures from the ground up in the midst of the crisis. Most major energy tribes had fledgling natural resources departments, usually established within the past one to six years, with a handful of employees who could be relatively quickly trained to conduct field inspections. None but the Navajos had royalty management or auditing divisions, and few tribes had within their staffs the educational background they needed for royalty management in fields such as finance, geology, data management, economics, marketing, and engineering. Thus building the government divisions and hiring the experts took tribes more time than it took the states.[55]

As Melcher pointed out, the rapid rise in oil prices resulted in royalty increases that made royalty accounting worthwhile. Several reservations—the Osage, Navajo, Kiowa, Comanche, Apache, Fort Sill Apache, Wind River, Wichita-Caddo-Delaware, Jicarilla Apache, and Uintah and Ouray—produced royalties of more than $10 million in 1981, according to a BIA task force on economic development. Several others—Cheyenne-Arapahoe, Five Civilized Tribes, Blackfeet, Southern Ute, Ute Mountain Ute, and Fort Berthold—produced over $1 million.[56]

These figures did not necessarily represent revenue available for tribal programs, however. For example, a substantial portion of the income in Table 5.1 actually belonged to individual Indian oil and gas owners (allottees), not to the tribes. The federal government does not distinguish between tribal and allotted income in such reports—in fact three of the reservations listed by BIA in its chart consist almost entirely of allotted oil and gas lands with limited tribal ownership, if any. A fifth, the Osage, is only an underground reservation because the 1906 Osage allotment act reserved minerals in communal ownership while the surface was divided. Although the Three Affiliated Tribes own oil and gas on the Fort Berthold Reservation, all production in 1981 was from allottee wells. Of the nine reservations listed on the chart, only the Ute Mountain Utes and Jicarilla Apaches had exclusively tribal production.[57]

The income that did go into tribal coffers was already stretched thinly to cover government services and was not readily available for new tribal programs. On some reservations, that tribal income was set aside for division among individual tribal members as their "dividends" from communally owned oil and gas, as distinguished from tribal members who received royalties from their individually owned oil and gas. For example, fed-

TABLE 5.1 Indian Reservations Receiving High Oil and Gas Revenues in 1981 (in dollars)

Revenue Level	Reservation	State	Oil and Gas Revenue	Per Capita Revenue[a]
Greater than 10 million	Osage	OK	71,332,912.69	15,020.62
	Navajo	AZ, NM, UT	39,065,047.00	372.13
	Kiowa, Comanche Apache and Fort Sill Apache[b]	OK	31,430,617.30	n.a.
	Wind River	WY	28,128,149.28	6,763.20
	Wichita-Caddo-Delaware	OK	23,680,172.90	n.a.
	Jicarilla Apache	NM	21,169,477.55	12,343.72
	Uintah and Ouray	UT	13,623,797.42	6,645.75
1 million to 10 million	Cheyenne-Arapahoe[b]	OK	8,684,238.62	n.a.
	5 Civilized Tribes Consolidated[b]	OK	7,424,806.73	n.a.
	Blackfeet	MT	7,276,076.07	1,316.94
	Southern Ute	CO	2,788,798.62	3,261.75
	Ute Mountain Ute	CO	1,715,387.23	1,544.00
	Fort Berthold	ND	1,218,752.00	461.65
0.1 million to 1 million	10 tribes	n.a.	n.a.	n.a.
160 to 99,000	12 tribes	n.a.	n.a.	n.a.

[a]Based on 1980 census data of the American Indian population living on reservations. Per capita figures are used for comparison purposes only; they do not mean that this amount is actually distributed to individual Indian residents on the reservation.
[b]These tribes do not have federally recognized reservations; thus there is no population data in the census.

Source: U.S. Department of Interior, Bureau of Indian Affairs (BIA), *Report of the Task Force on Indian Economic Development* (Washington, D.C.: GPO, July 1986).

eral law required that the Wind River tribes distribute 85 percent of their tribal royalty income to members in per capita payments. Westmoreland Resources's contract with the Crow Tribe stipulated that 60 percent of the coal royalties would go to annual per capita payments. Osage tribal members received 99 percent of their royalty income, leaving 1 percent for tribal use.[58]

The opportunities for state and tribal involvement offered by MMS and BLM all assumed availability of qualified personnel, of basic services (such as reliable telephone service), and of funding. During the first two years of MMS's training program, for example, only two tribes—the Northern Ute and the Navajo—sent tribal staff members to MMS offices in Lakewood for training, although five states used the program. Under the Intergovernmental Personnel Act, MMS employed tribal and state workers, preferably for a year or more, at its Lakewood headquarters. Despite the advantages of the program, some tribes felt they could not spare a college-educated tribal

Perry Shirley, a Navajo, received training from MMS in royalty accounting in 1986. Later he was named auditor general for the tribe. (Photo by Kenji Kawano.)

member to go to Lakewood for that long, especially given that he or she could only spend part-time on the tribe's own leases while there. The Northern Ute IPA employee stayed in Lakewood for only six months before his tribe called him home. On the other hand, the Navajo IPA employee, Perry Shirley, stayed in Lakewood for eighteen months before returning to his reservation where he became auditor general for the Navajo Tribe.[59]

It took several years before STATSS served the tribes as well as it served the states. Tribal STATSS computers sat idle until tribes geared up their staffs and MMS got the kinks out of its system, trained tribal staff members, solved telephone access problems in remote reservation locations, and replaced defective computers.[60]

As another example of the problems inherent in the reforms, tribes had to share costs, whereas states did not, if they wanted to take over responsibilities for royalty accounting. Congress specified in FOGRMA that tribes had to provide half the costs of audits and field inspections under cooperative agreements. In contrast, the law allowed 100 percent federal funding when states were delegated authority for inspection and auditing. Although the state and industry members of the royalty advisory committee joined with the tribes to request that the federal government provide both the states and the tribes with the same level of funding, the secretary of Interior refused. He said that because tribes received all of their royalties and states received only half, tribes should pay their own costs. The tribes argued that the federal government—including MMS and BLM—had a trust responsibility to them, not to the states. Meanwhile BIA refused to provide some tribes with contract funds to cover the tribal half of cooperative agreement costs.[61]

Although MMS and BLM continued to suffer from severe manpower shortages that resulted in audit and inspection backlogs, they resisted working with the tribes and states. The agencies took years to adopt regulations for cooperation or delegation, even though such agreements were intended to lighten the federal government's workload. MMS finally opened the door, and by 1988 eight states had signed up for delegated authority. In 1985 MMS signed its first cooperative agreement with a tribe, the Navajos, giving them sole responsibility for auditing their lease records for the previous six years. MMS was still responsible, however, for the actual collection of unpaid or underpaid royalties. By 1988 no other tribal cooperative agreements had been signed.[62]

BLM continued to discourage cooperative field inspections, even with tribes willing to share costs. The inspector general sharply criticized BLM for not even asking Congress for cooperative agreement funds and for not taking advantage of the opportunity of state and tribal inspectors' help. The agency did not recognize others' expertise unless they had taken BLM training. The inspector general found BLM inspectors did not communicate with tribes or with BIA inspectors to avoid duplicating efforts. By 1988 forty-five BIA and tribal inspectors had taken the BLM inspection classroom training, but none had completed the field training, which was necessary to take enforcement actions.[63]

Ironically, by certain delays MMS forced some tribes to take up mineral accounting. Until the spring of 1988 MMS refused to handle the day-to-day accounting for tribal joint ventures or other alternatives to standard lease agreements even though Congress authorized such agreements in 1982 and BIA approved them. MMS said they did not fit into the computer system for standard lease royalties, and each would require a separate program. The tribes knew that nonstandard agreement revenues needed to be scrutinized at least as much as standard ones. When MMS reluctantly agreed in 1986 to audit past revenues from such agreements, the audits yielded $250,000 in additional monies for tribes. MMS also delayed implementing its production accounting system, which tribes needed to verify payment of tribal taxes. This delay forced tribes to learn how to obtain their own production reports.[64]

Despite the resistance of the agencies, the major energy-producing tribes refused to put their trust entirely in the federal government ever again. Even if they could not get authority for auditing royalties, most audited their own tax revenues, and most also had at least a limited field inspection program. Although they could not issue their own citations without having completed BLM training, companies often responded voluntarily to tribal inspectors' complaints.[65]

The tribes used their own funds, audit proceeds, tax revenues, and user fees to pay their staffs and consultants for checking accounts and inspecting well fields. The Northern Ute Tribe, for example, funded its accounting and inspection programs with tribal funds except for one federal contract position. The Blackfeet Tribe paid the total revenue department budget, which amounted to $129,000 a year, out of tribal funds, including $45,000 for the inspection program. Several hired their own accountants. The Jicarilla Apaches' non-Indian accountant had a master's degree in accounting, and the Wind River Reservation Tax Commission director was an Arapahoe tribal member with a master's in finance. Others hired CERT or similar consultants to conduct audits, set up computerized oil and gas monitoring and accounting systems, and train employees. At CERT workshops, tribes swapped ideas on royalty management.[66]

Although neither the Apache nor the Southern Ute Tribes had a cooperative agreement with MMS, they did work with the agency informally on audits. In both cases, MMS remained responsible for collection. Without cooperative agreements, tribes were at a disadvantage because they theoretically could not get the data they needed for auditing and accounting. Companies submitted all production, sales, royalty, and other reports to MMS. Eventually some tribes persuaded companies to provide duplicate production reports to them. Others enacted ordinances or adopted contract provisions assuring access to industry records when necessary. Some tribes went even further. To ensure that they were being paid for actual production, they compared more data from sources other than those used by the federal government. The Southern Ute Tribe, for example, obtained its gas production data from gas pipeline companies rather than from the producers themselves, as BLM did. The Blackfeet Tribe's revenue department compared BLM and state production forms against tribal tax forms and verified that information through run tickets from oil tankers. For gas wells, tribal inspectors regularly checked gas meter charts. When they found any discrepancies, they notified BLM, BIA, and the companies.[67]

To avoid depending upon MMS, other tribes took their royalties as oil or gas. Even prior to the Wind River exposé, the Jicarilla Apache Tribe decided to get out of the federal royalty system as much as possible. Rather than continuing to get monetary royalty payments, the tribe began taking most of its royalty percentage "in kind," meaning the companies paid the tribe's share in oil or gas, which the tribe then sold. The Northern Ute Tribe, which also was not confident it was being paid for all the gas produced on its lands, in 1988 was considering the same course.[68]

For the larger tribes such efforts paid for themselves. Jicarilla Apache tribal accountant David Wong, a non-Indian, reported in 1987 that the tribe spent about $70,000 a year on salaries and overhead for its auditing and accounting work. In one four-month period the tribe brought in $500,000 through its audits alone. The Navajo Tribe and the Southern Ute Tribe also found that collections more than made up for the costs of running their auditing programs, and the Southern Utes said they also were able to pay for the tribe's oil and gas computer hardware and software.[69]

Although the smaller tribes with less production also had reason to be skeptical of the government's ability and willingness to protect them, they had fewer resources to address the problems themselves. For example, although the Jicarilla Apache Tribe had 1,500 producing tribal wells and the Navajo Tribe had 970 in 1980, the Fort Peck Assiniboine and Sioux Tribes had only 5. Only 3 percent of the $2.6 million in royalties from that reservation went to the tribes that year; the rest was from allotted wells. The larger tribes often spoke out on behalf of the smaller tribes, which sometimes could not even afford to send representatives to meetings with MMS and other tribes. These tribes and the allottees still had to depend upon federal protection.[70]

FIELDWORK NEGLECTED

In 1988 two serious royalty management problems continued to plague the states and the tribes, and even the most ambitious and independent tribes could not escape them. Despite Linowes Commission recommendations, the government had deemphasized its fieldwork, including production accounting, and had delayed its valuation regulations. MMS put off and tried to cancel plans for adopting the production accounting system, without which royalty accounting still relied upon the companies' honor. MMS said in 1985 that it would be too administratively difficult to begin production accounting, which required comparing sales data, lease use data, and production data, periodically verified in the oil fields, to assure that companies accounted for all oil and gas produced. The logistics were intimidating; the Interior Department received over 20,000 monthly production reports submitted by 1,800 oil and gas operators for 80,000 wells.[71]

No matter how difficult, however, implementing production accounting was essential. The tribes especially needed reliable production information to verify companies' payments of tribal severance taxes, but that information was crucial for federal leases, too, as congressional investigators and the Linowes Commission had emphasized. Without it, the agency and the tribes could not crosscheck information, and the system was little better than the old USGS unminded store. After considerable political pressure, MMS said it planned to implement the system in stages before the end of 1990.[72]

Despite the fact that the whole royalty reformation began because of field security problems, federal watchdogs practically ignored BLM's fieldwork in later years, turning their attention only upon MMS. When oil prices fell, the temptation for physical theft was reduced. Nevertheless, the reliability of the royalty accounting system depended heavily upon field inspections because otherwise it just compared industry-supplied data. The General Accounting Office conducted no studies of BLM fieldwork in the late 1980s. In the two Interior Department Inspector General reports that addressed BLM fieldwork, the department's inspector said that the federal government was not fulfilling the principal objective of FOGRMA—ensuring accountability. The reports criticized BLM management for not adequately supporting the inspection program and for not protecting against drainage. The drainage audit found that BLM had a fifty-three-year backlog of drainage cases, which could have resulted in losses to Indians and the federal government of as much as $6.4 million. Because federal agencies rarely verified production and sales volumes, the system still relied heavily upon operator integrity to report production.[73]

In 1988 these problems persisted. Although upper-management BLM officials said BLM had enough inspectors, others in the agency said it did not. BLM could not fulfill the minimum requirements of FOGRMA, which specified that every high-priority lease site should be inspected annually. Only 92 percent of the high-priority Indian lease sites were inspected in 1987. When BLM defined "high priority," it gave itself considerable leeway; some critics said its definition excluded the majority of Indian wells from the annual inspection requirement. If they were not included among the wells that required annual inspections, they might not get inspected at all, according to some BLM personnel who were concerned about the shortage of inspectors. Only 62 percent of Indian wells were inspected in 1987.[74]

VALUATION

Many of the tribes' frustrations stemmed from MMS's tendency to treat Indian royalties the same as federal. The Indians especially objected to MMS's methods for determining the value of federal and Indian minerals. Because royalties were figured as percentages of value, millions of dollars were at stake. In a valuation case against Supron Energy Corporation and three other companies, the Jicarilla Apache Tribe received over $1.6 million.[75] The Linowes Commission found that most underpayments of natural gas royalties resulted from undervaluation. For years Interior considered selling price as equivalent to value. The Linowes Commission said prices did not necessarily reflect value, especially with vertically integrated companies. For example, a company that produced oil might not charge full price for oil sold to its own refinery.[76]

As the owner of federal minerals the Interior Department's responsibility was to stimulate national economic development, encourage domestic energy production, and—incidentally—make some money for the public treasury. Its responsibility to the Indians, however, was different. Although the Interior Department sought only a "reasonable return" from federal minerals, CERT and NCAI argued that the department should guarantee the "maximum return" from Indian minerals.[77]

The courts agreed. In the 1970s the Jicarilla Apache Tribe filed three lawsuits against petroleum companies, challenging the valuation used by the companies and accepted by the Interior Department. The courts ruled that the Interior Department breached its trust responsibility to the tribe by allowing the companies to use values that were too low. In the *Jicarilla Apache v. Supron Energy Co.* decision, the Tenth Circuit Court of Appeals said, "Given two reasonable interpretations, Interior's trust responsibilities require it to apply whichever accounting method . . . yields the tribe the greatest royalties."[78]

Congressmen Mike Synar of Oklahoma and Sidney Yates of Illinois believed the Reagan administration sided with the oil and gas companies instead of defending the interests of Indian royalty owners. Because of the continued controversy over valuation, Interior did not implement its oil and gas valuation regulations until March 1988 nor its coal valuation regulations until March 1989. The Interior Department's inspector general chastised MMS for the delays, saying MMS spent over a million dollars a year resolving individual valuation disputes that would not have been necessary if valuation guidelines had been issued.[79]

Indian advocates usually focused on valuation battles concerning Indian resources rather than federal resources, saying if the federal government wanted to give away the public's oil, gas, coal, and uranium, that was its business. Occasionally CERT spoke out against the federal government's methods for valuing federal resources, too. In the West only the federal government owned more energy resources than the Indians. Consequently, when the federal government set the value for its coal or petroleum too low, it undercut not only resources owned by Indians but also those owned by states and private parties.

DILIGENCE AND DRAINAGE

The tribes' increasing involvement in royalty accounting benefited them far beyond merely receiving back payments and assuring more reliable royalty payments. Sometimes tribes acquired wells as part of their settlement for underpaid royalties. The biggest payoffs came when tribes discovered wells that were not producing. Under standard lease terms companies could hold leases as long as at least one well within the lease was producing in paying quantities. Soon after tribal members began combing through the records and crawling around the oil fields, they discovered leases with little or no production. Although companies sometimes needed to curtail production to work over wells or for other legitimate reasons, the leases required the Indians' consent for lengthy curtailments. In some cases, by not producing, companies were allowing non-Indian wells to drain tribal oil and gas. In other cases, the companies tied up acreage that tribes could have sold for much higher terms. In all cases, the Indians were deprived of needed income from wells when they were not producing.

The federal government was responsible for canceling leases that were not producing. Because of understaffing, the government's laissez faire attitude, and shoddy record keeping, however, federal agents had not done their job for many years. The Osage Tribe at the turn of the century and the Wind River Tribes in the 1920s and 1930s fought to have leases canceled for nonproduction (see Chapter 2). In some cases in the 1980s, such as on the Southern Ute and Ute Mountain Ute reservations, BIA did not even know whether lands were leased or wells were producing. In one case in Oklahoma, BIA extended a nonproducing lease without the consent of the Cheyenne-Arapahoe Tribes.[80]

Following several years of falling oil prices, the Interior Department adopted a policy in 1986 that allowed companies to stop production on "stripper wells" without losing their leases. According to BLM three-fourths of all Indian wells were stripper wells, meaning they produced ten or fewer barrels a day. The policy was justified to some extent to discourage abandonment of wells on federal lands. Indians, however, had different needs and different consent provisions built into their leases. But until CERT and tribal attorneys intervened, the federal stripper policy also applied to Indian leases, without the leaseholders' permission. Because of the resulting confusion, the federal stripper policy continued for several years to exacerbate the problem of companies shutting in wells without notifying the Indian mineral owners.[81]

Gradually tribes began to force companies to abide by development requirements, using lawsuits, tribal ordinances that required production, and innovative industry agreements. Recognizing the inadequacies of the federal record-keeping system, both the Southern Ute and the Ute Mountain Ute Tribes pieced together their own lease records, hiring consultants (including CERT) to check through county, state, and federal files at various locations. The effort paid off for the Southern Utes, who acquired several nonproducing wells. When BIA would not cancel a lease, the tribe enacted a tribal ordinance (approved by BIA) requiring production. After a lengthy evidentiary hearing, the tribe took possession of two oil and three gas wells relinquished by the company.[82]

The neighboring Ute Mountain Ute Tribe enacted a similar ordinance and also negotiated an agreement with Wintershall Corporation, which provided for the company conducting a $200,000 research effort to clean up tribal land records and find out which lands had been leased and which wells were still producing. During the market downturn, CERT helped other tribes to discover leases that were not producing and, in some cases, to take over leases from companies that no longer wanted them. Because they did not have to pay royalties or severance taxes, tribes often could make money or at least cover their costs at prices much lower than companies could afford to operate.[83]

CONCLUSION

Why did tribes experience such serious royalty accountability problems? The federal government, which was charged with minding the store, had consistently treated Indian resources the same as federal, to the detriment of the Indians. Reform resulted only after years of lawsuits and political pressure from the tribes. The Interior Department found itself, as any bureaucratic institution, unable to change quickly. Years of neglect resulted in a heavy toll on the Indian mineral owners and, to a lesser extent, on the public, both of whom would never recover millions of dollars because of inadequate records and the statute of limitations. During the reforms, Interior focused upon streamlining royalty accounting and reducing costs rather than recognizing its special responsibilities to Indians.

Recognition of royalty problems benefited the tribes. As a result of federal and tribal reforms, the tribes received millions of dollars in royalties and interest that otherwise would have been lost. These monies helped tribes reduce their dependency upon dwindling federal grants and contracts. Realizing the federal government's failings, tribes were forced to act, to exercise their long-dormant governmental powers. Initially the tribes became involved because they thought the federal reforms took too long, they wanted more job positions for tribal members, and they did not trust the federal government. Unlike federal officials, the tribes had a vested interest in seeing that their deals were enforced.

Soon, however, tribes learned that royalty management work had other advantages beyond money. It improved their staffs' and elected officials' overall understanding of the oil and gas business. As a result, they began negotiating contracts that provided for site security authority, access to records, and tribal monitoring of funding. They took more active interest in federal royalty administration, protecting themselves and other tribes from poorly thought-out initiatives. By computerizing their data bases they enhanced elected officials' planning abilities. With their increased understanding of the oil and gas business they could recognize when government and industry decisions jeopardized their interests. Finally, through joint royalty initiatives, such as the royalty management advisory committee, tribes forged bonds with the states. In many cases they also improved relations with industry as the result of better understanding and appreciation for one another's concerns. Such bonds helped tribes as they exercised their governmental powers in the regulatory arena.

At the same time, tribes encountered limits on their governmental authority, especially concerning the individual tribal members who owned minerals. Tribal progress in accounting for royalties and inspecting leases did not benefit the allottees. For many reasons, some of them justified, the federal government clung tenaciously to its responsibility for the allottees and turned down tribes' efforts to help their own allottees. Yet too often federal officials failed to fulfill those responsibilities.

CHAPTER SIX

The Forgotten People:
Indian Allottees

As a whole, they [allottees] are not highly educated. . . . They have not dealt with large sums. They have had no occasion to make business deals of any consequence. They are without information as to the value of their trust oil and gas. . . . These individual Indians are on the lower end of the economic scale. They tend to look on any sum of money offered for their oil and gas as a piece of good luck.

—Norman Hollow[1]

When Congress passed the Indian Reorganization Act (IRA) in 1934, hope abounded that the law would reverse the allotment process and reconsolidate Indian lands under tribal ownership. Yet today, allottees own more than nine million of the fifty-three million Indian trust acres in the lower forty-eight states. Most allottees on reservations still suffer from poverty, due largely to their lack of control over their resources and to the diminishing size of their estates, divided under inheritance laws.[2] On the energy reservations that were allotted, some of the allottees have caught a glimpse of the middle-class life style envisioned by Senator Henry L. Dawes in 1887 when he sponsored the General Allotment Act. Nearly forty thousand allottees—most of them in Oklahoma and New Mexico—received mineral income in 1988. A few oil-owning allottees became millionaires overnight.[3]

Generally, however, mineral owners have found that energy minerals have benefited them little. Instead of the American dream envisioned by Dawes, they have experienced a nightmare. Their ownership of land and minerals has been splintered. They have suffered under the Interior Department's trusteeship for the same reasons that the tribes have. The department has not involved them in development decisions, has denied them access to information that exists about their resources, and has not developed either the necessary resource information or expertise. After leases have been signed the department has failed to enforce federal regulations, thus allowing the theft of oil, underpayment of royalties, and ruination of water wells and pasturelands. Too often the department has sided with industry instead of the Indians. At the same time, the department has been hamstrung by a lack of funds and staff to tackle the difficult allotted mineral issues, exacerbated by complicated ownership. As with the tribes the department has been engaged in a constant balancing act between protecting the allottees and allowing them to determine their own destinies.

The advances made by the tribes during their renaissance of the 1970s and 1980s generally have not helped the allottees. Unlike the tribes the allottees have not had the money as individuals to hire their own attorneys, geologists, or accountants. They have lacked collective clout, which the tribes gained through the Council of Energy Resource Tribes (CERT). The Interior Department generally has discouraged tribes' attempts to assist allottees.

For the most part the energy mineral–owning allottees have been forgotten. When describing Indian minerals the Interior Department and CERT consistently have lumped tribes' and allottees' minerals together. BIA publications have referred to "annual revenues received by tribes" and then quoted figures that represented both tribal and allotted mineral revenue even though about one-third of the Indian mineral revenue produced in the lower forty-eight states came from allotted lands. Although a large percentage of the Indian lands on CERT reservations was allotted, CERT has glossed over the difference between tribal and allotted assets, inadvertently contributing to the illusion that energy tribes are rich.[4] After Indian Commissioner John Collier put the spotlight on the problems of the allottees in 1934, the nation turned its back on them. Yet in some areas of the country, allottees have found that by working together with tribal governments, organizing allottee associations, hiring attorneys, and enlisting help from Congress, they can effectively address the special problems of individual Indian mineral owners.

FRACTIONATED OWNERSHIP

For decades the allottees have been plagued by the ghosts of the nation's vacillating Indian policies. The allottees themselves exist only because Congress tried to destroy the tribes and assimilate the members into the mainstream through the General Allotment Act of 1887 and various other allotment acts, which divided many reservations into individually owned parcels. Allottee ownership rights suffer from two effects of earlier policies: split estates (when the mineral ownership did not go with the land ownership) and fractionation of their estates through inheritance. Because most of the original allottees have died, their allotments have been inherited at least once, usually as undivided parcels owned by multiple heirs, whose number grows at an exponential rate. Fractionation was already a problem when Indian Commissioner John Collier testified before Congress in 1934, advocating the Indian Reorganization Act. He told Congress, "For the Indian, the situation is necessarily one of frustration, of impotent discontent. They [allottees] are forced into the status of a landlord class, yet it is impossible for them to control their own estates; and the estates are insufficient to yield a decent living."[5]

The Indian Reorganization Act failed to reconsolidate Indian lands, mostly because Congress never adequately funded land acquisition. A generation later, over 40 percent of the realty work in many BIA offices consisted of heirship matters. A study in 1971 said BIA officials pored over complicated computations involving common denominators as large as 54 trillion when dividing an allotment's income. Millions were commonplace. Agricultural leases generated only pennies a year for individual allottees. In the 1980s BIA figured it spent $18,000 a year administering a certain agricultural tract in South Dakota, which was worth $8,000.[6]

Since 1934 Congress has attempted many approaches to reduce fractionation, without much success. Allottees themselves have fought many of the proposals, partly because they questioned whether they would be adequately compensated for land and resources. This problem is especially acute when minerals are involved. If allottees are to sell their minerals to the tribe or another person or bequeath them to an heir, the allottees should know the mineral's value. The federal government has failed to provide reasonable appraisals for known minerals and in many cases does not know whether or not minerals even exist under certain lands.[7]

Several South Dakota allottees successfully challenged the constitutionality of Congress's latest solution, the Land Consolidation Act of 1983. The challenged section of the law exemplifies the problems created by not knowing the value of minerals. The section provided for the tribes inheriting small, allotted land interests, incapable of producing $100 in a year. If mineral value had not been determined, however, no one could decide the land's potential for production. Another challenge was pending on a later amendment to the law. Despite the controversy, the law had spurred family members to consolidate their shares to avoid losing them to the tribes, according to BIA realty specialists.[8]

Because of fractionation, most allotted agricultural lands are leased to non-Indians, and thousands of acres are left idle. With possibilities for benefits from the surface thus stymied, using the subsurface estate—the coal or oil—can look especially enticing. Unfortunately, however, fractionation has made mineral-development planning impossible and consensus decision making difficult. For the potential developer the logistics of getting approvals can be costly and frustrating. An oil company landman in North Dakota said that he had to contact as many as a hundred owners for signatures on a 160-acre allottee lease. He traveled to Seattle, Dallas, and Oklahoma City, carrying lease forms for one lease and costing his company nearly $7,000. If an owner with a tiny fractional interest refused to sign, the lease could not be enacted at that time, and his efforts would have been for naught.[9]

By making land unusable fractionation has contributed to the loss of land and minerals as allottees continue to remove trust protection and sell their resources. When an allottee gets a fee patent for his or her land or minerals and takes it out of trust, his or her ownership is then just like a non-Indian's. The allottee can sell the land without restrictions. This situation combined with the loss of mortgaged trust lands contributed to serious land losses in the 1980s. If Indians use trust land as security for loans, they lose Indian ownership if they fail to make payments. By 1987 over 220,000 acres of allotted trust lands had been mortgaged to the Farmers Home Administration (FmHA) in Montana and the Dakotas. FmHA predicted that 132,000 acres would be lost by allottees. The land sale problem was especially serious on the Crow Reservation, where allottees had taken more than 4,800 acres out of trust in 1986 and 1987.[10]

At times the mineral rights were sold separately, with abuses reminiscent of Oklahoma at the turn of the century. During the oil boom in the early 1980s, some tribal members at Fort Peck were buying allottees' mineral rights "dirt cheap," according to the BIA superintendent and to Tribal Chairman Norman Hollow. Hollow said USGS contributed to the abuses by, for example, appraising oil and gas rights at $10 an acre, rights that already brought $325 an acre in bonus bids alone at oil lease sales. BIA at Fort Peck intervened to protect allottees and put a temporary moratorium on the sale of allotted mineral rights. Recognizing how difficult it was to evaluate the minerals' worth in untested areas, the regional office instructed BIA realty officers throughout Montana and Wyoming to discourage allottees from selling mineral rights even when they sold the surface. According to the BIA manual, however, BIA had no moral or equitable right to deny an application for a patent. So if allottees insisted, BIA realty officers removed trust protection from minerals. When the allottees retained their mineral rights, BIA's bookkeeping headaches continued, of course. Without adequate appraisals, however, allottees could not afford to sign over their mineral rights to anyone, including the tribe.[11]

SPLIT SURFACE AND MINERAL OWNERSHIP

Decisions to split mineral and surface ownership contributed to the growing number of split estates created earlier. Laws passed during the first three decades of the twentieth century separated the ownership of minerals and surface land in many ways among allottees, non-Indians, tribes, and state and federal governments. The severed estates created the same conflicts for Indians that they created outside the reservation where mineral and land ownership was split by homestead acts and railroad grants. The laws for federal and Indian minerals were designed to facilitate development. Consequently the mineral estate was considered dominant: No permission from the surface owner was required, and the lessee could use as much of the surface as necessary. The surface owner received less monetary compensation than the mineral owner, who sometimes displayed a cavalier attitude toward the rancher's or the homeowner's sacrifices.[12]

Such conflicts surfaced as early as 1909 in Osage County, Oklahoma. Allottee farmers objected when a drilling company cut fences and moved housing, cows, and horses onto drilling sites for the convenience of its employees. The allottees formed what may have been the nation's first allottee organization, the Osage Protective Association, to block renewal of the company's lease and to promote allotment of the tribe's mineral estate. Under the 1906 Osage allotment act, the tribe had retained an "underground reservation" owned communally by those on a 1906 membership list and their heirs. But because of conflicts with their rights, the allottees in 1909 wanted to reverse that decision. Soon after the formation of the allottee organization, however, the Osage Council decided to increase the minimum royalty, impose stringent safeguards against surface and water damage, and recommend putting wells on uncultivated land when possible. By thus addressing the landowners' concerns the council removed some of the impetus for mineral allotment, and communal ownership of minerals continues today.[13]

Gradually a system evolved over the years in this country for paying damages to surface owners and requiring minimal reclamation. When companies began strip mining coal and uranium in the West, however, the potential problems of the surface owners—both Indian and non-Indian—escalated. Although strip mining was much cheaper and safer for the miners than underground mining, it disrupted the historic means of balancing the needs of the surface owner with those of the mineral owner. Mining was no longer just an inconvenience. A strip mine could totally prevent use of the land by the surface owner, who often had to relocate.[14]

The threat of extensive disruption from coal strip mining loomed on three allotted reservations in the Northern Plains—the Fort Berthold Reservation in North Dakota and the Crow and Northern Cheyenne reservations in Montana—in the 1970s when BIA leased substantial portions of those reservations to coal companies with rubber-stamp approvals from the three tribal councils. On the Fort Berthold Reservation, 29 percent of the minerals proposed for leasing were allotted. The Northern Cheyenne Tribe owned all of the coal rights on its reservation, and the neighboring Crow Tribe owned most of the coal rights.[15]

Because BIA had violated its own regulations, the federal government voided all permits and leases at the tribes' request but not before creating an uproar on the Cheyenne Reservation. The Northern Cheyenne allotment act of 1926 said that mineral rights would belong to the tribe for fifty years and then go to the allottees. By 1968 the stakes were high. When testifying for continued tribal coal rights, the tribe told Congress that the coal's value had escalated to more than $2 billion. Concerned that the large number of allottees living off the reservation (one-third) could choose to mine, taking the use of the land away from the resident tribal members, Congress in 1968 extended tribal coal rights. Fearing future monetary claims by the allottees, however, Congress left a loophole. Congress said the tribe got the mineral rights only if the courts agreed; so the tribe sued the allottees to find out. The Supreme Court decision favored the Northern Cheyenne Tribe in 1976, a decision that agreed with a 1932 ruling on Osage tribal mineral rights.[16]

In the Northern Cheyenne case the tribe supported communal mineral rights because it wanted to be able to plan the reservation's future, not because the tribal leaders wanted coal royalties. The tribal council had made it clear in 1973—when the coal ownership was still up in the air—that it did not want coal development under the terms of the contracts approved by BIA. Subsequent councils still had not started coal mining on the reservation in 1988. The *Hollowbreast* court battle created some controversy on the reservation, but tribal members generally agreed with the decision to prevent coal development. Later, however, when the tribal council approved an oil and gas development proposal, the decision pitted surface owners against the tribal council.[17]

Neither the Crows nor Fort Berthold's Three Affiliated Tribes have suffered through a serious tribe versus allottee battle over coal rights yet. Congress in 1968 had similarly extended in perpetuity the coal rights of the Crow Tribe, which through the years has been far more interested in developing strip mines and power plants than the Cheyenne Tribe.[18] On the Crow Reservation, however, coal development would affect non-Indians more than Indian allottees because most of the Crow land has been leased to non-Indian ranchers. The Three Affiliated Tribes of the Fort Berthold Reservation have been more interested in small-scale coal development than in disruptive strip mining.

As a result of the growing number of conflicts between traditional uses of western lands and coal strip mining, Congress in 1977 passed a federal coal strip mining law. The law addressed the conflict for some landowners because it provided surface owners with veto authority over strip mining federal coal. The new law, however, specifically exempted Indian coal, did not apply to other minerals, and contained a loophole big enough for the whole Navajo Tribe to fall through. The split mineral and surface rights were so little understood that Senator James Abourezk of South Dakota, one of the senators most active in Indian issues, did not know that tribes could own minerals under allotments. Speaking on the federal strip mining bill, Abourezk insisted that only the allottee or the federal government could own the minerals, which was totally contrary to the *Hollowbreast* decision of the previous year.[19]

A background in land ownership is necessary to understand the problems of implementing the federal strip mining law. In the coal- and uranium-rich San Juan Basin of New Mexico, ownership of the surface and minerals is held by Indian tribes, allottees, Indian homesteaders, private parties, the state of New Mexico, and the federal government, creating what one federal study called a "jurisdictional no-man's land." Indian tribes and individuals own over 8 million acres—46 percent of the surface. At the turn of the century BIA tried to consolidate lands and to purchase land or obtain allotments for Navajo homesteaders, but the pressure on Congress from local stockmen was too strong.[20]

When the area was targeted for massive uranium and coal development in the 1970s and 1980s, the conflicts forced the Navajo Tribe, the federal agencies, and the landowners to again try to unravel the tangle of conflicting rights. Everyone shared an interest in resolving the ownership questions because they made development difficult and in some cases impossible. The courts and the bureaucracies inevitably took years to move, but a development hiatus (starting in the late 1970s for uranium and a few years later for coal) gave the surface occupants a reprieve.

Dinébeiina Nahiilna Be Agaditahe (DNA), the federally funded legal aid program on the Navajo Reservation, and to some extent the Navajo Tribe worked with the affected Navajo individuals to try to settle questions involving both ownership and the decision-making process for mineral development. In August 1983 DNA filed a class action lawsuit saying that 14,500 Navajo allottees—not the federal government—owned the minerals under their lands. As with the Cheyenne and Osage allottees who battled with their tribes, Frank Etcitty and the Navajo allottees wanted mineral rights for themselves—and the right to decide whether the minerals would be developed or not. Unlike the other cases, which involved tribal minerals, the *Etcitty* lawsuit asked the court to decide whether the allottees should have lost their mineral rights to the federal government. For unknown reasons the government reserved the minerals to itself in 1919 instead of to the tribe or the allottees, which it did on other reservations during the same period. The suit named several companies as codefendants, which could lose their coal leasing rights in the area if the allottees were successful. The Navajo Tribe shared in the costs of litigation. In addition to assisting the tribal members, the tribe wanted to protect its own interests in disputed tracts, which once were allotments that the tribe had purchased or inherited.[21]

The Navajo Tribe also tried to help tribal members whom the Bureau of Land Management (BLM) considered squatters. After being released from exile at Fort Sumner in 1868, many of the squatters' ancestors returned to the San Juan Basin area to settle and apply for allotments on public lands that their families had used for hundreds of years. In the 1910s the federal government had only completed about half of the planned land allotments when pressure from white ranchers forced it to stop. When BLM was planning the coal lease sale for the San Juan Basin in 1979, it took the position that many of the Navajos in the area did not qualify as landowners under the surface owner protection section of the new federal strip mining law.

Therefore, according to BLM, they had no right to stop strip mining of their lands for federal coal. BLM also said the Navajo Tribe had no veto authority over mining of federal coal under tribal lands because the tribe did not fit the definition of a "qualified landowner." Navajo occupants expected to soon see draglines on the horizon.[22]

A national controversy over the federal coal leasing program erupted in 1982, however, with several lawsuits and investigations. The conflicts were followed by a drastic slowdown of the coal market. During the interim the tribe embarked on a massive exchange to get tribal ownership of squatter lands in the San Juan Basin area. BLM relaxed its stance and began issuing land ownership documents to some of the remaining squatters, legalizing 500 homesites by 1987. The federal government insisted upon keeping mineral rights, but it provided allottees with the right to veto mining within 300 feet of their homes and required companies to compensate them.[23]

Meanwhile the tribes tried to agree upon surface owners' rights when the surface overlay tribal coal. The 1977 federal strip mining bill provided for a separate Indian strip mining bill, but Congress still had not passed such a bill by 1989. Although the coal tribes were able to agree on many thorny issues and present a united stand on an Indian strip mining bill, they remained divided upon the surface owner issue. Resolving the question involved a delicate balancing act of collective versus individual rights. Many tribal advocates said that Indian coal was like private coal, and Indians should be allowed to develop it. If surface owners had veto authority, one person could tie the hands of the tribe as a whole. In the Crow Tribe several members of the tribal coal staff believed, based upon basic principles of equity, that surface owners should have veto authority even though much Crow coal lay under non-Indian ranchers' lands. Other tribes' situations differed. The Navajos and Hopis shared coal ownership under some Hopi lands and some Navajo lands. In its study on the issue the Interior Department advised against providing consent authority to owners over tribal coal, saying it would unnecessarily impede development of Indian coal. Because Senator John Melcher of Montana felt strongly that non-Indian landowners over Crow coal should have veto authority, the surface owner consent issue contributed to delays in enacting the Indian legislation.[24]

Surface owners over other minerals had no veto authority because the federal strip mining law applied only to coal. Thus people living in the way of uranium open pit mines or oil and gas exploration efforts still had to depend upon other mechanisms, including litigation and tribal regulations, to protect their interests. During uranium's heyday in 1978, several allottees contacted DNA because they felt powerless in decisions being made by energy companies and BIA concerning their land. BIA had approved 303 uranium leases covering 250,000 Indian acres in the region, and the federal government estimated a total of 3.5 million acres, including federal uranium, were going to be developed. When the leases involved allotted minerals, many allottees had technically approved the leases themselves. Some appreciated the new income, the opportunities for jobs, and the pickup trucks that the companies offered in return for signatures. Others felt pressured by industry landmen and did not understand what they were signing. Some gave their consent by thumbprint to documents they could not read.[25]

When uranium exploration began, Walter Peshlakai and several other allottees realized that it could change their lives irreparably: They could not continue grazing their livestock on the lands because uranium exploration crews left poisonous chemicals around drilling pads that killed the animals. In addition, the companies admitted that the ground-water source supplying the local area would probably go dry. Even though this development offended the Indians' cultural practices and religious beliefs, the Bureau of Land Management completely ignored these impacts in its environmental studies for specific mine sites.[26]

In December 1978 one hundred Navajos—most of them allottees—joined with the environmental group Friends of the Earth to file suit against six federal agencies, demanding a regional study on the impacts of the uranium activity. The allottees' concerns were taken lightly by the federal court system. A federal judge refused them an injunction, questioning why they would want to continue their "pastoral way of life" when they could be miners or millworkers. The lawsuit represented one of the earliest attempts by allottees to act collectively on issues other than mineral ownership. The case was eventually settled out of court after poor uranium market economics made it moot. DNA attorney Alan Taradash believes, however, that the allottees' suit succeeded in slowing development and in reminding the federal government of the cultural impacts of such development. The government subsequently considered such matters in the San Juan Basin environmental study for coal development. BLM said, for example, that there would be no more unannounced mining; coal miners would notify the Navajo Tribe, the Navajo Medicine Men's Association, and the local Navajo government subdivision (the chapter) at least three months before beginning coal mining activities.[27]

Even oil and gas exploration with its relatively low impact could disrupt tribes when different parties owned the minerals and surface. A Northern Cheyenne dispute in 1980 over an oil and gas contract with Atlantic Richfield Company (ARCO) became one of the more notorious examples of the problems caused by split ownership. Some allotted landowners felt that because of the unresponsiveness of the tribal government, they had to go outside the reservation—to the American Indian Movement, non-Indian environmentalists, Congress, and the federal courts—for help in protecting their land rights against the tribe's mineral rights.

The council signed over oil and gas exploration rights for the entire reservation to ARCO, a major company that would not have been interested in small parcels. Thus if the allottees had retained mineral ownership, the contract most likely would not have been possible because of the difficulty in obtaining consent. Although the tribal council required the company to pay allottees for land disturbed during exploration, some landowners thought development violated the tribal traditions and threatened the agricultural and religious use of their lands. "That [oil and gas development] was not the intent of this reservation. It was intended as a home," said Wayne Little-whiteman, a member of the traditional Elkhorn Brother Society on the reservation.[28]

The conflict caused a constitutional crisis on the reservation, as the tribe used its governmental powers to protect its proprietary interests in the oil. The landowners filed a lawsuit in the tribal courts to keep the ARCO exploration crews off their lands. When the chief tribal judge ruled in favor of the allottees, the tribal president, Allen Rowland, fired her. Rowland, who once said there would be coal mining on the reservation only over his dead body, defended his decision to fire the judge and to expedite oil and gas exploration because he believed some kind of economic development was necessary for the survival of his people. He freely admitted that the tribe did not have separation of powers, judiciary from administrative. He saw the polling place as the forum for grievances, and he had won tribal presidential elections easily for sixteen years.[29]

The ARCO agreement also won support by a wide margin at the polling place when members voted at two local referenda on the matter. Opponents said off-reservation voters—who would receive benefits without suffering the impacts— had swung the vote, heavily influenced by the council's promise to distribute ARCO's $6 million bonus to the members. Although the tribe received the bonus and several other temporary benefits, it never received the economic development for which Rowland had hoped. After drilling seven dry holes and investing $28 million, ARCO left the reservation in July 1984. The controversy created a deep rift between some landowners and their elected leaders. Nevertheless, when the dust settled it looked as if Congress had been right to assume that decision makers on the reservation—the tribal council members—were more accountable to the tribe as a whole than off-reservation allottees might have been if the Supreme Court had awarded the mineral rights to Hollowbreast and the other allottees.[30]

PROBLEMS UNDER FEDERAL TRUSTEES

Non-Indians shared these ownership problems of split estates and, to a limited extent, fractionation. Industry landmen sometimes had difficulty finding all heirs to sign lease forms on non-Indian tracts and getting their consent. Surface owners' interests collided with those of their state and federal governments' when the governments owned minerals.[31] Private mineral owners and allottees also shared the difficulties of determining the worth of their minerals, arriving at fair lease terms, and assuring that they were receiving the proper royalties. However, the allottees, unlike the private mineral owners, were theoretically protected by the federal government, which had a trust responsibility for their land and resources just as it had a trust responsibility for those of the tribes. As with the tribes the federal government found it difficult to decide to what extent allottees should be allowed to determine their own future and how much protection they should receive from industry and from unscrupulous individuals, both Indian and non-Indian. As with the tribes, a guardian/ward relationship that had existed for years resisted change when the Indians grew more interested in participating in resource decision making. Some BIA officials were genuinely concerned about the allottees and worked hard and long to protect their interests. Others jealously guarded their responsibilities for allottees because they foresaw a time when their responsibilities for tribal affairs would be largely curtailed.

Despite the regulations implementing the government's trust responsibility and despite the desire of BIA to remain involved in allottees' affairs, mineral-owning allottees often lacked federal protection. The Interior Department did not have the staff and expertise to protect the allottees' interests. Just as with the tribes, the department was failing to assure equitable lease terms, enforce federal regulations, and provide adequate royalty and field monitoring. Yet the allottees did not have many of the alternatives that the larger tribes had, and their affairs—especially their royalties—were even more difficult for the department to administer. As individuals, the allottees usually could not hire their own geologists, accountants, and attorneys in order to know the value of their resources and achieve favorable lease terms. It was difficult for allottees to combine their often small, undivided interests in order to reap the greatest benefit for all owners. Even if they could get agreement among owners of a 40-acre tract, one tract was not nearly as attractive to industry as blocks of 160 acres or more. Allottees did not hire staffs to learn the regulations and lease terms and to assure compliance. Thus as individuals, allottees lacked clout and usually advocates.[32]

Carol Connor was an exception. Connor, an Assiniboine allottee and attorney, worked through the American Indian Law Center and the Native American Natural Resources Development Federation to request Interior's assistance for allottees. In 1978 she met with Assistant Interior Secretary Forrest Gerard, who agreed to fund her year-long study, which detailed serious problems of oil and gas allottees. She found BIA often had only one staff person at the reservation level assigned to assist and oversee leases covering thousands of allotted acres valued at millions of dollars. Allottees had no assurance that their interests would be protected by whatever lease compliance monitoring might be conducted by tribal or federal agencies. When one of Connor's reports was circulated in the Interior Department, the U.S. Geological Survey admitted that its personnel were not performing effectively. USGS, which was responsible for field and royalty monitoring at the

Carol Connor, an Assiniboine, served as a persistent gadfly forcing the Interior Department to address allottee issues. (Photo taken in 1988 by Stephen Konecheck.)

time, said it was "unable, with its present resources, to perform its regulatory responsibilities on Indian lands with what we would consider to be an acceptable level of competence."[33]

To address the disparity between protection for tribes and for allottees, Connor suggested making a publication in lay language available to allottees; holding workshops to provide legal and technical advice for allottees; and clarifying the responsibility of the federal government and the tribes to allottees. She said Interior should set up field offices where allottees could get legal and technical advice on their property, in much the way that the federal government assisted tribal natural resources offices and CERT. When some CERT officials blocked the appointment of Thomas Fredericks as the assistant secretary for Indian affairs in 1980, Connor's proposal became embroiled in politics and was never accepted by Interior. A persistent gadfly, she continued to pester the department with testimony on behalf of allottees, but Interior continued to ignore allottees' problems until she and other attorneys took the battle to the courts.[34]

The chairman of the tribes on the Fort Peck Reservation where Connor was born shared her concern for allottees. Assiniboine and Sioux Business Council Chairman Norman Hollow told a congressional committee that

BIA could not help allottees because it lacked the necessary expertise. Although the bureau had employees with appropriate graduate degrees for less-valuable resources, such as grasslands, farmlands, forests, and water, it had nobody with such credentials for oil and gas and other minerals. "No agency or area office, not even the central office in Washington, has a qualified petroleum engineer or petroleum economist or mining engineer," he told the committee. James Henry, chairman of the Turtle Mountain Band of Chippewas, said some of his tribal members who owned allotments on public lands in Montana and North Dakota received even less help from BIA, which had no offices nearby. Although these Indians had some producing oil and gas wells, they could get no information on the production and no data on the potential of the other lands.[35]

The Minerals Management Service (MMS—the royalty management agency created in 1982) and BIA, which in the mid-1980s could not handle tribal and federal lease royalties, found collection and distribution of the allottees' royalties completely overwhelming. They could not cope with thousands of allotted lease checks, some of which had to be divided among hundreds of allottees. In the Anadarko, Oklahoma, area, BIA served 7,500 allottees with 25,000 different accounts. In 1984 alone, that area distributed $26.9 million in Indian oil and gas royalties. In the Navajo area, BIA maintained 14,000 allottee accounts. Allottees' royalties there totaled more than $4 million in an 18-month period in 1983/84. When the number of producing leases doubled between 1980 and 1982, some of the allottees became millionaires overnight, adapting their life styles to suit the royalty checks. When the checks stopped coming or dropped dramatically without explanation, it devastated the allottees' pocketbooks even more than those of the tribes. Because of the federal government's ineptitude, allottees lost their homes, their cars, and their credit ratings. To pay their bills some Oklahoma allottees sold their mineral rights. Yet MMS put the highest priority on rectifying the higher yielding, federal, offshore royalty accounts.[36]

Employees at MMS were unaccustomed to dealing with Indians and initially referred all complaints to BIA, which in turn was accustomed to passive allottees who had generally allowed the trustee to manage their affairs for them. It was not until 1968 that the courts even recognized the right of allottees to sue to enforce an oil and gas lease. In that case, the Interior Department had allowed unauthorized flaring of natural gas owned by some Comanche allottees.[37]

The Federal Trade Commission staff found in its 1975 study that when allotted lands were nominated for leasing by industry, the Indian owners rarely exercised their veto power over the sales. Nor did they participate in other decisions after the sales were held. BIA typically erected barriers to allottees' participation in decision making while facilitating industry's coercion. Although the allottees normally had the power of consent, that authority was meaningless without information. BIA routinely allowed oil companies access to its records and data pertaining to allottees' lands—but denied access to the allottees themselves. BIA made no attempt to inform allottees of their right to develop their natural resources under their own terms and had no procedure to assist allottees in negotiations and planning.[38]

Norman Hollow blamed the companies and BIA for exploiting the allottees on the Fort Peck Reservation, where 97 percent of the oil was being produced on allotted leases. "The companies, with Interior's encouragement and blessing, hire an educated tribal member known to the Indians, and even popular with them, to solicit leases. Such an Indian has little difficulty in persuading aged and uninformed Indians to sign leases for sums that look very large to people in their income class." Testimony by other allottees, tribal leaders, and attorneys at congressional hearings detailed widespread problems: owners signing leases or rights of way when drunk or heavily sedated in hospital beds; company landmen coercing signatures by threatening condemnation or falsely saying other family members had signed; and BIA staff members conspiring with companies to defraud owners.[39]

Newton Lamar in 1982 asked the Senate Select Committee on Indian Affairs for a special investigation of fraud and incompetence in Oklahoma, where 95 percent of the remaining trust lands are held by allottees and where nearly half of the nation's Indian oil and gas leases exist. BIA employees there had helped the companies cheat the allottees out of millions of dollars, according to Lamar, who was president of the Wichita and Affiliated Tribes in western Oklahoma. He told the committee that two bureau superintendents (who "know all the loopholes") quit and set up their own brokerage firms to arrange leases. One misrepresented himself by using a business card that read, "Superintendent, Bureau of Indian Affairs," with tiny letters adding, "retired."[40]

In many cases BIA employees helped companies save millions of dollars by convincing unsuspecting allottees to agree to extensions of leases that were about to expire or to joining nonproducing leases together with producing leases through communitization agreements so they would not expire. In one such case in Oklahoma, it became painfully obvious that the allottees also lacked advocates at upper levels of Interior. When a $40 million communitization lawsuit was filed by an oil company against BIA, the Interior Department's attorney decided not to defend BIA and did not notify the allottees about the case. When the allottees found out about it and filed their own suit, the federal government opposed them and supported the industry position.[41]

The sordid stories resembled those from Osage County in the 1920s when an oil boom had left Indian oil owners prey to bandits and various white-collar crooks. Indeed, the problems of the allottees soared with the mineral booms of the late 1970s and early 1980s. When the federal government partially deregulated oil and gas and thus raised the selling prices, a lease that had been bought for a $100 per acre bonus might be worth $7,000 or $8,000 if it went up for competitive bidding again. Consequently, extending a lease through communitization or other means could save a company millions of dollars.[42]

One of the most blatant examples of BIA's catering to industry rather than Indians occurred in 1980 during the uranium boom in the San Juan Basin. BIA regulations required that all of the allottees holding an interest in an undivided tract must approve a mineral lease. BIA could sign for them only if they were minors, mentally incompetent, or could not be located. Through the years BIA had made several attempts to revise this cumbersome requirement, proposing at different times that only from 51 to 75 percent of the owners should be required to approve a contract. Some tribal leaders agreed that not all of the allottees should be required to sign the leases. However, others said that because of the value of the minerals and the threat to the surface involved, the 100 percent requirement should be retained.[43]

Both arguments were valid, and it was clear that a serious policy decision was necessary. In 1980, however, BIA tried to avoid a lengthy decision-making process. Some Navajo landowners did not want to sign a uranium lease renewal with Mobil Oil Corporation, which was paying a bonus of only $150 an acre when other companies were paying $15,625 an acre bonus for leasing state-owned lands nearby. Mobil, which had already invested $13 million in exploration, did not think it could establish production by 1982 when the leases would expire. Therefore it convinced BIA to waive certain requirements and allow it to negotiate directly with the allottees for lease renewals. BIA complied with Mobil's requests but refused the allottees' request to see the drilling logs, which indicated the value of their uranium and which would have bolstered their bargaining position. Nor would BIA get an appraisal. Mobil later admitted the Indian lands could yield at least 10 million pounds of uranium ore worth $350 million.[44]

One group of allottees, the McClanahan family, who owned 31 percent of the interests, enlisted the help of DNA-People's Legal Services. Mobil asked BIA to waive the 100 percent approval requirement because Mobil believed DNA was encouraging the McClanahans to make "unacceptable demands." Under the authority of an unpublished solicitor's opinion, BIA agreed to require only two-thirds of the allottees' signatures and proceeded to approve the leases, without notifying the McClanahans. Some local BIA officials objected to the decision, including the Crownpoint BIA superintendent, Ed Plummer, and Interior field solicitor Claudeen Arthur, both Navajos. The central BIA office overruled them. A federal judge in Albuquerque, New Mexico, later ruled, however, that BIA had acted illegally and unconscionably. U.S. District Judge E. L. Mechem said in his August 1987 ruling, "The BIA and Interior generally seem to have been more concerned throughout the leasing process with their relationship with Mobil than their relationship with the Indian owners." He chided BIA for not trying to get more money for the allottees, as DNA had tried to do. The judge said any

BIA shortcoming "pales before this arbitrary and continuing refusal to exact of Mobil one jot more than Mobil originally offered." After the court ruling, Interior attorneys decided they would need special legislation to change the requirement for 100 percent signatures instead of changing it administratively.[45]

With a few exceptions the allottees had been slow to rile until their royalty checks stopped coming and they discovered the communitization and uranium lease scams. When they did become concerned, some Interior employees demonstrated callous, patronizing attitudes toward their clients. After a long history of allottee passivity and disinterest, some BIA staff members seemed uncomfortable with the allottees' growing savvy. When allottees inquired about missing royalty checks, some employees hung up on them or threw them bodily out of their offices. BIA staff members told allottees that royalty checks from their own oil and gas were no different from welfare, and the allottees should be grateful for what little they received.[46]

Alan Taradash, who represented the Navajo allottees, vehemently criticized Interior's employees, one of whom asked Taradash why allottees should care if their payments were six months late so long as they were getting paid. Taradash did not accept the argument that it was too difficult for Interior to divide the royalties into the right accounts and send out checks promptly. He said banks routinely conduct such operations. He thought the federal employees should be held accountable: "If the Bureau of Indian Affairs and the Minerals Management Service and their predecessors were private trustees, they would have been fired or they would have gone to jail a long time ago for the mismanagement of the accounts."[47]

The investigations forced public attention upon the allottees' problems. Stories by allottee advocates horrified staff members of the Senate Select Committee on Indian Affairs, who said the competence of the federal trustee was "minimal, at best." The select committee staff said the Indian allottees seemed to be "at the mercy of the oil companies." Yet the loose leasing policies during the boom years invited abuse by all parties, not just the trustees and the companies. When allottees began negotiating their own terms after competitive sales, they commonly refused to execute documents unless they were paid a "signing bonus" or "gas money." Sometimes an individual would try to negotiate a whole new lease, which might not be approved by BIA. Tribal chairmen, as well as BIA officials, sometimes quit their jobs only to become oil company brokers, according to Newton Lamar.[48]

Federal employees sometimes seriously misjudged when trying to balance protecting allottees versus allowing allottees to determine their own fates. In 1981 BIA in Washington granted an extraordinary waiver to help an Indian firm, Fourstar Resources, develop blocks of allotted lands on the Fort Peck Reservation. John Gidley and three other members of the Assiniboine and Sioux Tribes had formed Fourstar Resources in 1981 in order to

Attorney Alan Taradash said that if federal officials had been private trustees, they would have been jailed for the way they treated allottees. (Photo taken in 1988 by Stephen Konecheck.)

help allottees develop their land. Gidley went to Washington and explained his ideas for getting better deals for his family's 3,000 acres. He was upset with the way oil companies had convinced inexperienced allottees to sign leases, and he wanted a waiver from competitive bidding for that acreage in order to negotiate better deals. His arguments apparently impressed some high-ranking Interior officials, who gave him much more than he requested—a blanket waiver so Fourstar Resources could negotiate oil and gas contracts on behalf of consenting allottees on the entire Fort Peck Reservation.[49]

At that time, prior to passage of the Indian Mineral Development Act of 1982, the department was not even allowing tribes to negotiate their own contracts without first having competitive sales. The waiver memo said BIA's central office supported the Fourstar Resources endeavor "in the interest of fostering the policy of Indian self-determination." It said, "Indians *must* be given the opportunity to make more decisions in the management and development of their resources" (emphasis in original). The waiver provided Fourstar with a tremendous advantage. If, for example, production were established in a certain area, Fourstar could acquire leases near that area without waiting for a sale, as every other company would have to do.[50]

A fiasco followed, which provided ammunition for those who opposed allottees' involvement in resource decisions. After two owners sold their interests in Fourstar, the third owner started brokering leases for 20 percent instead of 25 percent royalties, thus undercutting tribal leases and violating the waiver. The business soon threatened to escape the control of both Gidley and the remaining Indian owner. An oil and gas attorney in Los Angeles wrote a letter to several major oil companies offering them access to tens of thousands of allotted acres if they would join in a joint venture with Fourstar Resources. The letter misrepresented Fourstar's power, saying that Fourstar controlled and had exclusive rights to all allotted acres on the reservation and neglecting to mention that each allottee still had the power of consent under the waiver. The attorney threatened to exclude other companies from any Indian business in the area—tribal or allotted—saying that other tribal councils on other reservations were "considering designating Fourstar as their exclusive agent." At a congressional hearing Gidley insisted that Fourstar never authorized such a letter and said he wrote a second letter to the companies, rescinding the rash promises. Despite the scandal Fourstar continued operating on the reservation for some time, brokering leases for other companies.[51]

Gidley and Marvin Sonosky exposed the Fourstar fiasco at the congressional hearing, using it as evidence that both tribes and allottees continued to need more protection in handling their oil and gas affairs. Sonosky, a non-Indian attorney who had served the Fort Peck Tribes for more than twenty years, said that dispensing with the requirement for competitive auctions and allowing industry to negotiate with Indians would be "much like taking candy from a baby." Gidley advocated creating a new, special agency to assist Indian people with negotiations and with monitoring oil and gas leases, without which, he said, Congress would be "unleashing a pack of wolves on a helpless herd of sheep." A college graduate himself, Gidley said the Indian people were not unintelligent, just inexperienced.[52]

CAN THE TRIBES HELP ALLOTTEES?

When tribes realized they could not trust in the federal trustee, many wanted to help allottees get better leases and better enforcement. Sometimes BIA allowed the tribes to help, and sometimes it did not. For example, BIA joined with the Salish and Kootenai Tribes of the Flathead Reservation in Montana to present information to allottees on oil and gas leasing regulations and marketability. When in 1979 the Blackfeet Tribe imposed a moratorium on tribal oil and gas development so that the tribe could improve its information and expertise, the local BIA superintendent voluntarily stopped all lease sales for allotted minerals. Several local BIA agencies—including the Blackfeet—adapted the standard allotted lease form, raising royalties and making other changes to keep up with tribal improvements.[53]

Under BIA regulations and policies the bureau had a direct responsibility to protect the allottees, a responsibility that BIA officials insisted they could not delegate to the tribes. BIA said it could not allow tribes to have access to information about allotted resources, income, and lease terms, even at the allottees' request. To some degree BIA's reluctance to let tribes help allottees could be justified. There can be a conflict of interest between tribes and allottees if they both want to sell minerals in the same area. In fact, allottees and other individuals have objected to actions by their tribes on this basis. Federal Trade Commission investigators said some tribal members accused their leaders of giving inadequate and misleading advice to gain consent for leasing. Allottees do not necessarily feel that the tribal democratic process protects them. They can own land on the reservation without being enrolled and thus without being entitled to vote in the tribe. Sometimes they live on a different reservation. They cannot assume that tribal council members share their concerns; not all tribal members on allotted reservations own land.[54]

On the other hand, tribes often had both a selfish and a humanitarian interest in seeing the allottees obtain higher returns. When allotted minerals were available nearby for lower royalties and easier terms, it weakened a tribe's bargaining position for its own minerals. However, tribal staff members also resented seeing their neighbors exploited under BIA's protection when they could help. Blackfeet Natural Resources Director G. G. Kipp, for example, was able to increase the bonus paid to his grandmother by eight times, from $6,000 to $48,000, when he helped her negotiate. At the request of CERT tribal members, the CERT staff sometimes helped allottees negotiate better leases and right-of-way contracts.[55]

Although at times bureau staff members may have sincerely distrusted tribes' efforts to help allottees, at other times they seemed more interested in power plays with the tribes and in protecting their turf. Some tribes' mineral staffs said that BIA superintendents rejected tribal suggestions to increase allottees' standard royalty returns. On one reservation, tribal officials recommended holding an allotted lease sale because they thought nearby off-reservation wells were illegally draining allotted oil. The local BIA refused to hold a sale because the bureau staff believed the tribes just wanted to test the market for their own, tribal purposes. BIA's motives were not always easy to sort out. Regardless of intent BIA contributed to confusion and thus to industry's distrust of Indian mineral development, especially during the tumultuous boom years (1979–1982) prior to passage of the Indian Mineral Development Act.[56]

Many problems occurred when tribes tried to help allottees combine acreages, as John Gidley had tried to do, to get better terms than could be obtained for isolated, 40-acre tracts. The Assiniboine and Sioux Tribes on the Fort Peck Reservation succeeded for awhile in convincing BIA to combine allotted and tribal tracts in BIA lease sales. When a new BIA superintendent came, however, he insisted upon offering lands separately again, for administrative convenience.[57]

During the period that the central BIA issued its blanket waiver to Fourstar Resources to negotiate allotted leases, the Blackfeet Tribe, with CERT's assistance, was negotiating lease terms with Tesoro Petroleum for tribal and allotted lands on its reservation. As a result of vacillating federal policies toward the Blackfeet Reservation around the turn of the century, mineral and surface estates on the reservation are split, with the allottees owning most of the surface (76 percent) and the tribe owning most of the minerals (60 percent). In 1980, after several months of negotiation, Tesoro Petroleum agreed to most of the Blackfeet Tribe's demands for financial return and tribal control over oil development in return for the right to develop in a certain area of the reservation, which included both tribal and allotted lands. Although Tesoro officials knew they risked the wrath of other companies for setting such a precedent, they wanted the Blackfeet oil and also hoped to establish a reputation that would open other Indian resources to them. The Tesoro-Blackfeet deal was never consummated because on May 30, 1980, a week before it was to be signed, BIA announced an oil sale for nearby allotted lands without most of the tough conditions the tribe was demanding. Tesoro officials, infuriated, withdrew their offer to the tribe, saying they would wait until the tribe and BIA "got their house in order."[58]

BIA officials said that the tribe could not make promises for allotted lands and that BIA was obligated to hold a sale for those lands to assure allottees a share of the booming oil and gas income. The allottees had already waited out the tribe's moratorium. The sale brought extraordinary return for the allottees, with royalties as high as 25 percent and high bonus bids. Tribal officials, however, sought a congressional investigation of the allotted sale, saying the allottees could have received more if BIA had utilized the Blackfeet tribal contract requirements.[59]

In retrospect it seems clear that the tribe should not have been allowed to make promises for allotted lands without the allottees' involvement. Yet

BIA representatives present at early negotiation sessions apparently did not voice that objection. Thus the problems seemed to result from undefined authority as much as BIA jealousy of the tribe's or CERT's initiatives. During this same period, prior to the Indian Mineral Development Act of December 1982, Interior Department attorneys questioned tribes' authority to negotiate contracts for even tribal lands, much less allotted lands, without holding competitive sales. Offering large blocks of land rather than individual allotments was clearly preferable, as recognized by Congress when it finally passed the IMDA. The law provided for including allotted lands with tribal lands in negotiated contracts so long as allottees had the right of consent. Nevertheless, it took some time for BIA and the tribes to agree on how to implement that provision. The confusion was exacerbated by the fact that the Interior Department still had not issued final regulations implementing the IMDA six and a half years after its passage.[60]

On the Fort Berthold Reservation in North Dakota, for example, the Three Affiliated Tribes Business Council negotiated agreements after passage of the IMDA that involved allotted lands. BIA approval of one contract was delayed for seven months in 1984 as BIA officials exchanged memos about whether allotted lands could be included at all, what forms should be used to assure the allottees' involvement, and whether the tribes could have access to allottee mineral data for negotiation purposes. Tribal natural resources director Rich Schilf said some Interior officials mistakenly believed that the tribes wanted to exploit the allottees, not help them.[61]

In addition to trying to assist allottees in getting better lease terms, some tribes tried with varying degrees of success to help them monitor their royalties and their field production. CERT, at the request of the Navajo and the Southern Ute Tribes, launched an allottee royalty study in February 1985—its first major allottee project—that contributed to the political pressure on MMS to address allottee issues. On the Fort Peck Reservation the tribes filed a lawsuit on January 5, 1984, against the Interior Department and six oil companies, which, they said, failed to pay proper royalties to the tribes and allottees. The oil field included more allotted oil than tribal. Carol Connor, the Assiniboine allottee and attorney, commended the tribes for taking action on the problems, which affected her family's oil. However, Connor, who had consistently through the years pointed out the conflict of interest between tribes and allottees, told the council she objected to the tribes representing her and her family without asking them.[62]

Several tribes tried unsuccessfully to help allottees by including their wells in tribal inspecting and accounting systems. BIA at Fort Berthold eventually allowed the tribes to inspect allotted wells for surface violations—for example, spilt oil and security violations—using tribal funds. None of the other major energy tribes at that time included allotted leases in their inspections. Interior would not allow tribes to have access to allotted oil and gas production and accounting records, making it impossible for tribes to check royalty payments. Arguing that allottees had a right to privacy and that tribes' interests sometimes conflicted with theirs, the federal government said tribes would have to get all allottees to consent to tribal involvement. Thus the allottees still had to depend upon BIA and the Bureau of Land Management. BLM's inspection effort was especially limited in areas with substantial allottee oil and gas development, such as Oklahoma and Utah.[63] In some cases tribes were allowed to set minimum exploration fees, surface owner compensation requirements, and environmental standards for allotted lands. Yet these practices continued to be controversial.[64]

ALLOTTEE ORGANIZATIONS

When faced with the frustrations of trying to get assistance and protection from the federal and tribal governments, some allottees found that organizing allottee associations brought them the most relief. Through their organizations they obtained better lease terms, filed lawsuits, and won the ear of Congress. Some organizing occurred on a small scale. Families delegated one person as negotiator and consented to terms that he or she arranged, thus avoiding the disagreements that typically split families and infuriated industry landmen. When the Wind River Reservation royalty scandal rocked the Indians and the Interior Department out of their complacency, allottees saw the need to organize on a larger scale. After Chuck Thomas blew the whistle on the government's inadequate security, federal investigators paid particular attention to the Wind River Reservation. But once again, the allottees were forgotten. Investigators focused on tribal leases and neglected to even contact any of the allottees at Wind River. Realizing that if anyone was being shorted, it was probably them, Billy O'Neal (Shoshone) and several other allottees formed an association and hired Carol Connor to represent them. Connor teamed up with Robert Nordhaus, an attorney known among the Indians for his aggressive work to protect their interests, and asked the Interior Department for funds to investigate the allottees' accounts. Admitting that BIA had neglected allottees and concentrated on tribes, BIA Billings Area Director Richard Whitesell (Sioux) agreed eventually to provide $50,000 for Connor and Nordhaus to investigate the Wind River allottee leases. Whitesell had decided to fund the attorneys' study because he was not confident that either the USGS or BIA had enough expertise.[65]

The study found that Amoco Production Company made accounting errors, mixed tribal and allotted oil, and by-passed meters—the same problems alleged by the Wind River tribes when they sought cancellation of Amoco's tribal leases. After presenting the results of their investigation to BIA, the allottees filed a lawsuit in federal district court in Cheyenne, Wyoming, on October 18, 1983, seeking $41 million and charging that the company had not paid royalties on 1.3 million barrels of crude oil produced between 1971 and 1982. The allottees and the company reached an undisclosed, out-of-court settlement in September 1985 for money and future increases in royalties amounting to several million dollars.[66]

Other allottees decided to organize after they discovered that royalty reforms—such as the Federal Oil and Gas Royalty Management Act and the creation of the new Minerals Management Service—had not improved their situation. In fact, it worsened. Before the reforms, the oil and gas owners received checks regularly once a month, although they never knew whether the amounts were correct or whether a certain check covered an agricultural lease or an oil well. Then just after the peak in the oil boom, the checks in many areas stopped coming primarily because of computer problems. Although FOGRMA required royalty payments within a month after the company had paid MMS, the checks were as much as six months late, still without any explanation of what the money was for and with no way of checking whether the amounts were correct. Allottees in the Navajo area would travel as far as 240 miles round trip to pick up checks, only to find nothing in their accounts. BIA told Beatrice Saupitty, a sixty-eight-year-old Comanche widow in Oklahoma, that she owed an oil company $80,000 in overcharges, without trying to verify it. Berdena Holder, a Wichita allottee who had been receiving $3,700 a month, suddenly was getting only $132 a month.[67]

In Oklahoma and New Mexico, the two areas with the most allotted oil and gas leases, allottees formed organizations and hired help. The Navajo allottees' group, Shii Shi Keyah, hired Alan Taradash, formerly of DNA and then of the Nordhaus law firm in Albuquerque. Taradash filed a royalty lawsuit in October 1984 against Interior demanding prompt payments with explanations. A month and a half later, Oklahoma Indian Legal Services filed a similar suit on behalf of some Oklahoma allottees. The Navajo association in the meantime helped its members understand more about their oil and gas holdings. Using a grant from the Save the Children Foundation, the Nordhaus law firm—with cooperation from BIA and BLM—presented a four-day bilingual oil and gas workshop on the Navajo Reservation for allottees.[68]

Berdena Holder, sixty-three years old and armed with a sharp tongue and a rudimentary education, became a champion of hundreds of allottees in Oklahoma, most of them as elderly as she. She and Pressley Ware (Kiowa) founded the Oklahoma Indian Mineral Owners Association for allottees from several tribes in that area. To draw attention to their situation, the organization in 1983 erected a teepee encampment at Anadarko, directly across from the area BIA office, and tormented government officials with phone calls and letters. When it became clear the MMS computer system was not working, Holder called one of the top men at MMS, Robert Boldt, and told him "he bought a pile of junk for $15 million."[69]

Holder also took on the job of investigating where the other allottees' money went. With their written consent she scrutinized their files, some-

Berdena Holder, a Wichita, terrified BIA and MMS officials with her investigations of missing royalties. (Photo taken in 1988 by Sherry Brown, Tulsa Tribune.*)*

times traveling to the MMS offices in Lakewood, Colorado, in pursuit of records. MMS and BIA officials dreaded her appearances, but she recovered hundreds of thousands of dollars that had been deposited in the wrong accounts or inaccurately computed. She and her daughter, Frances Wise, taught allottees how to research their own files and in some cases advised them to hire outside help. The auditor for Saupitty, the Comanche widow and a member of Holder's organization, found that the companies owed her $64,000 instead of her owing the companies $80,000, as BIA had claimed.[70]

Following pressure from allottee associations Senator David Boren and Congressman Mike Synar of Oklahoma scheduled a public hearing in April 1985, the first dedicated specifically to allottees' concerns. At the hearing, Oklahoma state oil and gas officials testified that Indian allottees, unable to even get access to their records from BIA, had turned in desperation to the state. When the Oklahoma Conservation Commission investigated three cases, it found $71,613 that had not been paid to allottees as the result of Interior's—not the companies'—errors. Interior's own inspector general confirmed that the problems were widespread—BIA and MMS were holding money for months without distributing it, not paying interest, and withholding payments entirely when they figured that companies had overpaid allottees.[71]

Synar quickly won the hearts of the hundreds of allottees who filled the hearing room when his aggressive questioning made Interior Department officials squirm. Synar sharply criticized a policy adopted by MMS in February 1985 to put the highest priority on auditing and distributing royalties from the most lucrative federal leases rather than from tribal and allotted leases. Synar said priority should be given to allottees, who often depended upon royalties as their only source of income. The allottees had found that BIA and MMS responded slowly—if at all—to audit requests, even though such investigations were fruitful. Audits of six Indian leases resulted in returns of an average of $9,833 per lease. Synar asked an MMS official, "Now can you explain to me why it looks like the only time that you all are prepared to audit is when you have to be persuaded by a U.S. Senator to do it?" Synar told the Interior officials that he hoped they would be able to convince the allottees their problems were being solved so they would take down their teepee, which had been standing for seventeen months.[72]

Media coverage of allottees' oil and gas predicaments increased, as did their lawsuits against BIA and MMS. In one case BIA found a black man with the same name to sign an Indian's lease and receive his royalties and then sided with the oil company when the mistake was discovered. The federal judge said, "There is not justification for the Department of Interior to permit this type of procedural masquerade, wherein the government's trial attorneys know first-hand that the private attorneys filing these petitions are bought and paid for by the Indian's adversary." He blasted the federal attorneys for negligence and a breach of their statutory duty. After a series of articles exposing such swindles appeared in the *Tulsa Tribune* and the *Arizona Republic*, the Senate Select Committee on Indian Affairs in 1988 launched an investigation of the federal trustee, including its handling of allottee issues.[73]

The Navajo lawsuit and continued pressure from Congress gradually forced Interior to comply with the law. Partly as a result of Synar's admonitions, MMS gave Indian leases a higher priority, devoting 29 percent of its audit resources to Indian leases in 1986, although Indian lease income represented only about 2 percent of the total royalties handled by MMS. A federal judge set deadlines for Interior to meet each requirement of FOGRMA in the Navajo area. MMS and BIA computers began working better and issuing checks more regularly. MMS, which previously had referred all Indian calls to BIA, established ombudsmen to address allottees' complaints. BIA and MMS developed a joint strategy to expedite audits. BIA began providing explanations with royalty payments, telling allottees the amount of oil produced and for how much it sold. MMS established a royalty advisory committee, which included Pressley Ware of Oklahoma as an allottee representative as well as several tribal representatives. After the Synar hearing even Interior Secretary Donald Hodel expressed concern about allottees' royalty problems.[74]

Yet the bureaucracy moved much more slowly than the allottees would have liked. BIA automated its process one region at a time. Consequently some of the lower priority agencies still were delaying allottee royalty payments by as much as five months in 1988. Allottees feared that they might not ever recover thousands of dollars as a result of the shoddy record keeping, especially during the boom years. They continued to distrust the trustee's efforts and, like the tribes, explored ways to do the work themselves. In 1988 the Navajo allottees' association proposed that the Interior Department provide funds for the association to hire its own accounting firm, which Taradash said would cost less than using the federal system. An allottees' association on the Uintah and Ouray Reservation studied options for taking oil and gas in kind, by-passing MMS, to use in a local refinery.[75]

Although allottees when they were organized could act more like tribes, allottee groups lacked many of the tribes' advantages, such as tax revenues and formalized decision-making structures. Allottees who formed associations for a single purpose often were not interested in performing other functions. For example, the Wind River Allottees' Association, which formed to resolve royalty accounting questions, turned down a suggestion by Connor to tackle field monitoring.

CONCLUSION

In the 1980s mineral allottees have had some success at organizing, starting at the family level and sometimes encompassing members of several tribes. They have worked with tribal governments and have hired their own experts—accountants, geologists, and attorneys. Congress and the Interior Department have addressed some allottee issues by revising mineral contracting and royalty management laws and some regulations. In the spring of 1989, however, the department was still finalizing the tribal and allotted contracting regulations. Interior was seeking congressional action on the consent procedures so that less than 100 percent of the allottees owning fractionated mineral interests could approve leases.

Yet the department did not intend to take two critical steps to meet its responsibilities to mineral allottees. First, it had no comprehensive education plans for mineral allottees, such as a consulting process for specific deals, informational handouts, or workshops on allottees' general mineral concerns. The department also had no plans to include allotted lands in its mineral inventory, despite the obvious need to know what minerals lay beneath the surface—and despite the slow but continuous efforts to get that information for tribes. BIA argued that it would cost too much and be too difficult to get all of the signatures necessary for access. BIA also made an argument that was all too familiar to allottees: It was not worthwhile to look at small, isolated tracts, and allottees had not organized and made collective requests for mineral information. Thus the allottees had only the rudimentary information available to the tribes in the early 1970s; companies usually knew more about mineral resources than the owners or their trustees.[76]

Federal policymakers could resolve some of the debilitating confusion if they would, in consultation with the allottees and tribes, determine the limits of tribal and federal responsibility for allottees. Although the Interior Department would still have to be involved to some extent, its purpose should be to protect allottee interests, not federal jobs. The mineral allottees and their dilemmas are not going away. Neither land consolidation nor assimilation has eliminated the allotments, nor are they likely to in the near future, partially because of the resistance of the allottees themselves. In at least one way Senator Dawes and his compatriots succeeded when they passed the Allotment Act: They instilled the Anglo concept of private ownership in many Indian people to the point that allottees have described tribal land consolidation programs as "un-American."[77]

Unless the mood of the country changes radically—as it has in the past—and Congress acts unilaterally to take away the allottees' special status, there will always be individual Indians with their own separate relationship with the federal government. The allottees' situation will not be solved by ignoring it. Their interests will keep dwindling, becoming more and more infinitesimal and difficult to manage but never disappearing. Their political strength, on the other hand, seems destined to keep growing.

After the Contract Is Signed: The Tribe as Regulator

*Not all who speak of self-government mean the same thing by the term. There-
fore, let me say at the outset that by self-government I mean that form of government in which decisions are made not by the people who are wisest, or
ablest, or closest to some throne in Washington or in Heaven, but rather by the
people who are most directly affected by the decisions.*

—Felix Cohen, *The Legal Conscience*

During the renaissance of tribal powers in the 1970s, the energy tribes at first
concentrated upon their role as proprietors—improving their minerals' monetary returns. Gradually they realized the importance of their governmental
role, increasing their involvement with royalties as well as with permits, enforcement, standard setting, and taxation. The Supreme Court recognized
that tribes had inherent sovereign powers that had not been taken away by
treaty or statute. Through their government functions, tribes could exert
control over energy development's impact on their culture and on their reservation environment.

Early uranium mines, coal mines, and power plants had taken a heavy
toll on Indian lands and people, fouling the water and air and killing many
Navajo underground uranium miners. In many cases non-Indian lands suffered from the same neglect. Indian lands also became pollution havens because of unequal enforcement and jurisdictional conflicts. Recognition of
such problems convinced some tribes to halt proposed development, at least
temporarily, to protect their environments. Most tribes, however, did not
want to stop existing development because they depended heavily upon the
monetary income, however limited, and the jobs. Instead the tribes wanted
more control over setting the standards for their reservation environments
and assuring enforcement.

The tribes encountered significant problems as they flexed their regulatory muscles. Congress had not acknowledged tribal authority in the batch
of federal environmental laws passed in the 1970s. Non-Indians on reservations resisted being subject to tribal sovereignty and challenged tribal jurisdiction on all fronts. In addition to lacking recognized authority, tribes
lacked money for enforcement, which the federal and state governments derived from taxation. Under federal law, states generally had no jurisdiction
over Indian lands within reservations unless they had been specifically dele-

gated such authority; jurisdiction over non-Indian lands on reservations was less clear. Yet because the tribes in most cases had not been using their governmental authority, states sometimes had stepped in, further eroding tribal jurisdiction and making decisions that conflicted with tribal preferences. Because many of the uranium and coal mines and the oil and gas wells had been developed prior to enactment of the National Environmental Policy Act and other important environmental laws, even federal agencies sometimes lacked legal handles for control. The confusion often resulted in a regulatory vacuum on reservations.

Tribes also faced formidable internal hurdles. Only recently created in their modern form and lacking a financial base, many tribal governments had not formed the administrative or judicial structures needed for regulation. After the establishment of reservations, most had functioned in very limited capacities prior to the Indian Self-Determination Act of 1975. The young tribal governments, especially those on energy reservations, were forced to deal with complex and divisive issues. During the late 1970s and the 1980s, however, the tribes won many significant victories as they sought more control over both enforcement and setting of pollution standards. Court decisions in many instances recognized tribes' authority over that of the states. Congress's awareness of tribal issues grew, and with it came an increased willingness to treat tribes as states.

THE EXTENT OF INDIAN ENERGY DEVELOPMENT

As tribal governments gained regulatory authority, the change affected many existing and proposed uranium, coal, oil, and gas companies' operations. Beginning in the postwar uranium boom of the 1950s, uranium was developed extensively on three reservations—the Navajo and the Laguna Pueblo in the Southwest and the Spokane in Washington State. The four uranium mills that had been built on Navajo lands and the one within the boundaries of the Wind River Reservation all closed in the uranium bust of the 1960s. The two mills serving the Spokane Reservation mines closed in the uranium bust of the 1980s. If uranium markets improve, development potentially could occur on many other reservations where uranium is expected to exist, including the Ute Mountain Ute in Colorado; Wind River in Wyoming; Canoncito Band of Navajos, Zuni Pueblo, Jemez Pueblo, Zia Pueblo, and Acoma Pueblo in New Mexico; and Hualapai in Arizona. The Department of Energy estimated in 1979 that more than one-third of potential uranium reserves in the country lie beneath Indian lands.[1]

Although coal mining occurred on twenty-two reservations during the first half of the century, only four coal companies were operating on Indian reservations in 1989. These four companies held billions of tons of Indian coal. Most of the coal was mined in the Southwest, with the exception of

Westmoreland Resources's Absaloka mine on the Crow ceded strip in Montana, which produced 3.1 million tons in 1985. Pittsburg & Midway Coal Mining Company had operated the McKinley mine, which straddles the southern border of the Navajo Reservation in New Mexico, since 1962 and produced 4.9 million tons in 1985. Utah International (now BHP-Utah International) had operated the Navajo mine, which lay entirely on the Navajo Reservation, since 1963 and produced 7.0 million tons in 1985. For its 2.4 million ton per year Black Mesa mine in Arizona, Peabody Coal leased coal from the Navajo and Hopi Tribes. For its 7.3 million ton per year Kayenta mine, Peabody leased coal from just the Navajos. Peabody supplied both the Navajo generating station at Page, Arizona, on the Navajo Reservation and the Mojave generating station across the state near Bullhead City, Arizona. The San Juan and the Four Corners coal-fired power plants in New Mexico received their coal from the nearby BHP-Utah International mine. The Four Corners plant lay on the Navajo Reservation in New Mexico, and the San Juan plant was off the reservation.[2]

Coal regulatory issues could arise on several other reservations. Industry had shown interest in coal on the Northern Cheyenne Reservation in Montana, Fort Berthold in North Dakota, Southern Ute and Ute Mountain Ute in Colorado, and Uintah-Ouray in Utah. Three reservations—Fort Peck, Mescalero Apache, and Zuni Pueblo—had coal with more limited interest. Fourteen reservations—Cheyenne River Sioux, Standing Rock Sioux, Fort Belknap, Blackfeet, Rocky Boys', Wind River, Jicarilla Apache, Laguna Pueblo, Fort Apache, Acoma Pueblo, Kickapoo, Potawatomi, Isabella, and Iowa—had coal of marginal or uncertain value.[3]

THE LACK OF ENFORCEMENT

Despite legalistic haggling among the different governments asserting jurisdiction, their regulatory staffs rarely bumped heads in the field. Sometimes no one noticed the leaking mine wastes or the high radiation levels in houses. Too often tribes found that the Bureau of Indian Affairs, the Department of Energy, the Environmental Protection Agency, or state environmental agencies let energy industries break federal laws without rebuke. The federal and state staffs at times seemed to lack the will to battle recalcitrant companies and mountains of red tape on behalf of the Indians' environment.

On July 16, 1979, the tailings dam at United Nuclear Corporation's (UNC) Church Rock uranium mill in New Mexico failed, sending 100 million gallons of radioactive water into Rio Puerco. Navajos herding their horses and sheep along the banks said the water looked putrid yellow, like battery acid. A Navajo woman and several animals that waded through the river that morning developed sores on their legs and later died. For the next year the state told the Navajos not to eat their mutton, and butchers would not buy it. Friends and relatives shunned the Rio Puerco area people, refusing to shake their hands. A decade later the residents still could not use local water supplies for drinking or stock watering, partly because of the spill and partly because of years of contamination from several mines in the area.[4]

Despite the greater publicity surrounding the Three Mile Island nuclear plant accident in March 1979, Church Rock resulted in the nation's worst release of radioactivity. If political pressure had not interfered with the New Mexico professional staff's licensing process, the spill probably could have been prevented, according to the Southwest Research and Information Center, an Albuquerque-based environmental research group that worked often with Indian people. Although state officials disagreed with this assessment, the environmentalists' research shows that the Church Rock uranium mill had been built before it was licensed, and its license had been hurriedly issued three weeks before the state ground-water regulations took effect in 1977. New Mexico Governor Bruce King in 1981 told the New Mexico Environmental Improvement Division staff to allow UNC to continue an illegal water discharge, which the staff had been attempting to control for over a year. Chris Shuey of the research center said jurisdictional problems and the lack of continued state and federal interest weakened the investigation and the clean-up effort; the two EPA regions that shared responsibility for the Navajo Reservation refused to talk with one another for several years.[5]

The accident occurred at a dam site, licensed by the state and the federal Nuclear Regulatory Commission (NRC), in New Mexico just south of the Navajo Reservation boundary. The tailings flowed down the Rio Puerco, through lands owned by Navajo allottees, into Arizona and the Navajo Reservation. The most serious, long-term water contamination was caused by the UNC and Kerr-McGee mines in the Church Rock area—not by the mill spill alone. The Kerr-McGee mine was leased through the tribe in 1972, with BIA's approval.[6]

Thus the states of New Mexico and Arizona, the Navajo Tribe, and several federal agencies (Bureau of Indian Affairs, Environmental Protection Agency, Nuclear Regulatory Commission, Centers for Disease Control, and Indian Health Service) were involved. Yet they still lacked sufficient laboratories, technicians, or money to measure the effects of the spill or to reduce the impacts on local residents. UNC and the tribe provided alternative drinking water at first, but there was never enough for both livestock and human use. Local residents felt compelled to turn to an outside group, the Southwest Research Center, for help finding water supplies and monitoring water quality. With the assistance of DNA-People's Legal Services, they eventually won an out-of-court settlement of $525,000 in a civil damages suit, *Benally v. UNC*.[7]

The Church Rock investigation resulted in the different agencies pointing fingers at one another while revealing their own ignorance of what went wrong, according to the director of the Navajo Environmental Protection Administration, Harold Tso. Because UNC's Church Rock mill and dam were outside reservation boundaries, the tribe's only potential regulatory authority over the dam itself would have been under the Navajo water code, which was not adopted until 1984.[8]

The state agencies that had jurisdiction over water quality did not enforce their own standards, which differed from one another. New Mexico used water standards that assumed, incorrectly, that no one would drink surface water without treatment; Arizona's standards were stricter because they assumed that surface water could leach into ground water and pollute wells, as the Rio Puerco water in fact did. Thus Rio Puerco water that was acceptable under New Mexico standards was not acceptable under Arizona's. Both sets of standards were approved by EPA, but both were frequently violated by the Church Rock area mines and mill.[9]

The two states' water standards were adopted as a result of the Clean Water Act of 1972, which Congress intended to clean up the nation's waters to the point where they would be fishable and swimmable even if they could not be drunk. A different law, the Safe Drinking Water Act of 1974, was intended to protect drinking water. After 1974, any mines discharging water had to get permits under the National Pollutant Discharge Elimination System, which EPA left to the states to enforce. EPA's discharge permit system and the Clean Water Act were designed to clean up all the nation's rivers and lakes by 1983—not just the waters outside reservation boundaries.[10]

Yet Congress did not outline responsibilities for setting standards and enforcement on reservations, making neglect almost inevitable. For example, both EPA and the state of Washington ignored Dawn Mining Company's uranium mine on the Spokane Reservation. In 1977 BIA geologist Jim LeBret, a Spokane tribal member, discovered dangerous toxic wastes trickling from the mine at Blue Creek, a favorite camping and picnic spot for tribal members before uranium mining had begun. He was accompanied by his father and his uncle, who had discovered uranium on the reservation and previously owned interests in Dawn Mining. They left in tears after seeing the canary-yellow trickle of waste water and the destruction it had caused.[11]

The bureau forced Dawn to build a dam, which contained the toxic wastes for several years, but even more serious contamination occurred later after mining had stopped and the trickle grew to a 75 to 400 gallon per minute stream of wastes. The Indian Health Service said in 1983 that the heavy metal and acid contamination was "appalling" and recommended that BIA "prevent livestock and humans from consuming the water in question *by whatever means necessary*" (emphasis in original). When EPA tested the "seepage," the radiological lab chemist in Las Vegas said he had never seen such radioactive mine waste water before.[12]

Despite the initial urgency it was not until 1987 that EPA forced the company to stop the discharge—six years after mining stopped and ten years after the LeBrets noticed the discharge. By then it was too late for the reservation stream, Blue Creek, which previously had provided habitat for about thirteen thousand rainbow trout. In the spring of 1988 only five or six adults returned to spawn. EPA admitted that if the mine had not been on Indian lands, it probably would have come to someone's attention sooner. The experience convinced the tribe, which still hoped to establish its own fish hatchery, to develop its own water standards and contributed to a deep cynicism about the federal government's efforts to protect the Spokane Reservation environment.[13]

NO LAWS TO ENFORCE

Sometimes there were no laws to enforce. Congress took time to respond to environmental problems and moved much more slowly than some people—both Indians and non-Indians—would have desired, especially when it thought that regulation conflicted with other national interests.

Initially the federal government had no standards for reclaiming coal strip mines. In the mid-1970s the Navajos joined with non-Indians lobbying for a strip mining law by describing their problems trying to protect reservation lands and waters from mining impacts. Although the state of New Mexico had asserted jurisdiction over coal mines within its boundaries on the reservation, the state had not provided the enforcement the Navajos thought necessary. As for the federal government, "the Bureau of Indian Affairs was sound asleep," according to Leonard Robbins, a Navajo on the staff of the Navajo Environmental Protection Administration. Robbins and his boss, Harold Tso, told Congress that the Indian coal leases' paragraph-long reclamation requirements were totally inadequate. The Navajos had tried to enforce lease requirements themselves with mixed results. Because of the reluctance of some companies to accept tribal authority, Navajo officials sought Congress's backing. Congressman John Melcher of Montana held several meetings with Northern Plains tribes, and they, too, wanted tribal and not state regulatory control. After a decade-long battle Congress passed the Surface Mining Control and Reclamation Act (SMCRA) in August 1977. The landmark legislation provided for reclamation plans for new mines, funds for cleaning up old abandoned mines, and veto power for people who owned the surface over federal coal. The law delayed action on Indian coal, leaving it to the new Office of Surface Mining to enforce in the meantime.[14]

The reclamation legislation applied only to surface mining of coal, not uranium mining and not uranium processing. Navajos were instrumental in convincing Congress to address uranium mill tailings, the wastes that are left when uranium is processed and that still contain 85 percent of the radioactivity from the ore and thus remain hazardous for a thousand years. The Navajos were also among the first to start cleaning up tailings sites. In the

1950s investigators discovered that drinking water in Farmington, New Mexico, had high radium concentrations as the result of uranium mill tailings abandoned upstream. It was later learned that many of the mills in the West discharged their wastes directly into streams. The discovery of water contamination at Farmington, on the Navajo Reservation boundary, sparked the first research projects on the dangers of uranium mill tailings. In 1966 elevated radiation levels were also found in homes in Grand Junction, Colorado, where tailings had been supplied for construction purposes since about 1953.[15]

Despite these early studies, little action was taken on the tailings until the 1970s. Because uranium mills supplied the Atomic Energy Commission (AEC) with fuel for nuclear weapons, anyone criticizing uranium during the 1950s was thought to be criticizing the national defense program and thus was considered "disloyal, subversive, and perhaps a Communist," according to uranium analyst Justas Bavarskis. As the federal contracts began to be phased out prior to 1970, companies walked away with their profits, leaving communities to deal with mill buildings and tailings.[16]

In 1972 the Navajo Tribal Council established the Navajo Environmental Protection Administration, probably the first tribal agency on any reservation created to regulate energy development. Tribal members were concerned about oil spills, dust clouds from the power plant at Page, and proposals for injection uranium mining. The director, Harold Tso, was a Navajo nuclear chemist, and he was alarmed about the radiation from uranium mill tailings at four sites on the reservation, including his hometown, Shiprock, which lay down river from Farmington. Kerr-McGee Oil Industries had leased the mill site from the Navajo Tribe and built the mill on the banks of the San Juan River in 1954, and a later owner, Foote Mineral Company, abandoned the site in 1968. The wind blew the fine dust into Navajo houses; rain washed tailings into the river; children played on the piles; and, as Tso later learned, many Navajos constructed their houses with tailings material.[17]

Tso asked EPA to investigate the health hazards and help plan remedial action, and EPA asked Congress to act. Tailings had been abandoned at twenty-two sites around the country and used in construction in many places, and EPA was convinced they were not safe. Health studies later proved that EPA was right. They found three times the normal rates of Down's syndrome in Grand Junction and a definite correlation between elevated radiation levels in uranium mines and lung cancer rates among miners. Rather than waiting for the slow wheels of government to turn, Tso used a tribal construction firm and began work at Shiprock in 1973, the first large tailings decontamination effort in the country.[18]

Congress eventually passed the Uranium Mill Tailings Radiation Control Act (UMTRCA) in 1978. The law provided for cleaning up the old tailings sites, including contaminated houses, using 90 percent federal and 10 percent state funds, except on Indian lands where 100 percent federal funds would be used. States and tribes had a voice in the remedial plans through cooperative agreements. Unfortunately implementation was delayed by various interests' challenges of mill reclamation standards and by the continuing skepticism of some members of Congress that uranium caused health problems. Despite the tribe's first-aid work, DOE did not complete the Shiprock remedial work until 1986, nearly twenty years after the mill shut down and eight years after Congress passed the legislation. Nevertheless when the work did begin on the nation's abandoned uranium mill sites, Indian sites were among the first to receive attention, partly because of political pressure from the tribes. Shiprock was the second site in the nation to be worked on. Riverton, Wyoming, and the three remaining Navajo sites—Monument Valley and Tuba City, Arizona, and Mexican Hat, Utah—were scheduled to be completed by 1991.[19]

DOE ranked the Susquehanna-Western mill site—on the Wind River Reservation near Riverton, Wyoming—as a high-priority site because of serious ground-water contamination. Because it was located on non-Indian land within reservation boundaries, DOE did not consider it an Indian site, meaning that the Arapahoe and Shoshone Tribes had no official voice in the remedial action plan. The tribes fought hard to have the tailings removed from the reservation. The state of Wyoming agreed with the tribes and eventually convinced DOE to move the tailings despite the additional expense that the state had to share.[20]

Believing they had solved the environmental problems of uranium in 1978, members of Congress were incredulous a few years later when Navajos came to them for help with uranium-contaminated houses. The problems they had discovered indicated a serious omission in the uranium mill tailings law. Rural people had used the rocks around mines as well as the sandlike mill tailings for construction, but the 1978 legislation had provided remedies only for houses built with mill tailings near old mill sites, not with mine wastes.[21]

Once again the Navajo Tribe was at the vanguard for investigating a problem that others did not take seriously. The Navajo Tribe borrowed equipment from EPA to survey houses, used tribal funds, and spent years pounding on doors in Washington, asking various federal agencies and Congress for help in cleaning up the contaminated houses and in building replacements. A 1983 EPA study showed 3,000 uranium mine waste sites nationwide, about a third on Indian lands, but EPA considered the sites too remote to pose a national risk. Although some field EPA personnel disagreed with that conclusion, EPA said that if anything should be done, the states should do it. EPA did not say, however, who should be responsible for Indian sites, where the states had no jurisdiction.[22]

Congress had been much more thorough when dealing with coal mining than with uranium, probably because the coal mining problems were more widespread and better understood. The 1977 coal mining law had not only required reclaiming current mines; it had also provided for Abandoned Mines Land (AML) funds to reclaim old coal mines, and those funds could be used for other mines when the coal work was done. Thus the Navajos in 1987 could spend their AML funds on reclaiming old coal and uranium mines, using their own priorities rather than depending upon Washington to recognize the importance of a reservation environmental problem. In this case the tribe had an advantage over the state of New Mexico, which also suffered from abandoned uranium mines but which received much less AML money because it had less coal mining.[23]

Despite the requirements for reclaiming old and new uranium mill sites and despite the requirements for reclaiming old and new coal mines, Congress never required uranium mine reclamation. This omission plagued westerners near mines on both Indian and federal lands. Dawn Mining, for example, signed its leases with the Spokane Tribe and the allottees in 1964, before passage of the National Environmental Policy Act (NEPA) and before BIA added environmental protection regulations, both in 1969. Like other early leases, the Dawn leases said only that the company must leave the property "in as good condition as received, except for the ordinary wear and tear and unavoidable accidents in their proper use . . . [and] leave all the areas on which the lessee has worked in a condition that will not be hazardous to life or limb, and will be to the satisfaction of the superintendent." BIA only required Dawn to furnish a $15,000 bond. After the tribe became concerned about the company's plans, the Interior Department demanded a higher bond. Dawn Mining refused to comply and sued the department in 1982.[24]

Although the court said an adequate bond should be provided, Dawn Mining Company President Marcel DeGuire said the company could not afford to pay because its only assets were the mine and the off-reservation mill, which had been closed down in 1981 and 1982. DeGuire said the company had already spent $4 million reclaiming the mine and mill. Whether the Interior Department could tap Dawn's $1.6 million reclamation reserve fund or the vast resources of part-owner Newmont Mining was unclear. Thus the department potentially held only $15,000 to cover a mine reclamation project that it expected to cost nearly $10 million. Both the reservation land base and the federal treasury were likely to suffer from the lack of federal reclamation requirements and the lack of adequate bonding.[25]

The situation was much different for the nearby Sherwood uranium mine and mill. Although there was still no federal reclamation law when Western Nuclear built the mine and mill on the Spokane Reservation in 1978, the federal government by then required an environmental impact statement and in 1969 had adopted more stringent environmental protection measures. In its contract negotiations the tribe had protected itself by insisting upon a bond much higher than BIA regulations required.[26]

In 1989 Western Nuclear signed over ownership of the Sherwood mine and mill to the tribe, as well as $4.4 million for reclamation.[27] For a model in this innovative way of gaining control over environmental quality, the Spokanes had looked to the Laguna Pueblo in New Mexico. Although the uranium leases on Laguna Pueblo lands in New Mexico suffered from the same vague reclamation language as the Dawn leases, adequate reclamation was eventually assured, primarily because of the aggressive posture of the pueblo and because of the company's willingness to accept its responsibilities. If Anaconda had not agreed to pay for adequate reclamation, the consequences could have been disastrous not only for the Laguna Pueblo but also for downstream, non-Indian communities. Instead the company contracted with the tribe to do the reclamation work itself, thus turning potential disaster into at least short-term economic opportunity. The lengthy negotiations delayed reclamation for several years, however, and it was unclear what long-term health effects the pueblo people might suffer.

The Laguna Pueblo first signed its lease in 1952 with Anaconda Copper Company, now the Anaconda Minerals Company, a division of Atlantic Richfield Company. By 1980 Jackpile was the largest uranium open-pit mine in the world. The combined Jackpile-Paguate pits encompassed 2,800 acres of the 461,098-acre reservation. When the mines closed down in 1982, one commentator said it would take 400 million tons of dirt to fill the pits—enough to cover the District of Columbia to a depth of 45 feet. In 1977, five years before the mines shut down, pueblo officials began negotiating with Anaconda for reclamation as well as financial and training provisions. During its thirty years of mining, Anaconda had completely changed the local economy from agriculture to mining, employing thousands of tribal members over the years. With the assistance of an engineering firm and CERT, the Lagunas continued consistent reclamation negotiations under the administration of four different tribal governors until they were completed in 1987 by Laguna Governor Chester Fernando.[28]

When the mine closed in 1982 Anaconda provided a $175,000 grant to start a new tribal business. Pueblo officials and their consultants generally perceived Anaconda as a responsible company during the lengthy negotiations except in 1985 when the company—backed by other uranium companies—threatened litigation, saying that under the vague terms of the old leases, neither the pueblo nor the federal government had the legal clout to impose reclamation conditions. Some federal officials privately worried whether the courts would agree with Anaconda. They were well aware of the precedent a decision could set because Jackpile was to be the first uranium mine reclaimed on Indian land; other uranium mines on federal and state lands were abandoned without any reclamation unless states required it.[29]

Charges and countercharges flew at public hearings in 1985. Tribal attorney B. Reid Haltom said that Anaconda had made over $600 million dollars in ore value and should have spent some of it on reclamation. In an

Laguna Indian miners at the Jackpile-Paguate mine in New Mexico lost their jobs when the mine closed in 1982. The tribe began to reclaim the 2,800-acre pits in 1988. (Photo by Tom Barry.)

emotional speech that conveyed the Lagunas' sense of betrayal, CERT economist Ahmed Kooros said, "One of the things that I have learned from the pueblo people, Your Honor, is that when they pray, they pray for everybody. They pray for the well-being of humanity, for they believe that their well-being is not independent of the well-being of others." Anaconda said the pueblo demands were unrealistic. The Lagunas disagreed, saying they only wanted to be able to graze their livestock, use the water safely, and breathe the air without worrying about lung cancer. Unlike most uranium mines, which lie in remote, unpopulated areas, one pit lay just 1,000 feet away from the pueblo community of Paguate. Without proper protection of ground water, the mine pit area would be covered by toxic, saline wastelands, according to the Interior Department. DOI predicted that without reclamation, 95 to 243 additional radiation-induced cancer deaths could be expected within 50 miles of the mine. Anaconda said that DOI's figures overestimated the risk by 100 times or more.[30]

Despite encouragement from other uranium companies Anaconda never followed through on its threat of litigation, most likely because the company had posted a bond for $45 million and wanted to protect its corporate image. Anaconda agreed to pay the pueblo $43.6 million to take over the reclamation instead of the $17 million it had said that reclamation would cost. To conduct the reclamation the tribe planned to hire eighty people for about seven years, mostly tribal members who had served Anaconda as mechanics, heavy equipment operators, and so on. Although they had won some control over the future quality of their lands, the Lagunas joined with other CERT members to ask Congress to adopt radiation standards for reclaiming uranium mines and for safeguarding the health and safety of Indians. By 1989, however, the federal government had announced no plans to adopt uranium reclamation standards or to clean up abandoned uranium mine sites.[31]

CONTROL OVER STANDARDS

The federal government typically weighed lives against dollars and cents and set its standards by determining how many deaths would be considered "acceptable." Unlike the statisticians in Washington the local people could see the costs of the continual exposure to low-level radiation. Lung cancer reached epidemic proportions among uranium miners on the Navajo Reservation, and studies showed clusters of birth defects around the areas with the most abandoned uranium. A person who died of lung cancer was listed as a "negative health effect" in government studies, but to the Navajos he was a neighbor or a clan brother.[32]

Consequently the tribes wanted the opportunity, just as the states had had, to adopt federal minimums and then determine what impacts would be "acceptable" by setting standards to protect air, land, water, and cultural qualities. If a tribe were concerned that its rural members might drink stream water, it could decide for itself whether to use drinking water standards or less stringent requirements. If it did not want its members to be exposed to radiation in their houses, it could use stricter radiation level standards to initiate cleanup efforts. At the same time, tribes saw the adoption of standards and codes as a way of helping industry by increasing certainty about the rules of the game on a particular reservation.

In the late 1970s a few energy tribes established standards for cultural quality or air quality. Several tribes, including those from the Blackfeet and Fort Peck reservations, asked tribal elders to help tribal staffs develop tribal cultural codes by designating important cultural sites that should be off limits for development. In 1976 the Northern Cheyenne Tribe objected to Montana Power Company's and four northwestern utilities' plans to expand the Colstrip power plant by 1,400 megawatts and by the prospect of several other coal-fired power plants in the area. Tribal members learned that EPA had adopted regulations in 1974 that provided authority for states and local governments, including tribes, to redesignate their air as Class I, which allowed almost no degradation, or Class III, which allowed much more. Most air in the West was automatically designated as Class II, except for national parks and wilderness areas, which were protected by Class I. When the tribal council voted to demand protection for the reservation's pristine air, the tribe's insolence angered the state of Montana, neighboring rancher Marcus Nance, and especially the Montana Power Company (MPC). The Northern Cheyenne Tribe was the first government to make such a redesignation, which EPA approved after extensive public hearings and studies. The Cheyennes' surprising success changed forever the way that tribes looked at federal environmental laws.[33]

Nance and Montana Power Company challenged EPA's decision (*Nance v. EPA*). The Crow Tribe joined the lawsuit against the Class I redesignation on the neighboring reservation, although some Crow tribal members and nearby ranchers supported it. The courts upheld the Cheyennes, and the tiny tribe brought the $1 billion Colstrip project construction to a halt for three years, forcing the utilities to install better pollution control devices. The Cheyennes' air classification also contributed to the demise of the Crow Tribe's plans for a coal gasification plant and a coal-fired power plant by further increasing the already exorbitant costs of the facilities. The federal courts were convinced that EPA regulations provided such authority and that the tribe, through the Northern Cheyenne Research Project, had adequately studied the social, environmental, and economic impacts. Congress adopted amendments in 1977 that included tribal authority for redesignation, thus adding weight to the earlier EPA administrative decision.[34]

Despite their moral victory the Cheyennes' pocketbook was nearly drained by the cost of litigation. In April 1980 they dropped three lawsuits against MPC in an out-of-court settlement. Some critics questioned the decision, and a newspaper headline asked, "Is Cheyenne Class I Air for Sale?" Many years later, however, the value of the settlement was obvious, both in monetary and air-quality terms. It assured continued protection of the reservation airshed and benefits to tribal members for allowing the utilities to "rent" the air, including employment guarantees and job-training programs; air-quality monitoring stations and $25,000 annually for tribal air-quality programs; scholarships for air-quality monitoring courses; socioeconomic impact assistance; and an arbitration board to monitor compliance and handle complaints. Edwin Dahl, the tribal administrator for the agreement and a primary force behind the redesignation decision, said it had resulted in jobs for two hundred of the three thousand resident tribal members and had increased the standard of living on the reservation tenfold.[35]

Once the Cheyennes had breached the dam there was a flood of tribal initiatives to protect reservation airsheds. Two other reservations in Montana—Fort Peck and Flathead—obtained Class I redesignations, and the Spokane Tribe in Washington applied and was awaiting an EPA decision. The tribes decided to protect their air quality despite the constraints that redesignation imposed upon the tribes' own plans. Fort Peck's Assiniboine and Sioux Tribes and the Spokane Tribe were most concerned about coal-fired power plants, whereas the Salish and Kootenai Tribes of the Flathead Reservation were worried about sawmills.[36]

By 1986, nine reservations in the West had their own air-monitoring systems, allowing them to determine baseline conditions and recognize violations, which they then reported to EPA. The Wind River Environmental Quality Commission hired a consultant in 1988 to continue acid rain research on the reservation, which the federal government had abandoned for lack of funds. Several tribes hoped to obtain air regulatory authority from EPA in order to enforce federal and tribal air standards on their reservations. Geographical proximity gave tribes an advantage; some problems had occurred in the past because the reservations were so remote from federal agencies' offices. A tribal presence in the energy fields made a difference, even if the tribal monitors initially had no enforcement authority and just reported violations to federal authorities.[37]

HURDLES FOR INDIAN
STRIP MINING LEGISLATION

Despite such advances in shaping environmental standards, tribes continued to suffer from the lack of specific legislative authorization under the nation's environmental laws. It took years for Congress to adopt Indian amendments to those laws. In 1977 federal lawmakers took the first step by agreeing that tribes had a legitimate argument for regulating coal development on their lands. Although few of its members probably had thought out the full ramifications, Congress was prepared to provide tribes with regulatory authority under the federal strip mining law. The coal tribes asked for more time, however, to look at jurisdictional questions and design appropriate mining and reclamation programs for their lands. Consequently, rather than providing immediate tribal authority, Congress told the newly created Office of Surface Mining Reclamation and Enforcement (OSMRE) to study, in consultation with the tribes, how surface mining could be effectively regulated on Indian lands and to submit proposed legislation by January 1978 that provided Indian tribes the option of full regulatory authority. OSMRE was to regulate strip mining on Indian lands in the interim. Years later several of the tribes regretted their hesitation. During the study period Congress became more aware of the serious ramifications of tribal authority over coal and, in some cases, more reluctant to hand it over. Although it had looked as if coal legislation would be the first of many laws authorizing tribal regulation, the tribes still did not have the authority in 1989 that Congress offered to them on a platter more than ten years earlier—and no prospects of attaining it.[38]

On the other hand, the study was invaluable. Under a federal contract, Douglas Richardson of the CERT staff and staff members of the coal tribes worked together for a year, resolved many crucial issues, and produced a comprehensive document of conclusions and recommendations—the first of any significance on the topic of tribal regulation. Although completion of that study seemed to set the stage for quick action, the Interior Department did not submit its study until 1984, more than four years after CERT's study and six years after Congress's deadline. In the meantime CERT, in frustration, had its own Indian strip mining legislation introduced, but it was never acted upon.[39]

Why was the Indian strip mining bill delayed? Although Interior blamed the delays upon setting up a new agency and the change in administrations between Carter and Reagan, it seemed clear that the controversial nature of tribal regulation stalled the legislation. A look at the three main barriers provides insight into why various parties have opposed tribal regulatory authority in other areas, too. Two of the problems faced by tribes differed from those of states: jurisdictional ambiguities and lack of developed infrastructures and expertise. States' boundaries were clear, and their governments had better funding and more experience. The third hurdle—potential conflicts of interest—was shared by tribes and states.[40]

When OSMRE completed its 1984 study it did not even attempt to answer the most controversial question: To what lands will the legislation apply? Instead OSMRE offered six options for Congress to choose from for its definition of "Indian lands," ranging from the very narrow (on-reservation areas controlled by the tribe) to the very broad (all on- and off-reservation areas where tribes have some legal control of the surface or mineral estate). The tribes vehemently supported the broad definition, and the states just as vehemently advocated the narrow.[41]

Why should a definition be so controversial? The answer lay in the mixed ownership status of Indian coal and surface and in the tremendous quantity of coal and potential revenue at stake. Various allotment and homestead laws and land cessions had complicated the ownership of Indian coal and lands. The Office of Surface Mining's jurisdiction study required 273 pages just for the twenty-five coal-owning tribes. Jurisdiction over the coal then being mined was among the most complicated. In Arizona the Hopi and Navajo Tribes shared undivided ownership of some of the coal at the Black Mesa mine. In the other active Indian coal mining areas in Montana and New Mexico, Indians controlled both the surface and the coal for less than 2 percent of the land area, according to the General Accounting Office. Control over the rest was split among Indian, federal, state, and private interests. This mixed ownership created jurisdictional disputes, resulting in frustration and increased costs for the mining companies as well as lawsuits from the states. For a mine encompassing federal and Indian coal, a company had to meet both federal (or tribal) and state regulations, meaning multiple permit reviews, multiple inspections, and possibly different mining standards.[42]

Why were the states so eager to assert their jurisdiction? Like the tribes, the states were concerned about protecting lands and water within their boundaries; they lacked confidence in the tribes' interest in and ability to regulate. The states were also concerned about money. Millions of dollars in potential tax and AML fee revenue were at stake. For example, Congress provided in SMCRA for a 35-cent per ton fee on subsequent mining to create an AML fund for reclaiming old abandoned mines. When the U.S. Supreme Court in October 1985 upheld the Crows' rather than the state's right to the AML money from the tribe's coal, $4.6 million had accumulated.[43]

Such concerns were also involved in the intertribal jurisdictional dispute at Black Mesa, a dispute that also delayed action on the Indian strip mining bill. The Hopi and Navajo Tribes at the time were in the heat of battle over land at Black Mesa formerly known as the Joint Use Area. The controversy was so emotional that it led to violence and contributed to the defeat of Peterson Zah by Peter MacDonald in the Navajo tribal chairmanship election in 1986. There seemed to be no equitable solution to this problem, which like so many others was created by the federal government's vacillating Indian policies. In 1882 federal agents believed that the Hopis' and the Navajos' conflicting claims to the land would only be a short-term problem because the Indian people would vanish into the social mainstream. Thus although incomprehensible now, it had seemed reasonable to establish a reservation "for the use and occupancy of the Moqui [Hopis] and such other Indians as the Secretary may see fit to settle thereon." Instead of vanishing, the Hopi and the Navajo Tribes kept growing until all available lands had been assigned.[44]

Although the history and the specific laws pertaining to the Joint Use Area were unique, the joint ownership problem was not unique in Indian Country. Heirs of allottees owned undivided interests in surface and mineral estates throughout the West. Most Northern Plains reservation lands and minerals were owned by more than one tribal group, usually jointly. Wherever assets were undivided, conflicts were likely. The more valuable the assets and the more different the parties' value systems, the more arduous the decision making. On the Wind River Reservation, for example, the Arapahoe and Shoshone Tribes shared undivided interests in the surface and minerals after the federal government "temporarily" placed the Arapahoes on the Shoshones' treaty reservation in 1878. The U.S. Supreme Court in 1937 and 1938 said the two tribes jointly owned the land, minerals, and timber. To survive, the business councils of the two tribes learned to work together as a joint council. The court's solution made decision making difficult for the two traditional enemies but not impossible in most cases.[45]

At Black Mesa the U.S. Supreme Court ruled in 1963 that the Hopi and Navajo Tribes had joint and equal interests in both the surface and the minerals. Although it clarified authority and allowed both tribes to sign coal leases with Peabody in 1966, it did not resolve issues related to surface use and regulation, perhaps because the tribes had no mechanism for joint decisions. Congress therefore in 1974 passed the Navajo-Hopi Settlement Act, which said the courts should divide the surface property between the two tribes. Those who prefer simple, black-and-white explanations argue that the Navajos and Hopis could have continued to live peacefully in the Joint Use Area if the mineral companies had not insisted upon clearing title to the minerals. Navajo history scholar Peter Iverson argues persuasively that the Hopis would have demanded relocating the Navajos to make room for the Hopis even if minerals had not been involved: "Even though energy companies complicated the issue and had a stake in the outcome, they should not be used as a convenient explanation for the policy that was implemented." The division was inevitably controversial because it resulted in the biggest relocation of people since the internment of Japanese Americans during World War II.[46]

Although mining continued, the conflict directly affected the Peabody coal leases at Black Mesa because they included jointly owned coal, which lay partly under Navajo and partly under Hopi surface. The two tribes agreed upon the terms for the first Black Mesa coal leases in 1966 during the rubber-stamp era. When contracts were being renegotiated during the self-determination era, the tribes had more difficulty. They sometimes refused to meet with one another about either coal contract terms or strategy for Indian strip mining legislation. The attorney for the Hopi Tribe told Interior in 1982 that jurisdiction over the Joint Use Area must be determined before any Indian strip mining legislation could be introduced. He said the Hopis would not willingly give the Navajos jurisdiction over jointly owned coal so the only possible solution would be a joint Navajo-Hopi administrative agency.[47]

The Navajo Tribe for some time disagreed and insisted that it alone had jurisdiction over the jointly owned coal, and the Interior Department vacillated. When the Hopi Tribe tried to adopt a severance tax in 1983 for its portion of the coal, the Navajo Tribe objected. The Interior Department refused to approve the Hopi severance tax, saying that the Hopis did not have jurisdiction under the terms of the Hopi-Navajo Settlement Act. On the other hand, the Office of Surface Mining continued to provide funding and technical assistance to both the Navajo and the Hopi coal mining offices for enforcement at the Black Mesa mine, causing frustration for Peabody Coal officials.[48]

Because of its own and the tribes' concerns about tribal infrastructures and expertise, Congress in 1977 asked the Office of Surface Mining to study the jurisdictional issues and to suggest potential models for tribal regulation. Under the strip mining law, states had to choose between either assuming full regulatory authority or none at all. The coal-owning tribes suggested the alternative of partial authority, which later became the preferred alternative for tribal air and water regulation as well. Under a partial program tribes could, for example, develop a program for permitting an inspection and enforcement program while not administering it. Alternatively, the tribe and OSMRE could jointly administer all aspects of a program, with federal involvement decreasing as tribal capability increased.[49]

The coal tribes made it clear that they wanted the authority to assume full responsibility. They recognized, however, that they lacked the data, in-house technical expertise, and governmental structures needed to do so immediately. Cost factors alone would have been enough to discourage them. The CERT study estimated that for one large, 5 million ton per year mine, the regulatory program would cost as much as $88,900 for administration, $7,500 for inspection and enforcement, and $216,000 for permit review. The federal government subsidized state programs. Because of their special needs, the tribes suggested they initially should be eligible for more funding and training. OSMRE disagreed, saying it could provide more technical assistance to tribes but the same funding as that received by states.[50]

Many of the energy tribes had existing agencies and ordinances for protecting tribal resources, and most had been issuing tribal permits for some activities. Some had planning offices and research staffs. After passage of the federal strip mining law the coal tribes gradually improved their capacity to regulate strip mining with financial and technical assistance from Interior. The three coal-producing tribes—the Crow, Hopi, and Navajo—developed mining codes, reclamation codes, mining ordinances, and plans for reviewing new mine applications. They helped OSMRE enforce federal regulations, accompanying federal inspectors on visits to mine sites and learning how to analyze aerial photographs, survey vegetation, conduct enforcement proceedings, and manage contracts. The Abandoned Mine Lands program provided training and funding, as well as employment for Indian contractors and laborers, on twenty-two reservations. With help from coal income the coal-producing tribes expanded and improved their governmental structures and expertise. The Hopi Tribe, for example, in 1986 had more than four hundred government employees in four divisions with a budget of $5 million to $6 million, 60 percent of which derived from coal royalties, according to Don Ami, director of the Hopi Office of Natural Resources.[51]

Later, however, when Congress still had not acted on Indian strip mining legislation, the Office of Surface Mining reduced its spending on tribal regulatory programs. Why spend money to prepare tribes for power that they may never have? Cutting the training funds threatened to make the prophecy self-fulfilling, for with reduced funding, the tribes had trouble training and retaining capable staff. This paradox was reflected in an OSM evaluation in 1987, which said the Crows and the Hopis did not have enough qualified staff members. The Navajos, however, seemed able to meet inspection requirements.[52]

In energy tribes as a whole, tribal politics sometimes hampered efforts to increase tribal expertise. Tribal chairmen's terms were sometimes as short as one year, and there was often debilitating jealousy of one another's initiatives. When a new chairman was elected, he or she sometimes abandoned the predecessor's pet projects and replaced department directors. Although this also occurred in other governments, political bossism caused more problems for tribes, which often lacked a large staff to provide continuity. Department directors were told to train Indians to take over the positions of non-Indians, but the Indians often lacked the academic education necessary, especially in small tribes. Well-educated tribal members had to spread their efforts too thinly, and industry offered them more lucrative positions. Administrative and judicial decisions were often overruled by legislative bodies. In some tribes political factionalism became a tradition.[53]

On the other hand, some tribal leaders adopted personnel policies to provide autonomy and continuity. Tribal internship programs and scholarships encouraged members to get schooling in relevant fields. Tribal members who worked for their tribes generally did not page through job announcements every week, looking for positions in faraway places. Several tribal coal employees outlasted state and federal ones. Leonard Robbins from the Navajo Tribe, Nat Nutongla from the Hopi, and Ellis Knows Gun from the Crow each worked on tribal coal issues for more than ten years, during which time the Office of Surface Mining had seven different directors.[54]

Of all the tribal functions, tribal courts concerned industry the most. Coal companies argued, with cause, that most tribal courts had been established very recently, had limited experience, and lacked trained judges. The companies were also concerned about separation of powers, because judges often were appointed by and served at the pleasure of the tribal chairman or council. Tribes said regulatory conflicts should be resolved in tribal courts, which they felt were the only forums that could maintain balance between contemporary legal requirements and traditional Indian cultural values. Although tribal governments were not always organized the same as the U.S. government, with separation of powers, the tribes had many internal and external checks and balances for controlling power. Nevertheless some tribal spokesmen acknowledged that tribal courts would have to become more fair and judges would have to gain expertise in environmental law.[55]

Both environmentalists and industry groups were concerned about the tribes' potential conflicts of interest. Environmentalists feared that tribes would allow monetary considerations to override their concerns for protecting the land and individual surface owners. Industry feared that tribes might not enforce regulations equally if they were involved in one mine as equity partners and were not involved in another.[56]

It was obvious that tribal proprietary interests could conflict with some of their governmental interests. Adequate reclamation, for example, increased the costs of mining, thus reducing the potential profit to be shared between operators and mineral owners (the tribe). A federal trial judge found the Jicarilla Apache Tribe guilty of "unclean hands" because the tribe did not require as complete environmental studies for the oil and gas wells it shared ownership in as it required for others. Coal-producing tribes depended heavily upon their royalty income from coal mining. The Crow Tribe derived 61 percent of its tribal income from coal royalties and rents, the Hopi 57 percent, and the Navajo 19 percent. All three tribes used the royalties to provide government services. The Crow Tribe also distributed 60 percent of its royalty income among its members as coal dividends, thus increasing the possible personal conflict of interest for tribal members who worked in regulatory agencies.[57]

State and federal regulators sometimes suffered from conflicts of interest, but that had not stopped Congress from providing them with jurisdiction. For example, the Church Rock problems were blamed partly upon elected officials' interference with the New Mexico environmental staff's work. On the federal level the courts said the federal Atomic Energy Commission consciously decided for political (national defense) reasons not to tell workers of the dangers of radiation and not to force the companies to protect them. On the tribal level similar problems sometimes occurred. For example, Navajo Tribal Chairman Peterson Zah in 1986 fired the Navajo Tax Commission deputy director, reportedly because he had been overzealous in imposing tribal taxes on the reservation's biggest employer, Peabody Coal, while the tribe was renegotiating the company's coal contract.[58]

When discussing Indian coal regulation Congress had considered the potential conflict, and many lawmakers had decided that it would not be a problem. Arguing for tribal regulation Congressman Morris Udall of Arizona, one of the principal advocates of both tribal authority and strong environmental protection, told his fellow lawmakers that tribal conflict was like that of the coal-producing states, which derived millions of dollars in royalties from federal coal. Udall said he was confident the tribes and the states would operate in good faith. Congress in 1977 had not depended exclusively upon either the states' or the tribes' "good faith." Instead the law required public involvement, judicial and administrative review, and regulations and standards at least as stringent as the federal government's. The tribes agreed to accept the federal environmental restrictions as minimums despite the potential limitations they imposed on the amount of minable tribal coal.[59]

The environmentalists' conflict-of-interest concerns were important because of the groups' political clout. Although the Environmental Policy Institute, an environmental group, worked closely with the tribes for several years attempting to get an Indian strip mining bill passed, some observers blamed environmentalists for Congress's failure to pass such a bill. Most groups initially did not want to open the federal strip mining law up to possible weakening amendments. They distrusted local control by either states or tribes, and they insisted upon surface owner consent for stripping either federal or Indian coal.[60]

Although some environmental and industry interests tried during the debate to pigeonhole tribes as either anti- or prodevelopment, the tribes defied classification. Like the states, the tribes' primary interest was in protecting their options and their land base. However, tribes often would go further, given the opportunity, to protect their lands than either the federal or state governments. Tribes had a vested interest in protecting reservation resources because they possessed a limited land base, lived close to potential mining sites, and had cultural attachments to their reservations. They could not pack up and move elsewhere. Many also had a record of protection, passing environmental ordinances before Congress passed the federal strip mining bill. When reclaiming abandoned coal mines on reservation lands, the tribes often did more work than federal specifications required.[61]

CHANGING OTHER ENVIRONMENTAL LAWS

When faced with compelling reasons for tribal regulation of other resources, Congress was willing to overlook many hurdles that had stymied the Indian strip mining bill. EPA and Indian organizations conducted studies showing serious environmental problems caused by the regulatory vacuum on reservations. The Americans for Indian Opportunity (AIO) found, for example, that despite the Safe Drinking Water Act, many reservations depended upon untested and sometimes untreated drinking water supplies, which occasionally caused outbreaks of waterborne diseases. In a survey of hazardous wastes on twenty-five reservations, CERT found twelve hundred such sites, including several with potentially acute problems.[62]

In 1974 when the EPA wrote the clean air regulations, the agency suddenly realized that its twenty-seven regulatory programs each had three hundred holes because they delegated authority to the states, which lacked jurisdiction on the reservations. "It was a nightmare," according to EPA's Indian coordinator Leigh Price. EPA lacked funds to assume authority on those lands, according to Price. At the same time, EPA administrators could not give tribes the same authority and federal funding as states without changes in statutes, such as the Clean Air Act of 1963, the Clean Water Act of 1972, the Safe Drinking Water Act of 1974, the Resource Conservation and Recovery Act (hazardous wastes law) of 1976, and the Comprehensive Environmental Response, Compensation, and Liability Act (Superfund law) of 1980. Under these laws the federal government spent billions of dollars to support state environmental regulatory efforts while Indian environmental problems fell through the cracks. In the 1980s the agency commissioned the CERT and AIO studies and adopted a strong Indian policy. Relying upon

President Reagan's government-to-government Indian policy, EPA's Indian policy said tribes should set standards, make environmental policy decisions, and manage environmental regulatory programs on reservations.[63]

Tribes sought Indian amendments to the environmental laws in 1986. Indian lobbyists found that, although some members of Congress were outraged that tribes had been ignored in the early legislation, others wanted to protect states' interests, especially where a limited pot of federal money had to be split among fifty states and hundreds of tribes. Nevertheless, several factors in the 1980s added to the momentum for Indian environmental authority. At the suggestion of one of its members, Harold Tso, the National Commission on Air Quality recommended that Congress amend the Clean Air Act to provide for tribal regulatory programs. After revelations of Indian royalty abuses by industry, Congress passed the Federal Oil and Gas Royalty Management Act in 1982, which provided for tribal enforcement powers.[64]

In 1985 a federal appeals court gave the strongest push for the Indian lobbyists' efforts by ruling that the state of Washington did not have the authority to enforce the hazardous wastes law (RCRA) on reservations. The state had used reservation acreage to bolster its land base for federal funding purposes but had not tried to enforce its regulations on the reservation. The state said that because the hazardous wastes law was silent about Indians, Congress intended to sweep away tribal powers. The Ninth Circuit Court disagreed.[65]

Indian lobbyists succeeded in 1986 in convincing Congress to add Indian amendments to three laws—the Clean Water Act, the Safe Drinking Water Act, and the Superfund law—that recognized tribal governmental authority as being identical to or very similar to that of states. Congressional efforts to amend the Clean Air Act were stymied by controversy over acid rain, and thus Indian amendments were delayed. Under the other amendments, tribes could set their own standards, monitor their own waters, and be eligible for substantial enforcement and construction funds. Under the auspices of the new laws, some tribes, including the Navajo, began testing waters themselves and reporting violations. EPA embarked upon an education and training program that included loaning EPA experts and hiring the National Congress of American Indians and the Council of Energy Resource Tribes to help educate tribes about the laws.[66]

Tribes encountered problems implementing the new Indian amendments, however, partly because they had not been adopted until Congress was phasing out funding for the programs. Some EPA employees thought of Indians as another minority group instead of as governments. At the same time, federal policies treated tribes too much like states, without recognizing that tribes needed hands-on training and development money, much like states had received when the environmental statutes were first enacted. EPA tended to address Indian tribes as a homogeneous group without taking into

A seismic crew receives its morning instructions. (Ute Mountain is in the background.) (Photo taken in 1983; courtesy of Wintershall Corporation.)

consideration their different capabilities and objectives. Although tribes viewed the environment holistically and often could afford only one staff member responsible for several programs, EPA and Congress compartmentalized air, water, and waste programs.[67]

Like other agencies new to the Indian business, such as the Department of Energy and the Minerals Management Service, both EPA and OSMRE were accused of being ignorant of their special responsibilities to tribes and of ignoring tribes and dealing with the Interior Department instead. For example, when the states of New Mexico and Montana sued the Interior Department for jurisdiction over certain off-reservation Indian lands, the federal surface mining agency did not involve the tribes in the out-of-court settlements or even inform them about the results, according to the tribes.[68]

Later in 1985 Office of Surface Mining Director Jed D. Christensen shocked states and tribes by trying to take $24 million from the tribes' Abandoned Mines Land accounts to reclaim high-priority, non-Indian sites in the East. Christensen did not notify the tribes of the plan nor invite them to attend with the states the meeting where the reclamation was to be discussed. Saying the money was going to waste because the tribes had failed to win congressional authorization, Christensen promised his agency would clean up the worst Indian sites anyway, using the federal funds. To the surprise of Interior, states backed the tribes and unanimously opposed Christensen's idea, saying they considered the proposal an "absolute travesty" and "almost criminal." The General Accounting Office (GAO) later found that the surface mining office had the legal authority to take the money without congressional action. GAO said, however, that taking the money could prevent reclamation of many mines on Indian lands, both coal and uranium. Lacking support from the states, OSMRE withdrew its proposal, and the three coal-producing tribes convinced Congress to release the money, which had been accumulating for ten years, to clean up old Indian mine sites.[69]

Such a cooperative attitude was critical if states and tribes were to save the expense of litigation and adequately protect the environment. Dual sovereignty could paralyze all operations. Because water and airsheds cross reservation and state boundaries, differing standards could conflict, as illustrated by the Arizona versus New Mexico water standards in the Church Rock incident. Where small reservations were involved, the regulatory framework could be very fragmented. Some tribes and states recognized that cooperative management could solve such impasses. Despite the continuing legal tug of war between the Navajos and Hopis, staffs of the two tribes believed cooperative administration was possible even at Black Mesa.[70]

When jurisdictional questions still were being litigated or legislated, the various parties felt understandably reluctant to make many concessions. As more parameters were outlined, tribes and states began negotiating cooperative agreements that outlined responsibilities for regulating pesticides, air quality, and water quality.[71]

FUNDING TRIBAL REGULATION

Environmental regulation was costly, and money problems continually plagued the tribal programs. Some tribes found funds through innovative solutions, such as the Cheyenne Tribe's settlement with Montana Power Company and the Navajo Tribe's use of coal funds for uranium reclamation. Generally tribes had limited options: royalties, government grants, government contracts, fees, private grants, and taxes. A year's funding from federal contracts could not support stable programs; staff members had no assurance they would have a job the following year, much less a career. Department directors, hired for their scientific expertise, spent much of their time trying to raise money. While they waited for promised federal funding between fiscal years, some tribes had to borrow money just to meet salaries. Federal agencies inexplicably "lost" tribes' contract proposals. Federal hiring freezes and budgetary cuts during the 1980s exacerbated these problems.[72]

Tribes could not establish their own priorities but instead created offices that mimicked the federal agencies offering funding at that time—little

ANAs, OEOs, and OSMs sprang up on reservations. The demands of the different agencies frequently conflicted. Outsiders criticized federal programs for forcing non-Indians' ethnocentric concepts upon tribal government structures. The objectivity of tribal members warranted suspicion when regulatory staffs were on the payroll of the very companies they were assigned to police.[73]

The lack of independent funding took a heavy toll on tribal environmental regulatory programs, even on the energy reservations where tribes had some income of their own. Royalties, which were already stretched to their limits and declining, could not cover the costs of tribal government regulation and services. For example, the Fort Peck Tribes were forced to reduce their budget by $2 million in 1987 because of reduced federal grants and tribal mineral revenues. The Blackfeet Tribe's revenues fell from $8 million in 1982 to $4 million in 1988. Low royalty rates and poor environmental enforcement allowed the companies to externalize their costs; the royalties and salaries they paid to tribal employees did not cover the actual costs of their operations to the reservation environment and the Indian people.[74]

Given that the tribes were not likely to tear up the leases and the companies were not initially interested in renegotiating, the tribes had only one way to force the companies to internalize their costs: taxation. States relied upon sales, income, and energy taxes to support their governments. Local governments used property taxes. Although subject to the ups and downs of energy prices, taxation promised more stable income than continued dependence upon the whims of Washington. Taxation could also be used to facilitate tribal goals, such as reducing pollution and encouraging use of tribal employees and tribal products, thereby increasing the turnover of dollars.

In 1976 the Jicarilla Apache Tribe "innocently" adopted a severance tax on oil and gas production on its reservation in New Mexico. Because the tax amounted to only 29 cents per barrel of oil and because an economic impact study had shown it would have only a minuscule impact on consumers, the tribal leaders did not expect much opposition, according to Robert Nordhaus, the non-Indian attorney who represented the tribe. They could not have been more wrong. Newspapers in the state were filled with derisive letters and editorials criticizing the idea of a little Indian tribe taxing non-Indians. Nineteen major oil companies filed suit, saying they should not be subject to tribal jurisdiction and to double taxation by a tribe and a state. Some arguments against the Jicarilla tax were blatantly racist, as was attorney Bruce Black's when he said the tax would decrease oil production: "Indians do not require oil to carry on their traditional economic and social activities. . . . Oil conservation is the vital concern of the white public, not the Indian, and the accident of the oil's location beneath public lands set apart for Indians does not reduce the need for its conservation in the interest of white culture."[75] James Watt, then with a private, conservative legal foundation, argued that allowing the tax would expand tribal sovereignty without justification and would undermine the "overriding federal interest to become independent of foreign oil," especially if other tribes were to follow the Apaches' example.[76]

The companies were correct when they foresaw that the Apaches would open the door to more assertions of tribal regulatory authority and when they predicted that more tribes would adopt taxes. They vastly underestimated the strength of the Apaches' legal standing, however. On January 25, 1982, the U.S. Supreme Court affirmed the tax, using language that must have chilled James Watt's and Bruce Black's souls. In its *Merrion v. Jicarilla Apache* decision, the high court said: "The power to tax is an essential attribute of Indian sovereignty. . . . The power (derives) . . . from the tribe's general authority, as sovereign, to control economic activities within its jurisdiction, and to defray the cost of providing governmental services by requiring contributions from persons or enterprises engaged in such activities."[77]

The Jicarilla Apaches had a good case. They could demonstrate their fairness and their foresight; they used the revenue for government services and for a trust fund to provide for the tribe after the oil and gas were gone. For both Indians and non-Indians in the area, the tribe provided road maintenance, a recreational facility, and (except for one state policeman) the only ambulance service and police protection in the area. The tribal government of forty different departments was funded almost entirely by the tribe. The tax was backed by University of New Mexico impact studies, and it had been approved by the Interior Department, as required by the tribal constitution.[78]

Because the Supreme Court had implied that it relied upon the Interior Department to assure tribes' taxes were fair, industry turned its focus to influencing the guidelines the Interior Department was developing for tribal taxes that required the department's approval. The energy tribes objected to having guidelines, saying that although Interior required procedures that they were providing anyway, such as public notice and hearings, the guidelines represented an affront to tribal sovereignty. They especially objected to the Interior Department's asking industry for help writing the guidelines. Then in 1985 in its *Kerr-McGee Corp. v. Navajo Tribe* decision, the Supreme Court affirmed two Navajo taxes—a business activity, or value-added, tax and a possessory interest, or property, tax—that had not been approved by the Interior Department. The Navajo Nation was not organized under the Indian Reorganization Act and had no constitution that required secretarial approval, as the Jicarilla Apaches had. Because the decision did not rely upon the Interior secretary's approval, the decision was an even stronger affirmation of tribal sovereignty and included other non-IRA tribes, such as the Shoshone and Arapahoe Tribes of Wind River, in its scope.[79]

The *Kerr-McGee* decision left the door open to court challenges of taxes that were blatantly unfair and to congressional action, which was later

After winning a landmark severance tax case in 1982 the Jicarilla Apache Tribe presented its attorney, Robert Nordhaus, with an autographed and beaded cowhide. (Photo by Stephen Konecheck.)

attempted. It closed the door to challenges based on the general premise that the tribes had no jurisdiction under federal law to tax non-Indians. Susan M. Williams (Sisseton-Wahpeton Sioux) had put together the Navajo tax program as part of her senior thesis at Radcliffe College. The child of a BIA official, Williams had grown up on the Navajo Reservation and, after seeing the coal mining conflicts there, vowed to seek training in economics and law.

Her law firm, Gover, Stetson, and Williams, became a leading force in lobbying and litigation for Indian taxing authority and economic development.[80]

After nine years of considering various tax proposals, the Navajo Tribe in 1978 had enacted the three taxes that Williams and David C. Cole of Harvard had proposed: a sulfur emissions tax and the two taxes that were affirmed. The tribe had intended the two affirmed Navajo taxes to encourage use of Navajo products and employment of Navajo people through tax credits. With the emissions tax the tribe had wanted to clean up power plants affecting the reservation, especially the Four Corners power plant, which produced more particulate matter than all stationary sources in New York and Los Angeles combined, in clear violation of the contract signed by Arizona Public Service Company, the tribe, and the Interior Department. After the tribe's initiative the states doubled their pollution stringency requirements and the reservation plants cleaned up their emissions. The tribe then dropped its attempt to collect the emissions tax.[81]

Despite the multimillion-dollar *Kerr-McGee* victory, the Navajos' tax initiative was severely handicapped by the early leases for the Four Corners power plant at Fruitland, New Mexico, and the Navajo generating station at Page, Arizona. Those leases implied that the tribe could not collect taxes for twenty-five years from either the power plants or the coal mines that supplied them (BHP-Utah International's Navajo mine and Peabody's Black Mesa-Kayenta mines). Later, however, during contract renegotiations, the tribe was able to get other concessions in the form of monetary payments and employment of tribal members.[82]

After the *Kerr-McGee* decision, companies began to challenge state taxes on Indian resources, something that they had been mysteriously reluctant to do. Although tribal taxes tended to be lower than state taxes, companies that complained about double taxation previously had challenged only the tribal taxes, apparently believing them more vulnerable. By 1988, however, more than one hundred fifty cases involving Indian oil and gas alone had been filed against western states.[83]

In that year the Supreme Court handed the tribes another tax victory, *Crow Tribe v. Montana*, the third in the string with implications beyond taxation. The tribe had challenged Montana's 30 percent tax on tribal coal mined from the ceded strip. The high court refused to review the Ninth Circuit Court of Appeals decision, which said the state and local governments had no authority to levy their severance and gross proceeds taxes on the Crow coal. The circuit court said the state taxes interfered with tribal self-government. Quoting *Kerr-McGee v. Navajo*, the circuit court said "the power to tax members and non-Indians alike is an essential attribute of self-government." Because of the extraordinarily high Montana coal tax and because of jurisdictional questions, the Crow Tribe had been unable to levy its own tax. The Crow Tribe had enacted a 25 percent severance tax in 1976, a year after Montana enacted its 30 percent tax; the combined tax of 55 percent obviously would have been prohibitive. The Interior Department had refused to approve applying the tribal severance tax on the ceded strip, saying the coal lay outside reservation boundaries. The appeals court said, however, that the mineral estate was inside.[84]

Although the tribe had collected only $18 million in royalties from its coal, the state and local governments had collected more than three times as much ($62 million) in taxes from that coal. The appeals court said the state had not tailored its services to fit the money's origin. In fact, the state had discriminated against Indians in distributing the money by refusing to provide grants from coal impact monies to Indian communities in the midst of the impact areas. "It appears that Montana intended, at least to some extent, to use the taxes to profit from the Indians' valuable coal resources," the circuit court said.[85]

Westmoreland Resources, which was mining the coal, and Shell Mining Company, which had planned to mine other coal within the reservation boundaries, both testified against the state tax.

The decision was a clear victory for the Crow Tribe, which received $29 million in back taxes as well as the high court's affirmation that the ceded strip was part of the reservation.[86] In 1989, however, the Supreme Court ruled in favor of state severance taxes in another case involving the Jicarilla Apache Tribe's oil and gas.—*Cotton Petroleum v. New Mexico*. A 6 to 3 court majority said that Cotton had to pay New Mexico's 8 percent oil and gas taxes, which the court apparently considered much less onerous than Montana's coal taxes. The court majority implied that in some cases, a state tax could be invalidated if it interfered with the autonomy of an Indian tribe and its pursuit of reservation economic development. Nevertheless the *Cotton* decision represented a serious blow to tribes. The threat of double taxation meant that tribes would have to lower their own taxes or structure agreements to avoid state taxation.[87]

Up until that time, non-Indian westerners generally had believed that Indians did not pay taxes and thus the non-Indians had deprived them of many government services. Although Indians did not pay county property taxes on trust lands nor state income taxes, Indians usually did pay sales taxes, gasoline taxes, and federal income taxes. In addition, states and counties enjoyed considerable tax revenue from Indian resources, which tribes had neglected to point out. The resulting inequities shocked some investigators. Although the specifics varied from state to state, it soon became clear that Montana was not alone in profiting from Indian resources. For example, the Office of Civil Rights found that from 1976 to 1979 the non-Indian school board spent $38 per pupil for improvements at the Indian high school serving Laguna and Acoma Pueblo pupils and $802 per pupil for the high school in Grants, the mostly non-Indian town nearby. Although New Mexico had collected $40 million in severance taxes from the Jackpile uranium mine at Laguna, the school district refused to approve a bond issue for a new high school because residents believed Indians did not pay local property taxes.[88] A study in 1975 showed that Arizona taxing entities (school districts and counties and cities) received $24.7 million from the Navajo generating station, coal railroad, transmission lines, and Black Mesa mine and slurry pipeline. In contrast, the Navajos received $1.0 million from royalties, rents, bonuses, and rights of way and $4.0 million in wages.[89]

More recent studies showed similar disparities. CERT's chief economist, Ahmed Kooros, said the federal and state governments collected roughly $425 million ($221 million to the states and $204 million to the federal government) during 1985 from energy lessees on Indian reservations to finance activities everywhere, except for Indian economic development. Revenues from the Wind River Reservation in Wyoming exceeded state and county services by more than $56 million between 1979 and 1988, according to the results of a study commissioned by the tribes. A spokesman for Wyoming said the study ignored certain state expenditures, such as those spent on economic development, the state university, and community colleges. The state's own figures showed that the county government was heavily dependent upon the reservation, deriving 26 percent of its taxes from reservation resources.[90]

As the tribes' legal strength became clearer, state and local governments became more willing to talk. In 1987, after the Crow decision, the Arapahoe and Shoshone Tribes laid claim to all the taxes paid by oil and gas companies on the Wind River Reservation but delayed implementation to provide time for negotiations. Local officials initially reacted with alarm and distrust, refusing to attend the tribes' public hearing on the proposal. Some softened when the tribes said they planned to protect schools and services from a sudden loss of revenue and also said they would consider letting non-Indians serve with Indians on boards, such as sanitation boards, serving reservation residents. In 1989 the state and tribes negotiated a one-year tax and water agreement that provided part of the state's share of tax revenue to the tribes while protecting the county tax base.[91]

Nationwide, tribes had adopted more than eighty different tribal taxing ordinances, including several in Montana and Wyoming. Several states' and tribes' organizations, including the Western Governors Association and CERT, encouraged cooperation to avoid disruption of services and duplication of taxing infrastructures. At the same time, opponents continued their efforts to convince Congress to impose controls. Montana and Wyoming congressional representatives sponsored legislation in 1988 that would have imposed a two-year moratorium on tribal taxation, during which Indians were promised $500 million each year to "alleviate economic hardship." Tribal leaders who knew all too well the difference between "authorized" funds and "appropriated" funds expressed their skepticism that the tribes would ever see the money.[92]

Although most of the non-Indians presenting testimony at tribal and congressional hearings complained about double taxation, many also aired concerns about the adequacy of tribal courts, due process, and tribal services. Companies said, for example, that despite the taxes they paid to various governments, they had to provide their own road maintenance. Knowing that Congress could choose to limit tribal taxing authority at any time, tribal leaders and attorneys encouraged one another to assure due process and to demonstrate their actions were not arbitrary, and tribal testimony indicated that most complied. The moderate taxing levels chosen by most tribes proved they did not intend to adopt "confiscatory" taxes as had been feared. In the words of Wind River Tax Commission Chairman Wes Martel, "We do not intend to kill the goose that lays the golden egg." It remained to be seen, however, whether tribes would each be able to show that they used the new revenue to fill the services gap on reservations.[93] In 1905 non-Indians had succeeded in convincing Congress to abolish tribal taxes imposed by the Five Civilized Tribes. In 1988, however, tribal leaders succeeded in changing the tax hearing's focus to the problems with state taxation, delaying indefinitely any congressional action on antitaxing legislation.[94]

CONCLUSION

Tribes in the late 1970s and 1980s began to insist upon more local control, not just as a rhetorical exercise of sovereignty but because of the problems they experienced without local control. They believed as Felix Cohen—and Ronald Reagan—believed, that decisions should be made by those who had to live with their impacts. Rather than insisting upon absolute sovereignty, the tribes recognized the value of having minimum federal standards to back them up and in fact asked the federal government to establish standards in some cases—for example, uranium mine reclamation.

No one could assume that tribes would meet romanticized expectations and protect "Mother Earth" at any cost. With staggering unemployment and poverty rates, tribes had even more interest than states in encouraging economic development. Yet tribes had a unique relationship with their lands: They could not relocate. They knew that if they made mistakes or if they were influenced by economic considerations, their grandchildren would be faced with the resulting problems.[95]

Despite all the judicial and administrative victories that recognized tribal governmental authority, tribes were constantly aware that Congress could reverse them at any time. As CERT Board Chairman Judy Knight, vice chair of the Ute Mountain Ute Tribe, said,

No jurisdiction in the United States—not state government, not county government, nor municipal government—faces any more complex issues of enormous economic, environmental, and political consequences as do Indian tribal governments. Yet no other jurisdiction faces these issues in a more hostile environment. At no time during the frequent conflicts between other jurisdictions are they threatened with political, economic, or social annihilation as are Indian tribes. Tribal government walks a high wire, most often without a safety net should we lose our balance.[96]

Although increases in regulatory control improved their situation immeasurably, tribes still depended to some extent upon continued good will by the federal government. Yet the federal government's trusteeship suffered from a serious conflict of interest. In no area was this more evident than in water rights.

CHAPTER EIGHT

No Reservation Is an Island: Water and Off-Reservation Energy Development

*If our access to our water is taken away, our hope for the future is taken, too.
We are condemned to the status quo: a growing population mired in poverty,
with no way out, wards of the United States and a growing drain on the federal
Treasury, forever. . . . Any development at all will require additional water.*
—Peterson Zah, chairman, Navajo Nation[1]

Looking over the Mandan and Hidatsa Indian cornfields one simmering
summer day in 1811, botanist John Bradbury wrote, "I have not seen any
crop of . . . corn in finer order or better managed than the corn about the
three villages." The agricultural prowess of the Mandan and Hidatsa Tribes
had been known since Lewis and Clark had wintered with them and eaten
the products of the Missouri River bottomlands in 1804. German Prince
Alexander Philip Maximilian and George Catlin in the 1830s as well as
horticulturalist-anthropologist George Will, Sr., in 1930 had noted their fine
crops. Catlin described a "village of earth covered lodges, which were
hemmed in, and almost obscured to the eye, by the fields of corn and luxuri-
ant growth of wild sunflowers, and other vegetable productions of the soil."
Long before Will arrived, the Fort Laramie Treaty of 1851 had defined the
borders of the Fort Berthold Reservation, and in 1862 the Arikaras had
joined the other two tribes living along the Missouri River. Unlike the many
tribes that lost their economic autonomy soon after they were confined to
reservations, the Three Affiliated Tribes thrived. Only 6 percent of tribal
members' income in 1945 was from public assistance.[2]

In 1953 a giant reservoir drowned those rich bottomlands and the tribes'
dreams of economic self-sufficiency. Eighty percent of the tribal members,
who had lived amid the cottonwoods along the river, had to relocate. As the
waters backed 200 miles upstream, they buried 155,000 acres of Indian land,
nearly all the reservation's commercial timber, most of the wildlife habitat, and
many of the accessible coal seams that local Indians had used for fuel. Tribal
leaders say that some of their people died of broken hearts. Today the people of
the Arikara, Mandan, and Hidatsa Tribes live off the reservation or in isolated
communities, nearly cut off from one another by the five giant arms of the reservoir, which the army—not the Indians—chose to call Lake Sakakawea. Some

George Gillette, chairman of the business council, wept in 1948 when Secretary of Interior J. A. Krug signed the contract for selling the best land on the Fort Berthold Reservation for construction of the Garrison Dam. (AP/Wide World Photo.)

Indians raise cattle on the short-grass prairie. A few have gardens, where remnants of the famed old seed stock survives. Most are dependent upon jobs with the tribal government and welfare from the federal government. Nearly half are unemployed.[3]

Lake Sakakawea is a constant irritation to the Three Affiliated Tribes, not only because of the buried farms, forests, coal, and berry bushes but also because they must drive hundreds of miles around its shores to reach tribal headquarters at New Town or to visit their relatives. Its shimmering surface taunts them with its unrealized potential: the tourists it has not attracted and the Indian fields it has not irrigated. The reservoir seems to benefit only Basin Electric's gasification plant and coal-fired power plant near the south shore. Garrison Dam stands as a symbol of the federal government's disregard for the Indians' water rights. The reservoir behind Garrison is only one of many federal reservoirs that inundated Indian lands at great cost to Indian people and benefit primarily to non-Indians.[4]

The federal government built its dams to stop flooding and make hospitable a region that explorer Stephen H. Long once called the Great American Desert. In some areas of the region, especially the Southwest, reservoirs' waters and rivers were soon completely allocated without considering Indians' needs and rights. Then and now irrigation consumed most of the water in the West. Yet other interests threatened farmers' grip on water supplies just as the Indians increased demands for their share. In the early 1970s the federal government joined with coal companies in plans to industrialize the West to provide energy. When those plans were put on hold, western urban areas kept growing and buying water extravagantly, ignoring the constraints of an arid region and of Indian water rights.

Already in 1975, without the intensive development foreseen, the towns, farms, and existing industries used most of the available water. More than 80 percent of the normal flow of western streams—all of it in some areas—was being used in average years. The National Academy of Sciences in 1974 said that the water supply in Montana was "completely committed, perhaps over committed"; Wyoming's supply was nearly accounted for; and the Colorado River Basin was clearly over committed. Underground the situation was no better in some places. Ground water was being withdrawn faster than it was being replenished in wide areas of the West, especially in New Mexico, Texas, Oklahoma, Utah, and Arizona.[5]

Because of increased tribal awareness and involvement no one could quietly take Indian water in the 1980s without loud tribal protests, as had been done for irrigation during the first quarter of the century. Four key questions remained in the century's fourth quarter: How would Indian water rights be quantified—by litigation, negotiation, or legislation? If Indian water was used by others, would the Indians gain money for their own economic development, jobs, or irrigation projects? What controls could the tribes exert over how their water was used on the reservations? Could they lease water for use off the reservations? The viability of the Indians' reservations as permanent homelands depended upon the answers.

WATER THREAT FROM THE COAL BOOM

When the Army Corps of Engineers and the Bureau of Reclamation (BuRec) forced the tribes of the Upper Missouri River Basin to sacrifice their lands for Pick-Sloan reservoirs in the 1940s and 1950s by authority of the Flood Control Act of 1944, the tribes lacked the political strength to defeat the projects. Confident of its ability to condemn Indian lands, the corps began construction of Garrison Dam before ever contacting tribal leaders at Fort Berthold. The water had begun to inundate their lands before the tribes had accepted the government's terms. To accommodate non-Indian interests, the corps changed its plans, but it rejected alternative sites proposed by the tribes. From that experience the tribes of North and South Dakota learned the need for political vigilance and collective action, just as all tribes learned those lessons from the termination legislation of the same era.[6]

By the 1970s western tribes were better prepared to fight for their rights, and the public was more inclined to respond. Congress, no longer in a "termination" mode, had grown to accept the idea that Indian tribes were not likely to vanish. The tribes of the Upper Missouri Basin banded together in 1974 to form the Native American Natural Resources Development Federation (NANRDF) of the Northern Great Plains and issued a "Declaration of Indian Rights to the Natural Resources in the Northern Great Plains States." The nation's effort to reduce dependence upon foreign oil rested heavily upon western fuel resources and power plant sites, particularly those in the Upper Missouri Basin. The tribal federation recognized that this threatened the tribes' survival and it asked the president of the United States to call a moratorium on allocation of waters by federal agents until Indian rights were recognized and protected. The National Congress of American Indians called the Missouri Basin energy plans the "sharpest attacks" against Indian water rights anywhere in the nation. Tribes claimed ownership of all the waters that flowed under, around, or through their reservations.[7]

The tribes' alarm was justified and was shared by many others throughout the West. Some states and rural communities looked forward to the increased jobs and tax revenue that energy activity would bring. Others, however, were concerned, especially about the potential demands on the region's limited water supplies. They believed massive changes in water use in the West threatened farmers and ranchers and the western way of life. Although mining and mine reclamation required little water, converting coal into electricity or synthetic gas or oil used mammoth quantities. A 1,000-megawatt coal-fired power plant with an evaporative cooling system required about 15,000 acre-feet of water, enough water to irrigate nearly 10,000 acres in intermountain states. To gasify 12 million tons of coal required 15,000 acre-feet per year. Liquefaction (converting coal into an oil-like substance) required about half that amount, depending upon the technology used. Oil shale extraction and processing required the most water—about 200,000 acre-feet a year to support a 1 million barrel per day shale industry.[8]

Because of the expected social and environmental impacts of large coal conversion facilities, rancher conservationist groups advocated exporting coal out of the region. If coal slurry lines instead of railroads were used to move the coal, however, water had to be mixed with coal particles in the pipeline. To transport 35 million tons a year required 20,000 acre-feet—an amount that otherwise could supply a western city of 150,000. In 1974 the Western States Water Council commissioned a study that confirmed the governors' worst fears: It projected that as much as 3 million acre-feet of water would be needed each year for power plants, slurry pipelines, and mining in the member states. Other studies indicated that the water used by the power plants would reduce the flow of the mighty Yellowstone River by half in dry years.[9]

Federal energy plans focused most heavily upon the Northern Plains and the Southwest. In the Northern Plains, there was—theoretically—extra water that could be devoted to energy uses. In the Southwest the picture was much different. Except for the deserts of the Great Basin, the Colorado River Basin had the worst ratio of rainfall to evaporation of any basin in the country. Within that larger basin the San Juan coal and uranium basin was an especially dry area, getting fewer than 10 inches of precipitation a year in some places. No perennial streams flowed south of the San Juan River, where most of the coal and uranium lay.[10]

Indian tribes shared the belief held by many farmers and ranchers that the water would be taken from them. The agricultural water users had some assurances they would be compensated for water sales, but the Indians had no such faith. Indian tribes legally had rights to water in nearly every coal- and uranium-producing river basin in the West—the San Juan, Green, Colorado, Bighorn, Yellowstone, Missouri, and Columbia. Yet in most cases, the extent of their rights had not been determined. The energy crisis forced some tribes to realize that until their rights were quantified, they could be taken with impunity.

Some farmers and ranchers saw the Indians' demands for water as extensions of the federal government's. Unfamiliar with the concept of Indian self-determination, they feared the federal government was using Indian water rights as a vehicle for getting more water for energy development. Montana rancher Ellen Cotton said, for example, "I would be happy as a lark to have the Northern Cheyenne have the water—even if they claim all of it. What bothers me is that the federal government would be using Indians to get water for industry." The federal government had at times fed those fears by saying that treaties reserved water for the United States, not the tribes, and that the secretary of Interior should control Indian waters.[11]

Because the Bureau of Reclamation was one of the most enthusiastic boosters of western energy development, Indian tribes also saw the federal government as the biggest threat to their water rights. In the San Juan Basin, tribal and industrial conflicts surfaced early. As far back as 1948 the Navajos had asked Congress not to approve conflicting uses of the San Juan. As the result of federal legislation in 1962 Congress had to study water availability and specifically authorize all BuRec water contracts from the Navajo Reservoir, which held most of New Mexico's share of water under Colorado Basin compacts. Thus Congress had provided tribes in the San Juan Basin with more legal handles for affecting federal water sales than those in the Missouri Basin.[12]

In the late 1960s BuRec began leasing water from federal reservoirs in the Upper Missouri Basin to energy firms. By 1975 BuRec had sold contracts for 658,000 acre-feet a year and accepted industrial applications for an additional 1.7 million acre-feet of water, mostly from Boysen and Yellowtail reservoirs in the Bighorn River Basin but also from Fort Peck and Lake Sakakawea reservoirs on the Missouri mainstem. BuRec planned massive diversion projects from four other dams in South Dakota to serve the coal fields in Wyoming. The application list read like a who's who in western coal development and included many companies with ambitious plans for Indian reservations—Peabody Coal Company, Gulf Oil Corporation, AMAX, Shell Oil Company, Texaco, Exxon Corporation, Kerr-McGee Corporation, Western Energy Company, Consolidation Coal Company, Atlantic Richfield Company, Conoco, Mobil Oil Corporation, and WESCO.[13]

Environmentalists, rancher conservationists, and Indian advocates challenged the decision of the bureau to turn itself into an industrial water marketing agency, to the detriment of irrigators, without authorization from Congress. States questioned the bureau's jurisdiction to market the water within their boundaries. Although the Upper Missouri had more unappropriated water than any other basin, critics argued that the massive industry contracts would harm the region's farmers—even without considering potential Indian water rights. They feared the Missouri Basin waters would become polluted and depleted by inadequate controls and planning, as the Lower Colorado and the San Juan basins of the Southwest had.[14]

The Bureau of Reclamation had never built most of the irrigation projects promised to Upper Missouri Basin states and tribes under the Pick-Sloan legislation. When BuRec claimed in 1975 that Pick-Sloan provided authority to build an industrial empire, the agency in effect was saying that Army Colonel Lewis A. Pick and BuRec Assistant Regional Director William G. Sloan were going to exploit them again. Those were fighting words to the states and tribes. Most of the BuRec and Army Corps of Engineers reservoirs involved in the tug of war lay within Indian reservations—Boysen Reservoir on the Wind River Reservation in Wyoming, Yellowtail (also known as Bighorn Lake) on the Crow Reservation in Montana, and Lake Sakakawea on the Fort Berthold Reservation. The South Dakota mainstem reservoirs involved Sioux Indian lands of the Standing Rock, Cheyenne River, Lower Brule, Crow Creek, and Yankton reservations. Although Fort Peck Reservoir lay outside the Fort Peck Reservation in Montana, the Assiniboine and Sioux Tribes claimed much of its waters. The Bureau of Reclamation and Army Corps of Engineers took no notice of Indian water rights, not consulting tribes and not even mentioning them in the 1975 memorandum of understanding upon which the marketing program was based.[15]

Indian advocates said that if Indian water rights were considered, there would be no excess water to market. Reid Peyton Chambers, an Interior associate solicitor, estimated that Indian water rights exceeded available supplies above Garrison Dam in average years and required 83 percent of the "uncommitted" water below Garrison. If water was to be marketed the Indians wanted a part in the decision making and the benefits.[16]

Although the states and environmentalists won part of their battle, the tribes initially had less success. The states convinced BuRec to give them first option to market the water. The environmentalists and agriculturalists successfully sued to force BuRec to prepare an environmental impact study before proceeding with marketing water. The tribes succeeded in convincing Interior to invite them to later meetings on marketing, but BuRec and the states ignored the tribes' requests for representation on the Missouri River Basin Commission, the principal agency coordinating federal, state, regional, and local plans for water development in the region.[17]

The water marketing conflict between the Bighorn Basin tribes and BIA on one side and the Bureau of Reclamation on the other is a classic example of the inherent conflict of interest in the Interior Department. It also illustrates the federal agencies' habit of excluding tribes from decisions about their resources. Thus the details of the conflict warrant close attention. In 1970, before the environmentalists' suit against BuRec, the tribes on both the Crow and the Wind River reservations had objected to BuRec selling industrial water from Yellowtail and Boysen reservoirs. Despite a previous resolution by the Shoshone and Arapahoe Tribes claiming "all waters arising upon, flowing through, or adjacent to the Wind River Reservation," BuRec did not even notify the tribes when it leased water from a reservoir within their reservation. When the tribes learned of the contracts and protested, Interior Secretary Walter J. Hickel agreed to delay approval of industrial contracts until after a study of the tribes' needs.[18]

During the same period BuRec contracted almost all of the water from the Yellowtail Reservoir without notifying the Crow Tribe. In the ensuing debate a federal attorney leveled a direct attack upon Indian water rights. The Interior Department field solicitor (who served both BIA and BuRec) argued that when the Bureau of Reclamation condemned the Crow land for the Yellowtail Reservoir site in the late 1950s, it also condemned the tribe's water rights. "No opinion could be more damaging to the interests of the Solicitor's Indian clients," Chambers later said.[19]

The Bureau of Reclamation's effort to reduce the tribes' water rights and thus increase the water available for industry continued a BuRec versus BIA dispute that began in 1902 with the birth of the reclamation agency. The battle between the warring siblings came to a head in 1975 when Acting Reclamation Commissioner E. F. Sullivan asked the secretary of Interior for permission to proceed with water marketing. He said the secretary should not recognize the larger Indian claims because they would preclude using the water for meeting "the Nation's energy needs." Commissioner of Indian Affairs Morris Thompson strenuously objected. He said BuRec's plans would not leave enough water to satisfy the rights of the Shoshones, Arapahoes, and Crows. William H. Veeder, an attorney within the Interior Department, said BuRec's plans would "throttle all reasonable economic development of either the Crow or Wind River Reservations."[20]

If the power plants and slurry pipelines had been built as expected, BuRec might have succeeded in robbing the tribes of the water necessary for their economic survival. In its 1977 EIS on Upper Missouri Basin water

Although western tribes expected to lose water to dozens of coal gasification plants in the late 1970s, all but one were canceled. The Great Plains plant (in background) produces synthetic gas using water stored behind Garrison Dam. (Courtesy of the Department of Energy.)

marketing, BuRec said 1 million acre-feet were available. Responding to criticism from several Sioux bands, BuRec said it was "beyond the scope of the EIS to predict the outcome of Indian water litigation" but promised the marketing program would be modified after Indian water rights were quantified. BuRec ignored the critical question: Who would lose their water if a power plant or a city were already using water that courts later said belonged to a tribe? Despite all the threats, neither the Interior Department nor the tribes challenged any of the BuRec water contracts in court. However, they did file general suits to quantify Montana tribes' water claims and defended against the state of Wyoming's attack on Indian water.[21]

Soon, however, energy consumption dropped, and so did oil and gas prices, completely changing the economics of coal gasification. Only one of the thirty-five giant gasification plants expected—the Great Plains gasification plant—was built to commercial scale. The plant received state and federal—but not tribal—permits for 17,500 acre-feet from Lake Sakakawea. The permits for the coal-fired power plant, which supplied power to the gasification plant, provided 15,000 acre-feet a year. Despite a threat to sue, the Three Affiliated Tribes of Fort Berthold insisted only upon a provision saying the contract was subject to court-recognized claims by the Missouri Basin Indian tribes. Despite the high unemployment rate on the neighboring reservation, only 5 percent of the gasification plant's work force was made up of tribal members. Great Plains Gasification Associates defaulted on its $1.5 billion federal construction loans, forcing the Department of Energy (DOE) to take over ownership in 1986. When DOE sold the plant in 1988, the Fort Berthold tribes submitted a bid to try to increase tribal employment and to sell tribal coal. DOE accepted the bid of the Basin Electric Power Cooperative instead.[22] The only other Missouri River water actually marketed by BuRec was for the Colstrip coal-fired power plants. Montana Power Company received 6,000 acre-feet of water from Yellowtail, also through both state and federal—but not tribal—permits. BuRec charged a minimal storage and handling fee of $30 per acre-foot for each of its contracts, none of which benefited tribes.[23]

South Dakota, on the other hand, could have received $9 million a year for use of water from Oahe Reservoir if Energy Transportation Systems, Incorporated (ETSI) had built its coal slurry pipeline to transport coal from Wyoming coal fields to Arkansas. Because of the railroads' political clout, however, the ETSI companies abandoned their project in 1984, and no other coal slurry pipelines were built during the boom. The Black Mesa pipeline, using water sold by the Navajos in 1964, remained the only example of the technology that was expected to crisscross the West. ETSI failed to convince Congress to give it the power of eminent domain, and thus railroads could prevent ETSI from getting the necessary rights of way. A federal court awarded the state of South Dakota $600 million in an antitrust lawsuit against the railroads based upon its lost water-marketing revenue. Tribes with rights to the water wanted a portion of that judgment if it was confirmed by a higher court, saying they should have been part of the marketing agreement. The tribes did not join Lower Missouri Basin states that successfully challenged BuRec's authority to market water from Oahe to ETSI. All of BuRec's other industrial option contracts from the Upper Missouri expired. In December 1988 Interior announced a water marketing policy under which BuRec would facilitate between willing parties leases of waters from federal projects.[24]

In the Southwest the slowdown in the energy market gave tribes more time to try to resolve BuRec marketing conflicts. In 1989 the Jicarilla Apache Tribe was trying to settle a lawsuit challenging BuRec's decision to sell 44,000 acre-feet of water from Navajo Reservoir to BHP-Utah International. The company first acquired the water in 1967 and had planned to sell about half of it to Public Service Company of New Mexico (PNM) for use in a coal-fired power plant in the Bisti area south of Farmington. The Apaches opposed BHP-Utah International's contract because it interfered with their rights to San Juan water. Use of ground water instead would have made the power plant even more objectionable to tribes because it would lower water tables on Navajo and Jicarilla Apache lands and affect wells on Ute Mountain Ute, Southern Ute, and Laguna Pueblo lands as well.[25]

The PNM plans took on an ironic twist in the late 1980s when the Navajo Tribe joined with PNM to plan the power plant, potentially pitting the Jicarillas' water rights against those of the Navajos. Southwestern tribes held many contesting claims to the San Juan waters. For example, the Jicarillas sued the city of Albuquerque when it accommodated the interests of pueblo tribes in its water acquisition from the upper San Juan Basin but ignored those of the Jicarillas. For the city of Albuquerque's future use the San Juan–Chama Project diverted water from the Navajo River in Colorado, upstream of the Jicarilla Apache Reservation, destroying the Apaches' fisheries and reducing their domestic water supplies.[26]

Luck, as much as legal or political clout, had protected most Indian water rights during the energy boom. The boom had demonstrated the potential extent of conflicting water demands and the stakes, giving tribes incentive to pin down their claims before markets for electricity and coal improved and before the urban demands escalated as much as expected. By mid-1989, however, the Indian water rights had been determined for only two reservations in the Upper Missouri Basin—Fort Peck and Wind River. The Fort Peck–Montana compact of 1985 provided the tribes with 1 million acre-feet. The U.S. Supreme Court in June 1989 confirmed an award of 500,000 acre-feet for the Wind River tribes. Although in the San Juan Basin most water rights were not expected to be quantified before the turn of the century, the state of Colorado and the Southern Ute and Ute Mountain Ute Tribes reached a settlement, confirmed by Congress in November 1988, for 87,000 acre-feet of water from two proposed water projects and 42,000 acre-feet from various streams.[27]

How to quantify Indian water rights turned out to be more complicated and time consuming than BIA, BuRec, or the tribes could have guessed in 1975. The squabbles of the 1970s only hinted at the warfare to follow, in which the Bureau of Reclamation continued its efforts to reduce Indian claims.

QUANTIFICATION

Tribes had the most at stake in the battle for water because the extent of their water rights had not been determined. Indian rights were different from other western water rights. Western states allocated water on the basis of the prior appropriation doctrine, a law under which water that was used earliest had a higher priority—"first in time, first in right." Water rights were assigned priority dates based upon when they were first put to use. The prior appropriation doctrine also dictated that a person must relinquish his or her water right if he or she did not continue to use it.

Indian and federal reserved water rights, on the other hand, were based upon a court decree known as the *Winters* decision, not upon laws passed by state legislatures or Congress. Under the *Winters* doctrine, the priority date for water rights was determined by the date that the Indian or other federal reservation was set aside, and the water need not be put to use for the right— theoretically—to continue. Indian rights almost always predated all others, meaning they had higher priority in times of shortage.

The evolution of the *Winters* case is important for both legal and historic reasons because it illustrates the crisis orientation of BIA and the importance of actually developing Indian water. In 1888 pioneers, at the invitation of the federal government, settled former Indian lands around the Fort Belknap Reservation in Montana Territory and started diverting water from the Milk River, upstream from the Indians' reservation. Fort Belknap Indian Agent William R. Logan did not object until a drought caused a crisis that threatened his success at turning "his" Gros Ventre and Assiniboine Indians into farmers. In 1905 he begged his superiors to make the settlers respect the Indians' rights, saying, "To the Indians it either means good crops this fall or starvation this winter."[28]

In response to his plea the federal government in 1905 sued the non-Indian irrigators, including Henry Winter. (A clerical error resulted in Winter's name being recorded for posterity as "Winters.") In 1908 the U.S. Supreme Court recognized that the federal government created the problems by giving the water away twice, once to the Indians and again to the pioneers. Nevertheless the high court ruled in *Winters v. U.S.* that the Indians' rights came first. The court said that when an Indian reservation was set aside, the action also set aside sufficient water for the Indians for "all their beneficial use, whether kept for hunting, and grazing roving herds of stock, or turned to agriculture and the arts of civilization."[29]

Although important to other Indian water claims, the victory was only a paper one to the Gros Ventre and Assiniboine Indians of Fort Belknap. Two years after the decision they were starving. Hawk Feather, an Assiniboine, said in 1910, "There was a man here once and he told us that this water right belonged to us. He said we would use the water first and raise a crop and make a living on that, but we do not get enough water yet." If Hawk Feather had been alive eighty years later, he would not have seen much improvement. The tribe's irrigation project, begun in 1903, still had not been completed by the 1980s, and some of it was in an "advanced state of deterioration." Non-Indians were still pirating water that the *Winters* court said should be irrigating Indian fields. During the 1985 drought the Bureau of Reclamation shut the tribe's floodgates to protect non-Indian irrigators. That crisis forced the Interior Department to improve its water measuring and its surveillance technology in preparation for another lawsuit to protect Indian farmers.[30]

Throughout the West during the years after *Winters*, BIA consistently defended Indians' rights only after water was taken by someone else. Just as at Fort Belknap, that "someone else" was often the Bureau of Reclamation. Senator Edward Kennedy of Massachusetts said in 1971 that for the Indian people, "reclamation might just as well be the cavalry all over again." Neither BIA nor BuRec fully believed in the legitimacy of the *Winters* ruling; until the late 1950s BIA continued to apply for state water rights for Indians instead. The states and the federal government ignored the implications of vast Indian water claims, just as they ignored the constraints of living in the arid West, until most of the water was allocated. The federal government never integrated the two conflicting principles in its policies.[31]

Thus the water crisis of the 1970s found the tribes' water rights unprotected. "It's like inviting everyone to come and feast when you only have a 10-pound turkey. Pretty soon, everyone starts looking at the next table," said Mel Tonasket, chairman of the Colville Tribes in Washington and president of the National Congress of American Indians, where he was one of the most outspoken critics of federal water policy. That the Indians' "turkey" barely had been nibbled upon encouraged the appetite of would-be water marketeers and eager industry users. The differences between the prior appropriation rights and the Indian reserved rights would have been of little importance in the 1970s if Indian water had already been put to use. Instead water development decisions had been allowed to determine water rights. A national water policy had never been developed and so market and political forces determined the de facto policy: The federal government subsidized development around Indian reservations for non-Indians using Indian water and, in some cases, Indian funds. By minimizing development of Indian lands, Indian water rights were suppressed.[32] When Indian rights conflicted with those of BuRec irrigators, Interior sometimes instructed BIA employees not to present evidence that would damage the BuRec position.[33]

The Indians were decades behind their non-Indian neighbors in putting water to use in domestic, municipal, agricultural, or industrial ways. Overall, 25 percent of Indian houses lacked plumbing; on the Navajo Reservation, it was more than 50 percent. On the Pine Ridge Reservation, 60 percent did not have running water or even wells. Indians along the Missouri dipped water from reservoirs for their families to drink.[34]

Although the lack of domestic water had a great impact on the Indians' daily lives, the lack of irrigation had an even greater impact on their water rights. Because 90 percent of the West's water goes to irrigators, farmers and ranchers own more water rights than anyone else. The Bureau of Reclama-

Even tribes that live at the headwaters, such as the Blackfeet, have had to wage lengthy court battles to confirm their rights to water for irrigation and other purposes. (Photo by Marjane Ambler.)

tion provides nearly half of the surface water used. The federal government failed to develop irrigation projects on Indian lands, just as it failed to build drinking water and disposal systems for Indian people. As Indian legal expert Charles Wilkinson said, "The reclamation program proceeded on the backs of the Indian people." The National Water Commission recognized this fact in 1973:

> In retrospect, it can be seen that this policy was pursued with little or no regard for Indian water rights and the *Winters* Doctrine. . . . With few exceptions, the projects were planned and built by the federal government without any attempt to define, let alone protect, prior rights that Indian tribes might have had in the waters used for the projects. . . . In the history of the United States government's treatment of Indian tribes, its failure to protect Indian water rights for use on the reservations it set aside for them is one of the sorrier chapters.[35]

How much water *Winters* provides for is not clear. In most of the cases that have been decided, the Indians were determined to have rights to as much water as they could use for the principal purpose for which their reservation was established, usually agriculture. Such measures determined the quantity of the right but did not necessarily dictate how the tribe could use the water.[36]

Determining the amount of land that could be irrigated required costly, technical studies of soils, hydrology, geology, economics, and engineering. The Interior Department said in 1987 that the proper studies cost $1 million to $4 million for each reservation. The Indian water study program suffered from a chronic lack of funding and from crisis-oriented prioritizing; manpower initially was allocated only after reservations became involved in litigation. Because of unresolved questions about the definition of "practicably irrigable," the studies' conclusions involved subjective judgment. Thus, who conducted the studies became a critical consideration. The Bureau of Reclamation came up with much lower figures than BIA on several reservations, including the Wind River, Crow, Jicarilla Apache, and Navajo. BuRec officials tended to protect "their projects" from Indian claims. Battles ensued when tribes wanted to hire their own consultants.[37]

As water demands expanded, complaints about the uncertainty of Indian water rights escalated. Non-Indians perceived Indian rights as a "time bomb" because they could not evaluate the value of other rights until the Indian rights were settled. The uncertainty also discouraged on-reservation development—a 1984 study on economic development found that tribes considered water rights disputes a significant hurdle to development. Non-Indians increased pressure to quantify Indian rights more quickly than the Indian tribes thought was possible or desirable. Historically speaking, however, the complaints themselves signaled success. No longer were the irrigators and the industries of the West allowed to ignore Indian rights. A string of major water rights victories had forced the people of the West to acknowledge the legitimacy of the *Winters* ruling, no matter how inconvenient and disruptive it might be.[38]

Indian water rights could be quantified by three different means—legislation, litigation, or negotiation. Although several attempts had been made through the years to convince Congress to unilaterally take away Indian water rights or to give tribes a deadline for quantifying their water needs, none had succeeded. In most cases tribes and states resorted to lawsuits at great cost to all parties. By 1989 the Wind River tribes estimated that they had spent more than $9 million defending their water rights from the state of Wyoming, in addition to $1.9 million spent by BIA and $864,000 by the U.S. Justice Department on the lawsuit filed by the state against the tribes. Wyoming figured it had spent $9.9 million. The money had gone either for consultants or toward attorneys' fees, clearing records, and administration; none had gone toward the storage and construction that would be necessary to satisfy the conflicting rights.[39]

As demands for water increased, tribes realized they might win only paper victories. Because litigation often took ten to thirty years, all the water the Indians wanted might be allocated by the time lawsuits were settled. When President Carter came into office in 1977 the time was ripe for a new approach. The energy crisis had become a water crisis, lending urgency to the drive to settle Indian rights. When Carter proposed reducing the federal government's subsidies of water projects, he raised a storm of protest from non-Indians. Many Indian leaders and tribes also objected to the Indian water policy that Carter proposed because it required tribes to quantify their water needs and it encouraged tribes to negotiate with states. Indian groups and their advocates objected to quantification because they believed tribal rights under *Winters* were open-ended and could be expanded as tribes' needs changed. They said states had no authority to negotiate binding contracts with Indians. Much of the opposition was based upon the tribes' losses through other "negotiations" for land, minerals, dam sites, and water and upon the assumption that tribes' negotiating skills and knowledge might not suffice. One article in *Akwesasne Notes* about water negotiations phrased its criticism in language that now sounds incredibly patronizing: "Negotiations are reduced to ludicrous sessions between experienced, strong corporate legal staff and the inexperienced [tribal] delegates."[40]

Despite the objections, some Indian interests recognized that Carter's Indian policy offered several advantages over past practices. It promised money for studies—up to $200 million for the ten-year study period—and called for federal agencies to encourage new Indian water development. Through the ensuing years the U.S. Supreme Court removed some of the other objections to negotiations with states by giving state courts jurisdiction in general stream adjudications and by saying that *Winters* rights were not open-ended. Gradually tribes became more confident because numerous court decisions recognized both the legitimacy of large *Winters* claims and the expertise of tribal advocates, such as Native American Rights Fund attorneys. Many tribes and tribal organizations came to accept the concept of negotiating with states so long as tribes were free to choose tactics without coercion from the federal government.[41]

Negotiations could shape individual agreements that litigation could not and could address important transboundary issues affecting Indian and non-Indian lands, such as water pollution and administration of water codes. Congressional ratification of a negotiated agreement could give tribes powers that existing laws—and thus courts—could not provide, such as authority to market water outside their boundaries. Sometimes tribes combined negotiations with litigation when the dispute involved unresolved legal questions, such as, Can a tribe quantify its right on the basis of future mineral development, recreation development, or instream flows? If it can, how will those rights be measured? What are the economic limitations of a *Winters*'s claim? Can a tribe claim reserved rights to ground water? Who will administer water allocations? How much water is an allottee entitled to? Can his or her water right be transferred?

Although the Reagan administration endorsed Carter's preference for negotiated settlements, the federal government undermined them by refusing to provide the necessary financial support for studies, water planning, inventories, or construction. In the Upper Missouri Basin, for example, where all Montana tribes except the Blackfeet were negotiating with the state in 1986, only 50 percent of the baseline studies needed to begin negotiating had been completed. Funding for water studies decreased dramatically during the 1980s because of budget cuts. Settlements involving two Arizona tribes—the Tohono O'Odham (Papago) and the Ak-Chin—and urban users were not funded until after the unfulfilled promises threatened to jeopardize other negotiations in the country. The Ak-Chin water was critical to the tribe's efforts at self-sufficiency. The community gained jobs for all its members and netted $1 million annually from farming, half of which was reinvested and half used for tribal services. Despite such payoffs, Indian advocates feared for some time that the federal administration would accept only settlements that involved no federal expenditures, such as that reached in 1985 between the state of Montana and Fort Peck's Assiniboine and Sioux Tribes. In 1988, however, Interior endorsed and Congress approved the Colorado-Ute settlement, including $60.5 million from state and federal sources for the two Ute tribes' economic development.[42]

Settlement skeptics did not forget broken promises. In New Mexico the federal government had sabotaged a negotiated settlement and the Navajo Tribe's attempt to increase its self-sufficiency. The Navajo Tribe agreed to the construction of the San Juan–Chama Project (a diversion and storage project benefiting non-Indians) if the federal government would build the Navajo Indian Irrigation Project (NIIP) delivering 508,000 acre-feet. The tribe agreed to share shortages, in effect relinquishing its priority date for that water in the hope of attracting industrial as well as agricultural development. After Congress authorized the construction of the two projects in 1962, however, the federal government tried to break its side of the bargain. BuRec sought to reduce the Navajo water allotment and make more water available for industry, saying only 370,000 acre-feet could irrigate the same Navajo acreage if conservation measures were used. Although it completed the non-Indian project on schedule, the federal government dragged its feet building NIIP, and at one point the Reagan administration tried to defer funding altogether.[43]

The NIIP controversy once again involved the Bureau of Reclamation competing with tribes to market water. The Navajo Tribe said any NIIP water saved by conservation should be available to the tribe for other uses, such as the proposed power plant. BuRec meanwhile had leased some of the San Juan River water to BHP-Utah International for use in the same power plant. Ironically, the federal government's interpretation of the NIIP settlement was more hostile to tribal interests than was New Mexico's interpretation, according to tribal attorneys.[44]

Despite its rhetoric encouraging negotiations, the federal administrative branch by its actions usually seemed to tell the tribes to litigate rather than negotiate, to use as much water as possible, and above all not to trust the Interior Department to defend their interests. Nevertheless Indian and non-Indian cooperation grew, partly because the crisis atmosphere of the 1970s was removed. Although fast-growing urban areas continued to demand more water in the late 1980s, those demands were not combined with huge industrial ones upon limited water supplies. When an industry group, the Western Resources Council, sought Congress's help in 1982 to force quantification of Indian water rights, Indian groups thwarted that threat by opening negotiations over more palatable solutions. As a result industry, western governors, and three Indian groups (NARF, NCAI, and CERT) formed an ad hoc Indian water rights group, which was still active in 1988.[45]

Because of improved communications brought about by such efforts, the Western Governors Association in 1987 adopted a surprisingly pro-Indian policy resolution that said the federal government should make a "fair and just contribution" and should not coerce tribes to negotiate. The western governors established improved state-tribal relations as their top priority and planned a series of meetings with tribes in 1989. On the state level the Montana state compact commission was negotiating with all but the Blackfeet Tribe to resolve quantification and administration questions. The Northern Lights Institute, a private, nonprofit group in Montana, brought states and tribes together to discuss basin water allocation at symposiums in 1986 and 1988, an effort that represented a marked departure from earlier years when tribes had been deliberately excluded from interstate negotiations, such as the Missouri River Basin Commission and the Colorado River compacts.[46]

States and tribes could reach settlements easiest when both their interests were served. For example, the states of Colorado and Montana benefited by water allocations to the Ute and Fort Peck tribes because the reservations lay on state boundaries; if not used by the tribes the waters would flow out of state. Settlement was more difficult for the Northern Ute Tribe and Utah because the reservation lay upstream from Utah's urban demand areas. Although the Wind River Reservation lay in the center of Wyoming, non-Indians nevertheless benefited by a one-year settlement negotiated in 1989. The settlement guaranteed irrigation water for the non-Indians; it provided the tribes with money to improve their irrigation system and to relieve drought-stricken farmers.[47]

WATER FOR ENERGY ON THE RESERVATION

Beyond the question of quantification, tribes also focused on gaining control over how water would be used for energy. Water had been used for energy development on Indian reservations for several decades. Throughout the early years, energy companies routinely used water available at the sites without cost and in most cases without even acknowledging the privilege. An 1896 oil and gas lease between the Osage Tribe and Fosters Brothers provided that the leaseholder could use as much timber, building stone, water, and wood as needed.[48]

By the early 1970s little had changed. Although tribes objected to BuRec's marketing their water without their consent, companies during the same period convinced tribes to give away thousands of acre-feet of water. The companies' arguments were bolstered by standard mineral lease terms, which assumed that mineral leases carried with them implied water leases, free of charge. Finally tribes and legal experts began questioning these assumptions, especially for water-intensive uses such as secondary recovery of oil, oil shale treatment, oil refining, and coal conversion.[49]

Many oil and gas fields and some mines lay within reservations. In most cases the energy conversion plants, which used larger quantities of water, were located outside reservation boundaries. Although industrial facilities outside reservation boundaries still sometimes used Indian water, the tribes had only limited possibilities for influencing decision making on them. Within reservation boundaries, only six uranium mills had operated through the years—four on the Navajo Reservation in the Southwest, one on the Spokane Reservation in Washington, and one on the Wind River Reservation in Wyoming.

The only other significant on-reservation energy conversion facilities were two coal-fired power plants built in the 1960s on the Navajo Reservation. These early power plants were important not only for the amount of reservation water supplies they consumed but also for the decision-making process involved in their planning. The unfavorable water contracts that resulted from the negotiations have haunted the Navajos and other tribes throughout the nation ever since. The Navajos' claim to water necessary for their own economic and agricultural development was severely undermined. The obviously one-sided terms contributed—as the Pick-Sloan dams had done—to Indian aversion to negotiations and continued distrust in federal agencies. Although the Navajos in the 1960s possessed more potential political clout than the Three Affiliated Tribes had in the 1950s when they signed the Garrison agreement with the waters licking at their heels, the Navajos lacked the information they needed to protect their people and their resources during the negotiation process.

In 1961 the Navajo Tribe leased land for the Four Corners power plant near Farmington, New Mexico, to Arizona Public Service Company and Utah Mining and Construction Company (now BHP-Utah International). The companies rushed the negotiations, saying delays "would be fatal" to the project, which they said would benefit the tribe with revenue, employment, and "desired industrialization." Utah Mining acquired water rights for the plant from the state, not the tribe. The next power plant, the San Juan, also was built near Farmington but across the San Juan River, off the reservation, by New Mexico Public Service and Tucson Gas and Electric companies.[50]

Water did not become an important part of Navajo power plant negotiations until 1968 when the Navajo generating plant was planned. Congressman Wayne Aspinall of Colorado, chairman of the House Interior Committee, recognized the threat posed to his state by the potential extent of Navajo rights to the Colorado River. He told the Bureau of Reclamation and the Salt River Project to convince the tribe to sign away most of their rights to Upper Colorado Basin water before he would approve the Central Arizona Project, a massive water project.[51]

Faced with threats that the plant would be built elsewhere, the Navajo Tribal Council agreed so that the tribe could gain jobs, land lease revenue, and a market for its Black Mesa coal. Tribal officials passed a water resolution, drafted by the Interior Department, that donated to the plant their rights to 34,100 acre-feet, leaving less than 16,000 acre-feet of Colorado River water for their own future needs. Tribal economist Phil Reno figured that the water donated for that power plant alone was worth $6.8 million a year at $200 per acre-foot.[52]

Guilty of innocence and driven by economic imperative, the Navajo officials failed to realize the strength of their bargaining position. Interior Secretary Stewart Udall admitted later that the Navajos suffered from the department's conflict of interest. Tribal officials relied upon data supplied by those on the other side of the negotiating table—the Salt River Project, the Bureau of Reclamation, and the Upper Colorado River Commission—who bribed them with jobs and misled them about the value of the Navajos' water and about their future water needs. The advice from the tribe's side of the table was no better. On the eve of the nation's biggest coal boom the local BIA superintendent said, "If you're going to sell all this coal, you've got to have both of these plants. . . . We don't find anything else you can do with the coal." One of their own attorneys told the Navajos that there was no better use for their water. Although attorneys with DNA-People's Legal Services told them to protect their water rights, tribal officials ignored the advice. Other uses for their water rights seemed to the Navajos ephemeral, whereas jobs were immediate.[53]

Unlike the power plant contracts, which involved no payment for water, Peabody Coal Company agreed in 1966 to pay the Navajos and Hopis for water used in the slurry pipeline from Black Mesa. Because money changed hands and water was used off the reservation, the deal represented the first marketing of Indian water. This slurry water, which is now recognized as the best of reservation waters, was sold to convey pulverized coal through a 275-mile-long pipeline to the Mojave power plant on the banks of the Colorado River in Nevada. The Navajos agreed to provide more than 3,000 acre-feet of ground water each year for just $5 each. The Hopis, who shared owner-ship of the water, received $1.67 per acre-foot.[54]

Recognition of such inequities gradually changed the way that Indian tribes across the West looked at their water, just as the unfavorable coal leases changed the way that they looked at minerals. Tribes began recognizing water's monetary value and how much tribal water rights contributed to the value of tribal minerals. Rather than kowtowing to energy companies to convince them to locate on reservations, the tribes began setting tougher terms. In 1987 the Hopi and Navajo Tribes during coal contract renegotiations succeeded in increasing the price of the slurry water from $5 an acre-foot to $600, by far the highest published prices paid for leased water. To discourage overuse, Peabody must pay twice as much when it uses more than 2,800 acre-feet a year. Peabody also agreed to provide half the funds for a study of the aquifer that will determine what must be done, if anything, to protect it. After the renegotiations Peabody sold the pipeline to Williams Technologies.[55]

Under the terms of the original contract and federal leasing regulations, Peabody or the new owner must replace water or provide compensation if production wells significantly affect other users. Several Navajo and Hopi communities get their water from the same aquifer that Peabody uses. The water table has dropped by 10 to 70 feet in those communities and is projected by federal scientists to drop as much as 175 feet by the year 2032 under the worst case scenario. Tribal and federal geologists at the time did not consider this impact significant, however, because they expected sufficient water pressure for community wells. The Hopi Tribe was more worried about ground water and successfully insisted that the federal coal mining agency prepare a full environmental impact study, including a water impact study. Although neither the Hopis nor the Navajos were happy with sending so much high-quality water off the reservations, their negotiating strength was limited by the fact that the old coal leases did not require renegotiation. Because Peabody Coal is the biggest employer of Navajos, the tribe did not consider revoking the contract. Nor did the tribe insist that Peabody build a second pipeline from the power plant back to the reservation so that water from the Colorado River could be used for the slurry. Instead the Navajos received monetary compensation and promises of replacement water.[56] In return for these concessions from Peabody and other coal companies during contract renegotiations, the Navajo Tribe agreed to allow some companies to mine more coal and guaranteed certain quantities of water to them. This was particularly important for BHP-Utah International, which supplies coal and water to the Four Corners power plant. If tribal ownership of the water were proven through adjudication, however, BHP-Utah International agreed to pay the Navajos for the water.[57]

Tribes in the 1980s did not always trade water for money in negotiations. At times they insisted upon recognition of the value of their water while bartering for jobs, irrigation development, and other concessions. Because the Navajo Tribe was constrained from changing terms for water use at the two on-reservation power plants, the tribe focused upon employment, with mixed success. The Salt River Project employed only 42 percent Navajos at its plant in Page, Arizona, whereas Arizona Public Service Company employed 65 percent Navajos at the Four Corners power plant. By a lawsuit the Navajos also succeeded in increasing the number of tribal members employed at the San Juan generating station off the reservation.[58]

The Three Affiliated Tribes of the Fort Berthold Reservation also were able to win more benefits by tough, assertive demands in the 1980s than they got in the late 1940s and early 1950s when "negotiating" with the Army Corps of Engineers. In 1984 Congress passed legislation restoring mineral ownership under the lake and lakeshore to tribal ownership. The restoration was a windfall to the tribes, more than quadrupling their communal mineral base and resulting in more than $240,000 of revenue within the first two years. The tribes also won congressional authorization for $62.5 million in irrigation development as compensation for the Garrison taking.[59]

TRIBAL WATER MANAGEMENT

Water quantification and water management were often related. Both states and tribes wanted a voice in deciding how much water was surplus and thus available for other uses. They both also wanted authority over how water was used. States insisted they had the right to administer the water rights of non-Indians who had purchased former allotments or homesteads within reservation boundaries. State-tribal battles over administrative authority sometimes created rifts that prevented amicable negotiations about quantities. In Montana, for example, where the state had challenged the Blackfeet tribal water code in 1980, the Blackfeet Tribe steadfastly refused to negotiate eight years later.[60]

The Interior Department hesitated to side with the tribes in the frays. In response to non-Indians' growing opposition to tribal regulation of their water use within reservation boundaries, Interior Secretary Rogers C. B. Morton imposed a moratorium in 1975 on approving any tribal water regulation ordinances or codes. He adopted the temporary moratorium to give his department time to develop uniform guidelines so that subsequent tribal codes would treat Indians and non-Indians fairly. Morton said he feared that independent tribal water codes "could lead to confusion and a series of separate legal challenges which might lead to undesirable results."[61]

After several aborted guideline proposals, however, the moratorium was still in effect thirteen years later. Tribes objected to the guidelines; they did not want the department to arbitrarily decide on jurisdiction over non-Indian landowners before the courts had ruled. The Ninth Circuit Court of

Appeals had recognized tribal authority in the *Colville Confederated Tribes v. Walton* case, in which the waterway lay entirely within a reservation. The same appeals court had recognized state authority in another case, *United States v. Anderson*, however, saying state control would not infringe upon the tribe's right to self-government nor its economic welfare. In *Walton* the circuit court recognized the importance of water jurisdiction: "Regulation of water on a reservation is critical to the life style of its residents and the development of its resources . . . [and] an important sovereign power." Realizing the need to decide jurisdiction over non-Indians' water once and for all, the appeals court pleaded with the high court to review its decision. In 1981 the Supreme Court let the ruling stand.[62]

Other critics, both within and outside the Interior Department, objected to Morton's adopting the moratorium to protect non-Indians because his responsibility was to protect Indians. Despite the moratorium, some tribes adopted codes or tried to manage their water without codes, usually affecting waters of non-Indians. Interior granted the Fort Peck Reservation a special exception in 1986 because the tribes and the state of Montana had reached agreement upon a water compact, which required a tribal water code. When the department heralded the Fort Peck code as a model that should be followed by other tribes, it neglected to mention the moratorium. Nevertheless Interior continued to block tribal codes, saying it would consider them only if a tribe's water rights had been quantified.[63]

Why would tribes want to enact codes? Would these codes be fair? How would they affect mineral developers on reservations? How would tribal regulation differ from state regulation? The water codes adopted by the tribes offered some answers. By using their governmental powers tribes could protect their proprietary rights to certain quality and quantities of water. To many Indian people, water played an important part in their spiritual beliefs. Unfortunately state and federal regulations did not take individual cultural practices into consideration. A tribe whose culture depended upon fish did not want its fish streams dried up, which many states allowed. As Shoshone councilman Wes Martel said, "Culturally we view fish and wildlife as life, too. They're our relations. We have laws to protect people, and we need laws to protect them." States used water administration to put limits on transfers of use, with which a tribe may or may not agree. For example, North Dakota forbade transferring water from agricultural to industrial use. However, most states did not have such prohibitions. Under Montana law, for instance, water rights could be transferred with some limitations from agricultural to energy use, but water could not be used for coal slurry pipelines.[64]

On reservations where jurisdiction had not been decided, the uncertainty about administration was nearly as debilitating as the uncertainty about quantification. Some people applied for state permits, others for tribal, and still others for none at all. No one knew who was using water for what purposes, and there was no way to prevent inequities, waste, and over appropriation. To chart its future a tribe needed information (about current water usage and future potential) and control.

The Navajo code deserves attention because it was one of the first adopted and because it affects energy producers. It demonstrates how one tribe incorporated its own priorities, which differed from state laws, and how it addressed past mistakes in water for energy decisions. The Navajo Tribe used its code to protect ground-water resources important to Navajo sheepherders and other rural residents. The state of New Mexico allowed companies to mine ground water—that is, to use it faster than it could be replenished naturally—under certain conditions. Ground-water depletion from coal and uranium mining, both on and off the reservation, threatened Navajos in many areas. To protect themselves, Navajos filed formal objections to many mining plans and projects outside reservation boundaries, including the Escalante power plant at Prewitt, New Mexico, and federal coal and uranium leasing in the San Juan Basin. For water usage within its territory the tribe included a prohibition on mining ground water in its 1984 water code. Under this code the Navajo Tribe could deny new permits if necessary to protect existing stock wells. The tribal code was not affected by Interior's moratorium because Navajo ordinances did not necessarily require the Interior Department' approval.[65]

The code was designed to help the tribe receive fair compensation and make decisions based upon the Navajo people's long-term goals. A tribal spokesman said that if it had been in force in the 1960s, the code probably would have blocked the slurry pipeline water lease. Although the code did not specifically prohibit use of tribal waters off the reservation, tribal officials would have had their own data upon which to rely and probably would have realized it was not an appropriate use of the best reservation waters. By the 1980s the Navajo Division of Water Resources had developed extensive information about existing and future water use on the reservation. With a budget of $7 million, the division employed 200 people, including ground water geologists, soil scientists, hydrologists, and water-quality specialists.[66]

Although the state of Utah raised objections to the Navajo water code, no formal challenges had been filed by 1988. Various mining interests said soon after it was adopted that they initially saw no major problems with it. They agreed to comply with both the Navajo code and with state regulations. The code gave the tribe certain prerogatives, such as charging a water administration fee, but also authorized coal mining companies to use water for reclamation and dust suppression. The companies felt reassured by this firm recognition of their water prerogatives and by the code's standard provisions for notice and appeals—some state-tribal jurisdictional questions had not been resolved in 1988, especially in the Eastern Agency.[67]

The Fort Peck–Montana Water Compact of 1985 represents one of the few examples of state and tribal agreement over water management. The

Fort Peck settlement divided administration of water rights at Fort Peck among the United States, the tribes, and the state. The federal government continued to administer the water from the federal irrigation project. The Assiniboine and Sioux Tribes administered uses of the tribal water right (including both Indians and non-Indians with such rights) and reported uses to the state. The state administered state rights on the reservation and reported them to the tribes. To resolve disputes between users of tribal and state rights, the compact established a quasi-judicial board with a state representative, a tribal representative, and a neutral representative. Appeals were made to any court deemed to have jurisdiction—federal, state, or tribal.[68]

The complicated compromise on water administration pleased the participants but was criticized by outsiders from all sides; it was too soon in 1988 to evaluate its success. Although some tribes believed the Fort Peck tribes gave up too much in jurisdiction, some states thought Montana gave up legitimate jurisdictional authority and created an "administrative nightmare" at Fort Peck. It was clear, however, that both Montana and the tribes benefited from reaching agreement on the quantity of the tribes' water rights.[69]

As with the Navajo code, the forty-eight-page Fort Peck water code included standard provisions for notice and hearings. The real test of tribal codes would come when reservation water supplies became overextended and choices had to be made between competing uses. Then differences between state and tribal approaches would become apparent. The Navajo code, for example, listed stock watering as the second highest priority use, above agriculture and industry. The Fort Peck code placed a higher priority upon instream flows for fish and wildlife than upon agricultural, large stock watering, industrial, or power uses. Neither the Montana nor the New Mexico law specified priorities by uses, only by dates, with the earlier users having the higher priority.[70] In the 1990s more tribes could be expected to take an active role in administering water on their reservations. Participation from both states and tribes continued to grow at water management conferences sponsored by organizations such as the American Indian Resources Institute in Oakland, California.[71]

USE IT OR LOSE IT

Non-Indians remained concerned about how tribes would use their water. On the one hand, they feared Indians would use too much water and not conserve it. On the other hand, states resented tribes that did not use water and let it go to waste; the prior appropriation doctrine and competition with other states encouraged states to use as much water as possible to prevent its passing outside their borders. Would the tribes forego conservation? Technically, reserved water rights continued whether they were used or not, but Indian leaders became increasingly aware that their rights were at risk until they were used. If all the available water were put to use by someone else first, the Indians were likely to win only financial compensation, not "wet water." Always living with the fear that Congress could step in and quantify their rights for them, the tribes had added incentive to utilize their water quickly.

In 1985 Peterson Zah, chairman of the Navajo Nation, told a room full of state and tribal leaders at an American Indian Resources Institute conference, "I hope we will be able to say soon that the quarter million acre-feet now running off this reservation have been stopped." To accomplish this goal Zah urged his people to take up their shovels to put water to use. During his tenure as chairman the tribe spent $3 million a year developing water, drilling water, building dams, installing stock tanks, maintaining canals, and refurbishing ditches. The Presidential Commission on Indian Reservation Economies recommended marketing Indian water as a means of protecting rights. Indian water attorney Robert Pelcyger told his clients that if they could not market their water, it was to their benefit to build projects on the reservations that consumed as much water as possible.[72]

State and federal officials tended to reinforce that thinking when they indicated that the benefits of conservation would go to non-Indian interests. For example, when the Bureau of Reclamation imposed conservation measures on the Navajo Indian Irrigation Project, state and federal officials said the excess waters should go to other parties, not the tribe. And the state of Wyoming and its high court tried to impose stricter limits of conservation on Indian irrigators than state law required for non-Indians.[73]

Theoretically conservation could reduce conflicts between Indians and non-Indians by making more water available to share. Economist Philip Reno estimated that 4 million to 5 million acre-feet a year might be saved in the Colorado River Basin through conservation. As long as the tribes could share in the benefits of conservation, they had as much impetus to support it as anyone, and tribal water codes, which discouraged waste, indicated that they did.[74]

How would tribes use their water rights? For industry or irrigation on the reservation? For leasing off the reservation? Would energy tribes use water for energy projects just to put the water to use and guarantee that tribes would benefit from the water? Judging from some tribes' arguments in adjudication cases, one would have thought that the Indians would cover their lands with sprawling energy industrial centers. Wanting the highest quantification possible—and wanting to provide for future tribal leaders' choices—the tribes listed all possible future water uses, including heavy industry, whether or not they desired it.[75]

However, the courts tended to say that the quantity of the tribal rights could be determined by just one use—serving the primary purpose of the reservation when it was set aside a century ago. That purpose was usually agriculture. Commentators said future courts would probably not allow quantification on the basis of water needed for coal gasification or oil shale processing because those uses could not have been foreseen when reservations were established. Irrigation provided more water than other measures; so if only a single use were allowed, tribes generally preferred to quantify their rights using irrigable acreage as a measure. The courts did not limit how that water could be used once the quantity was determined.[76]

Once beyond the broad, hypothetical claims for adjudication purposes, some tribes still indicated they considered certain projects at least partly because they would put tribal water rights to use. The Northern Cheyenne Tribe in 1988 considered participating in building a coal-fired power plant outside its boundaries, in part to assure the tribe's rights to the Tongue River. When the Crows announced in 1986 that they were considering building a hydroelectric plant at Yellowtail Dam, Tribal Chairman Richard Real Bird said their first purpose was to "enhance the posture of our inherent rights to the use of our water."[77]

Both the Crow Tribe and the Navajo Tribe used their water rights as incentives for coal conversion on their reservations, especially because uncertainties about Indian water rights were a major barrier to building outside reservation boundaries. The Crows sought partners to build a coal-fired power plant and a synthetic fuels plant on the reservation, saying the tribe could provide the land, the coal, and the water. The Navajo Tribe in 1983 joined with Public Service of New Mexico and two suppliers to promote a 2,000-megawatt power plant, the Dineh plant, that would be built in northwest New Mexico. When PNM initially proposed the power plant in 1980, Navajo water claims posed a serious threat to it. Thus PNM and the project's other partners insisted upon a firm water commitment from the tribe when the Navajos joined the project.[78]

When making water-use decisions, tribes necessarily considered economic as well as legal factors. Western water law and tradition discouraged any sector from spending disproportionate amounts of money to conserve water. Because of the low prices charged for water, energy companies tended to use evaporative rather than dry cooling towers for their power plants. Dry systems consumed only 2,000 acre-feet of water for each 1,000-megawatt unit compared with 17,000 acre-feet required for evaporative cooling. Just as with other power plant owners, the Navajos and PNM planned evaporative cooling for the Dineh power plant. Thus unless there were overriding legal or political pressures, economics dictated the decision. A 1977 study predicted that thermal plant designers would select evaporative cooling systems until water costs exceeded $660 per acre-foot. After that, a system combining wet and dry methods would be chosen until water costs reached $4,200 per acre-foot. Only then would plant designers choose all dry systems.[79]

Why would these tribes use water for energy rather than for other, potentially less disruptive kinds of development, such as irrigation? In the first place it was not an either/or question—most tribes wanted both agricultural and energy development. Two reasons encouraged decisions to use some of the water for energy: the money needed to develop water and the money that water development would produce. Tribes generally lacked capital for investments in any kind of water development. While the tribes had futilely waited for their turn for irrigation funding, the pendulum had swung, and by the 1980s the federal government no longer heavily subsidized irrigation development in the United States. Congress and the administration cut funds for BIA irrigation construction and maintenance even more than for BuRec projects.[80]

Many tribes persisted in asking Congress for irrigation funds, but those who had alternatives—such as the energy tribes—sought water development funds from industry for industrial development, too. Tribes also made their decisions on the basis of the economic development that would be produced. With Indian and non-Indian farmers and ranchers going bankrupt throughout the region, many tribes believed agricultural development held little promise in the 1980s for producing revenue. The lower agricultural prices also skewed cost/benefit ratios against proposed irrigation projects. The Interior Department shared the tribes' dim outlook for agriculture. In 1988 the department encouraged the Colorado Ute tribes and the Fort Berthold tribes to use their water rights to obtain economic development funds instead of irrigation projects.[81]

MARKETING INDIAN WATER

Some tribes wanted to lease their surplus water for use outside their boundaries, at least temporarily. Tribal water marketing held many advantages for both Indians and non-Indians who had to share limited water resources. Without marketing, tribes faced the probability that their water would be used by others without compensation to them, as the Bureau of Reclamation Upper Missouri marketing program in the 1960s and 1970s would have done. Some, such as the Sioux tribes in South Dakota, had few resources to exploit other than water. Others with energy resources preferred not to have power plants or gasification plants on their lands because of environmental and social impacts. By leasing water they could get money for economic development on the reservations without suffering the impacts of the plants.

Given that tribes wanted to lease, not sell, their water, they could not expect the thousands of dollars being paid for sales in some areas. Even so, depending upon their proximity to areas of need and the unavailability of agricultural water, they could sometimes expect substantial sums. Tucson offered the Tohono O'Odham Tribe $100 per acre-foot for a 99-year lease, and Phoenix offered $16 million for a 99-year lease of the Salt River Pima-

Maricopa Indian communities' 13,300 acre-foot entitlement from the Central Arizona Project. In other areas, however, subsidized agricultural water leased for as little as $2.50 per acre-foot.[82]

Because of the potential for revenue and for protecting rights by usage, many tribes—including those of the Fort Berthold, Northern Cheyenne, Crow, Fort Peck, Cheyenne River Sioux, Northern Ute, Southern Ute, Ute Mountain Ute, Jicarilla Apache, and Wind River reservations—expressed interest in water marketing through the years. In 1981 Navajo Tribal Chairman Peter MacDonald proposed creating a national or regional trust to pool Indian water rights for trading or leasing.[83]

Some non-Indians saw benefits to tribal water marketing, too, because it was clear that when Indian water rights were recognized, there was not enough water to go around. Although both energy and urban demands decreased between 1980 and 1988—driving water sale prices down by a third or more—some urban areas were still actively buying agricultural water rights for use after the turn of the century. Leasing Indian water had many advantages over buying farms and ranches to obtain their water rights, a practice that disrupted families and the rural agricultural social structure. The states of Montana, Colorado, and Wyoming recognized advantages to tribal water marketing and agreed to it under certain conditions. The Tucson urban area in Arizona benefited from an arrangement with the Tohono O'Odham Tribe, in which the Indians leased their water.[84]

On the other hand, leasing Indian water had its disadvantages, both philosophical and practical. Marketing of Indian water violated nonintercourse acts—first passed in 1790 to limit non-Indians' trade with Indians—and thus required special congressional authorization, which had only been granted in two instances: to the Tohono O'Odhams and to the Colorado Utes. Even if a tribe just deferred its water use, it might still require congressional authorization. Many negotiations, including those with the Northern Utes in Utah and the Shoshones and Arapahoes in Wyoming, involved such deferrals. Although Congress had willingly changed the Nonintercourse Act to allow widespread mineral leasing, Indian law experts did not expect Congress to pass a national water leasing law because it would not benefit non-Indians as much as mineral leasing. Action on individual marketing proposals was more likely. Some states believed that Indian *Winters* rights could not be used outside reservation boundaries, although that objection apparently had not been raised during negotiations over the Navajo irrigation project or the Black Mesa slurry pipeline.[85]

Some states and interest groups opposed any transbasin diversions; thus many state laws and interstate compacts prohibited sale of water outside state or basin boundaries. It was unclear, however, whether tribes would have to abide by those restrictions.[86] When individual water leasing proposals were brought before Congress, they met with opposition from some states. South Dakota objected to marketing by tribes within its boundaries because they would be competing to sell the same water. Downstream states objected because without tribal marketing, they could use the excess water without charge. For example, Arizona, Nevada, and California initially objected to marketing by the Southern Ute and Ute Mountain Ute Tribes, which wanted to make short-term leases while retaining long-term rights to use the water for coal development on the reservations. The Interior Department supported the Ute marketing proposal, saying it would enable the tribes to repay the costs of their water project. The downstream states withdrew their objections only after the tribes agreed to accept strict limits on the quantity of water that could be leased and on the terms.[87] The law was called an antimarketing law by some critics because it subjected the tribes' rights to the same restrictions as other rights in Colorado.

Beyond the legal questions, some Indians opposed exchanging water for money for spiritual, cultural, and historical reasons. They saw it as selling the future of their children because without water, Indian cultures could not survive. Indian people had exchanged "surplus" land for money before and had lost potential for self-sufficiency. By imposing strong contract stipulations, by having reliable information on water needs, and by leasing for shorter periods, tribes could protect themselves, however. Careful marketing of water represented one of the tribes' few opportunities to finance economic development on the reservations and to participate in decisions on the region's future rather than being victimized by those decisions.[88]

OFF-RESERVATION ENERGY DEVELOPMENT

Water was only one example of the important energy development issues that crossed political boundaries. Sulfur dioxide, construction worker poachers, radiation, and water-borne uranium tailings did not stop at the reservation line. On the other hand, federal and state impact assistance usually did not reach across reservation borders to help alleviate the impacts of energy development there. Just as federal agencies had ignored tribes' water rights in their water allocation decisions, they often had disregarded potential impacts on Indians when making energy decisions. Working with states and industry, the federal government planned development outside reservation boundaries that directly affected Indian resources. Smaller reservations and pueblos were especially vulnerable to outside development. Through litigation, legislation, and negotiation, tribes sometimes succeeded in affecting such energy decisions.

When federal agencies prepared regional environmental impact studies (EIS), they usually referred to Indians only in the past tense. Archaeological sections described use of the area by Indian peoples hundreds of years earlier. The studies generally omitted references to present-day tribal boundaries, communities, and resources within a few miles of proposed coal and uranium leasing sites, pipelines, and power plants. In 1982 Interior Secretary James Watt ignored complaints of the Northern Cheyenne Tribe and proceeded with the Powder River Basin coal lease sale, the largest federal coal sale in history. The leases surrounded the reservation with the farthest only 17 miles away. After a federal judge voided the resulting coal leases because of the tribe's lawsuit, the Cheyennes could no longer be overlooked.

U.S. District Court Judge James Battin in Billings, Montana, said the Interior Department's Bureau of Land Management had "systematically excluded" consideration of Indian people when it prepared the EIS for leasing federal coal around the reservation. In court the department made the ridiculous argument that it had consciously decided to deal with Indians "simply as people affected by the sale" and their reservation as any other real estate in the sale area. Battin rejected that argument and said the EIS should have recognized the reservation as culturally distinct and considered the tribe's lack of a tax base, which affected its ability to mitigate impacts. In fact, the EIS did not treat the Indians as equals. Although the authors thoroughly analyzed impacts on non-Indian towns, they did not even mention reservation towns. Battin especially criticized Interior for downplaying its responsibilities to the tribe compared with the national interest in leasing coal.[89]

Although National Environmental Policy Act regulations require agencies to seek comments and involvement from Indian tribes, agencies ignored impacts on contemporary Indians in most environmental studies, including the Northern Tier oil pipeline, the White Mesa (Utah) uranium mill, and the Susquehanna-Western (Wyoming) tailings cleanup. When they did mention living Indians, such as in an environmental study for the Dalton Pass (New Mexico) uranium mine, the studies usually focused upon economic advantages to local people, with no mention of the cultural impacts of losing their traditional livelihoods. At Dalton Pass none of the residents, including government and religious leaders, was interviewed for the study.[90]

Battin praised a synfuels EIS's treatment of the Northern Utes in Utah, but a geologist who worked for the Northern Ute Tribe said it initially was difficult to convince the BLM in Utah to even include the reservation on its maps. As a result of the tribe's protest, BLM incorporated tribal issues throughout the final Uintah Basin EIS, from socioeconomic impacts to air quality. The tribe's efforts resulted in a commitment from industry and state representatives to help mitigate impacts on the reservation and provide impact funds, but the oil shale was never developed so the promises were never tested.[91]

The San Juan coal study could serve as a model for thorough consideration of Indian cultural, social, economic, and environmental concerns. Tribes were not happy with the consequences it projected, however. Both the Jicarilla Apache Tribe and the Navajo Tribe threatened lawsuits to stop the San Juan Basin coal lease sale planned in March 1984 because of the risk to their ground water and people. BLM estimated that as many as twenty-two thousand Navajos would have difficulty pursuing their traditional life styles if federal coal leasing proceeded as planned. The tribes charged that BLM had no comprehensive water management plan. Interior canceled the San Juan sale after a request from the governor of New Mexico and after several challenges of the Powder River Basin sale. Still pending were lawsuits by the Navajo Tribe and by Navajo allottees challenging the ownership of coal and surface in the San Juan coal area.[92] Tribes frequently asked to be represented as members of regional coal teams so that they could participate with state and federal representatives in decisions on selection, delineation, and ranking of coal tracts—and on scheduling coal lease sales. Finally, as a result of the Cheyenne lawsuit, Interior appointed a Cheyenne and a Crow as ex-officio members of the Powder River regional coal team.[93]

The Northern Cheyenne Tribe set an important precedent in 1978 when it became the first tribe in the nation to utilize the Clean Air Act to shape off-reservation coal development (see Chapter 7). The Yakima Nation opened the door for tribal participation in decisions on nuclear waste disposal sites. In 1982 tribal leader Russell Jim, working with Senator Gary Hart of Colorado, convinced Congress to give tribes the same authority as states to participate in decisions on siting the highly radioactive wastes from nuclear power plants. Under the Nuclear Waste Policy Act (1982), tribes' authority extended to any lands over which they had treaty rights to hunt and fish. Because of their concern over contamination of Columbia River Basin waters, the tribes of three reservations—the Yakima, Umatilla, and Nez Perce—won affected tribe status. One of the three sites considered was at Hanford, near the shores of the Columbia in Washington State and within the treaty area of all three Indian groups. As affected tribes they received $12.8 million over five years to help them participate in the decision making; the state of Washington had received $11.2 million as an affected state. The money was spent on environmental, social impact, and cultural studies. Under provisions of the law, CERT provided environmental and computer training while the National Congress of American Indians helped potentially affected tribes across the nation with educational workshops and policy issues. In 1988 Congress decided tentatively upon a Nevada site and discontinued funding for studies at the other proposed sites.[94]

The Nuclear Policy Waste Act gave tribes a mechanism for affecting decisions concerning areas of religious significance to them that lay outside their boundaries. Others have relied upon the American Indian Religious Freedom Act of 1978 to protect religious sites from federal energy decisions. The tribes met with little success either on or off reservations except for Navajo ranchers, who convinced a federal judge that tribal religious leaders should be consulted before a coal-hauling railroad was built on the reservation. Some Blackfeet tribal members fought oil and gas development in a national forest area east of their reservation. Several pueblos fought a geothermal plant in the Jemez Mountains. The Zuni Pueblo, Acoma Pueblo,

Russell Jim, a Yakima, led the battle in Congress in 1982 to give tribes equal status with states in making nuclear waste storage decisions. (N.E.W.S. Photo N.W.)

and Navajo communities opposed mining that would affect Zuni Salt Lake, which was on state land, surrounded by BLM coal lands. In the Black Hills a group of Sioux Indians fought uranium exploration.[95]

Despite the religious freedom law, the courts had as much trouble as the federal agencies being challenged in relating to the esoteric concerns of Indians. The Blackfeet were concerned that seismic work would drive the spirits away from the Badger-Two Medicine region. The pueblo people feared the geothermal plant would cause an imbalance of nature and that "the energy produced by the plant would be used against the earth," leading to human anxiety and deterioration in health. In traditional pueblo belief, health lies in the flow of life forces, which would be interrupted by drilling rigs and "technical gadgetry." Ultimately economic rather than judicial forces blocked some of the projects.[96]

CONCLUSION

Tribes in the 1970s and 1980s became both more aggressive and more innovative in affecting decisions concerning water and energy development. Well aware of the neglect and abuse they had suffered at the hands of the federal government in the past, they turned to the courts to protect their rights. The downturn in the energy market gave tribes breathing space, time to address water quantification and administration so they would be better prepared when the next energy boom hit. The relaxing of time pressures encouraged negotiations.

Tribes became aware that they needed to cooperate with other political entities to resolve difficult transboundary issues. To do so required Indian people, while maintaining a healthy skepticism, to abandon some of their suspicions and become more confident in their own ability to protect themselves. It required non-Indians to respect the tribes' legal strength and to recognize that Indians would be business owners, miners, and power plant operators—not just farmers and ranchers.

Despite the reasons for optimism, progress on quantification and administration proceeded excruciatingly slowly, handicapped by the continued conflict of interest within the Interior Department and by the lack of federal money. Water policy expert Charles Wilkinson predicted that only water projects with an Indian component would be built in the future. Yet instead of jumping on that bandwagon and assuring a role for itself, the Bureau of Reclamation continued its long tradition of sniping at BIA and the tribes.[97] Clearly tribes' ownership of vast energy and water supplies could no longer be denied. In the next boom the Indians hope they will not be drowned by others' ambitions. Instead they plan to be among the developers themselves, a force to be reckoned with as both governments and proprietors.

CHAPTER NINE

Into the Boss's Seat:
The Tribe as Developer

We will make mistakes. But they are less painful than suffering the conse-
quences of other people's mistakes. We're trying to learn, so that those who fol-
low us won't be as helpless as we've been in the past.

—Earl Old Person[1]

"I'm the great grandson of Cochise, and I'm here to help you." "I have a magic black box that can tell you where the oil is—just watch these dials, gauges, and flashing lights." Such promoters' spiels echoed in the council chambers in the remote reservation capitals. Some handed out fistfuls of cash; others sent letters to tribal members promising checks before the tribal fair. One man who billed himself as an "international financier" turned out to be a hotel operator secretly offering tribal coal at cut-rate prices for a quick commission. Another who claimed to be a Chippewa Indian said he had found investors to provide $1.2 billion for a tribal power plant. Many were fly-by-night shysters spending someone else's money. Some displayed impressive portfolios, filled with figures showing potential profits for tribes. Others had scribbled their get-rich-quick schemes on cocktail napkins. Nearly all of them displayed the "Cherokee syndrome"—they could trace their heritage to a Cherokee princess just a few generations ago. Like ghetto drug pushers they depended upon glitter and giant cars to sell themselves and their schemes.[2]

Prior to 1982 tribes and individual Indian mineral owners for the most part were protected from such wheelers and dealers by specific requirements of the 1938 Indian Mineral Leasing Act. The tribes began objecting to the law's restrictions, however, because they also protected them from sharing in profits. As the result of the tribes' insistence upon negotiating their own mineral contracts, Congress passed the Indian Mineral Development Act in 1982 to authorize tribes to join with industry as mineral developers and choose which development schemes to pursue. By the time Congress finally acted, the energy boom was fading and the Interior Department delayed the regulations implementing the law for several years. Some said the law was passed too late to benefit the tribes. Others, looking at the limited bidding at oil auctions, said Indians had bargained themselves out of the energy business by maintaining high royalty rates and other demands despite the slump.[3]

Some companies did turn their backs on Indian resources. Yet in the tribal council chambers and the corporate board rooms, others took their places. Negotiators continued to cut Indian mineral deals. With the depressed prices some companies took more interest in innovative contracts with tribes, hoping to reduce their costs. They talked of jointly owned refineries, coal drying plants, synthetic fuel plants, and coal-fired power plants. On two reservations the tribes went beyond partnerships to drill new oil wells using 100 percent tribal funds. Far from being helpless victims of exploitation and colonization, tribes were becoming major players in the energy game, sharing the profits and the perils of a highly speculative field. Several agreements negotiated by tribes illustrate the risks tribes and industry face and how they can be avoided.

RISKY BUSINESS

Passage of the 1982 Indian Mineral Development Act marked a major turning point in Indian mineral history, potentially exposing the Indian mineral owners to great financial risk and dramatically limiting the role of the federal government in protecting them. The 1938 Omnibus Tribal Leasing Act had required competitive lease sales to protect the tribes from abuses prevalent up to that time. Under the 1938 act, leases for coal and uranium could be negotiated but only after prior written permission of the secretary of Interior. The competition requirement for oil and gas in the act was even stricter—negotiation could only follow an advertised sale. Congress decided in 1938 to require competition, based upon the premise that companies would bid higher bonuses for tracts that showed the most promise. The bonus bids paid to Indians consequently would track market forces, whether or not BIA or the tribes knew what they had or what the market was doing. The advantages would be similar to those of an unsophisticated person buying into a mutual stock fund instead of buying individual stocks. In addition to the competition requirement, leases under the 1938 law required certain rents to be paid so that if no minerals were found, the Indian owner would at least get the bonus and rental money. If minerals were found, he or she would be paid a royalty, based upon how much was produced and at what price.

In practice, however, competition for Indian leases was often limited for several reasons: Sales were inadequately advertised; geologic information about the proposed tracts was not widely known; large tracts were offered when industry was only interested in parts of tracts; and isolated tracts were offered instead of more attractive contiguous tracts. In addition, the federal government set rents and royalties lower than market conditions justified, and rents could be subtracted from royalties. Thus the protective measures guaranteed only minimal levels of income.[4]

When Congress held hearings on the proposed legislation allowing negotiation, some old-school Indian advocates objected, implying that the Indians—and especially the allottees—were not ready to protect themselves. Attorney Marvin Sonosky, who had represented the Fort Peck Tribes since 1962 and the Shoshone Tribe at Wind River since 1955, testified against dispensing with the competition requirement, saying,

> How much nicer for industry to negotiate a deal without competition. Much like taking candy from a baby. There are places where industry charmers and persuaders would bear down on the key council members and on individual allottees. The way would be opened to side money, bribery, graft, and corruption—a return to precisely the conditions that led Congress to write into the 1938 act the competitive lease sale provision.[5]

Sonosky went on to suggest that the Interior Department would not adequately protect the tribes because "industry muscle men" would exert pressure against the department. Although he agreed that changes were needed, he wanted Congress to preserve the safeguard of competitive sales. Sonosky was not alone in his caution. Several Indian advocates and two tribal chairmen said at the hearings that the proposed legislation did not provide enough safeguards, especially for allottees. Several others objected to specific contracts that would be affected. Nevertheless the vast majority of the tribes testifying in 1982 favored the changes, and Congress decided to allow tribes to negotiate contracts, subject to the Interior secretary's approval.[6]

Under the new legislation the department had only 180 days to approve or disapprove. Allottees' minerals could be included in a tribal minerals agreement if the allottees consented and the Interior Department found that such participation was in the allottees' best interests. As a further step toward tribal authority, Congress adopted a controversial provision that said the federal government could not be held liable for a tribe's or an allottee's financial losses. Congress also imposed restrictions that made it more difficult for the Interior secretary to disapprove than to approve alternative contracts that had been negotiated by tribes. However, these moves did not absolve the federal government of its trust responsibility to assure that the agreement had adequate provisions for economic return, for controlling social, environmental, and cultural effects, and for resolving disputes.[7]

The lack of implementation funds reduced the Interior Department's protective role even further than Congress intended. Although legally the federal government still had the responsibility to provide expertise to tribes during negotiations and to evaluate proposed contracts, Congress never provided the funds necessary to fulfill those responsibilities. The BIA Energy and Minerals Office in Lakewood, Colorado, which reviewed all mineral contracts, had only three people to evaluate agreements and in 1988 was still trying to get funding to hire an economist. When the Interior Department evaluated alternative contracts, the principal criterion used was whether the agreement was likely to produce at least as much revenue as a standard lease. Because of the many variables involved and because of its lack of expertise, however, BIA could not always predict the results. Although the law presupposed that tribes would have substantial mineral information, funds had been cut for BIA's mineral inventory work to only $5 million a year, leaving a backlog of $18 million in 1988. The lack of readily available expertise increased the risk to the tribes of the alternative contracts.[8]

WHY NEGOTIATE?

Despite the increased difficulty and risk, negotiated contracts offer many advantages over standard BIA leases for both Indians and industry. Under a standard lease a company has to pay a bonus before it makes any money off its properties. Theoretically it has to continue production in "paying quantities" or risk having the government cancel its lease, thus losing its whole investment because of one poor profit year. Even if it is just barely breaking even, it still has to pay royalties to the tribe. On the other hand, with an alternative contract a company can offer a tribe no bonus or a minimal bonus prior to production in exchange for a bigger share of the profits, which represents a significant advantage for smaller companies. A company and a tribe can negotiate an agreement that has different diligence requirements and that provides the tribe a share of net profits instead of royalties. If there are no profits, the tribe does not get paid. Rather than paying a tribal royalty rate that companies consider unrealistic, they can negotiate terms that suit the land, whether wildcat or proven acreages.

Tribes also can benefit substantially. Under a standard lease much of the profits goes into tax payments to federal and state governments. Tribes have little voice in development decisions once the lease is signed. In years past, tribes had watched corporate coffers grow during boom times, disproportionally to the royalties paid to tribes. The revenue tribes did receive leaked off the reservations, benefiting border towns instead of reservation economies. In contrast, with alternative contracts money that would have gone to taxes can instead be split between the tribe and the company. Tribes do not have to pay state sales, severance, or property taxes or certain federal taxes, and they rarely tax their own interests. Contracts can be structured to take advantage of the exemptions. Rather than depending upon federal regulators and federal courts to control developers, the tribes can negotiate for access to records and for environmental protection.[9]

The slump in energy markets caused many companies to plug their marginal oil and gas wells and close their mines, leaving potentially valuable minerals in the ground. This disruption in income hurt tribes and allottees, who depended upon the royalties as well as the companies. With alternative contracts, however, tribes can discourage premature abandonment by allowing companies to pay lower royalties when they produce less. Tribes can encourage companies to develop low-grade as well as high-grade minerals by charging lower royalties for the less valuable ores.

The potential difference between being passive royalty beneficiaries and partners in development is substantial. In 1987, despite continued low oil and gas prices, the Jicarilla Apache Tribe made 25 percent of its mineral revenue from less than 2 percent of its wells—those in which it held an ownership interest. The lease wells (98 percent of the total) produced much less revenue proportionately for the tribes.[10]

TYPES OF AGREEMENTS

By 1988 tribes and companies had negotiated sixty-seven alternative contracts with different levels of risk and return. All of them covered exploration and development of oil and gas, except one for coal (Crow-Shell), one for uranium (Navajo-Exxon), and one for an oil refinery (Northern Ute–Charging Ute Corporation). Eight of these had been negotiated prior to 1982. They covered both wildcat and proven acreages. Some tribes' contracts involved allottees, including those of the Southern Ute, Three Affiliated Tribes, and Blackfeet.[11]

By 1988 Indian lease sales were held regularly in only a few places, partly for industry and partly for Indian reasons. Both parties usually preferred the flexibility of negotiations. Most Indian lease sales that were held involved allotted lands. Although the 1982 law provided for negotiation of allotted minerals, they had to be included with tribal minerals, which was difficult in most places and impossible where there were no tribal oil and gas lands, such as most of Oklahoma. Industry showed little interest in bidding at state and federal lease sales because of the low oil and gas prices. Unless an area was especially promising, companies were even less interested in Indian sales if tribes insisted upon higher minimum royalty, rental, and bonus levels than those for state and federal leases. Industry preferred to negotiate acceptable terms to match the area's potential. For example, on the Fort Peck Reservation, where two sales were held each year, the tribes required 25.0 percent royalties, $5.00 an acre annual rents, a minimum bonus bid of $25.00 an acre, and a primary term of 5 years, during which production had to be established. In contrast, the federal government after December 1987 required only a 12.5 percent royalty, a $1.50 per acre minimum rental, a $2.00 an acre minimum bonus bid, and a 10-year primary term for federal oil and gas.[12]

Mineral contracts fell into four basic forms: service contracts, joint ventures, production sharing contracts, and leases. Tribal contracts were difficult to categorize, however, because most agreements were hybrids, combining several different elements, and because their true nature was often disguised, either for tax purposes or—before the 1982 act—to avoid competition requirements. Just as with an investor's choice of stocks versus certificates of deposit, the choice of the best type of contract depended upon the tribe's objectives and level of expertise as well as tax considerations. How much capital or resources does the tribe have on hand? How much data and experience does it have? How much risk does it want to take? What are the relative tax advantages of the company or the tribe taking the risk?[13]

Risk generally correlated directly with the potential for profits. The most risky venture for tribes was the service contract. Under a service contract a tribe hired a contractor to drill wells and it paid all the costs. In return for the substantial jeopardy the tribe received total control and all the revenue produced. The contractor, who had no property interest in the minerals, got paid for his services by the tribe. Because of the risk and because tribes could not write off the cost of drilling for tax purposes, only two tribal governments—the Jicarilla Apache Tribe of New Mexico and the Fort Peck Assiniboine and Sioux Tribes of Montana—had drilled their own wells.

Under pure joint ventures, tribes shared in both the costs and the income by a predetermined percentage. For example, they would pay 40 percent of the costs and get 40 percent of the revenues. Exploration was very expensive; a well could cost from $50,000 to tens of millions, and nine out of ten exploration wells drilled in wildcat areas were dry holes. Tribes would not necessarily have to contribute their share in cash; they could contribute resources instead. Most joint ventures were not pure joint ventures. Rather than helping to pay the cost of exploration, most tribes backed into the agreements, acquiring an ownership interest (sometimes called a "working interest") only after exploration was completed. (For the sake of convenience, any agreement in which a tribe held an ownership interest at any point will be referred to here as a joint venture.) Even if a tribe waited until after wells were producing to acquire its ownership interest, it faced some risk. If sand got into the pumps or the equipment broke down, for example, a workover rig had to be brought in, which cost thousands of dollars a week.[14]

Several tribes acquired total ownership of already developed wells, many from companies that wanted to dispose of them because of low oil and gas prices. The Jicarilla Apaches were the first to do so when in 1980 they purchased Palmer Oil and Gas Company's operation on the reservation, including six producing wells. The tribe and Palmer had had a joint exploration and development agreement since 1976, and Palmer wanted out of the oil business. Among the other tribes that acquired complete ownership were the Southern Utes in Colorado, the Arapahoes in Wyoming, and the Blackfeet in Montana. Others held a smaller ownership interest in wells, including the Fort Peck Assiniboine and Sioux Tribes with U.S. Energy, the

Ute Mountain Utes with Wintershall Corporation, and the Jicarilla Apache Tribe with ARCO and Odessa Natural Corporation. Several others negotiated contracts that provided for a possible equity share later, at the tribe's option; these tribes included the Navajos with Exxon, the Northern Cheyennes with ARCO, and the Crows with Shell.[15]

Even when tribes did not share costs, alternative agreements were often labeled as joint ventures. Tribes could receive a share of either the gross proceeds or the net proceeds. A 16 percent share of gross proceeds was practically the same as a 16 percent royalty because the percentage was computed on total sales; such agreements differed little from leases. However, a share of net proceeds was very different—if there were no profits after costs were subtracted from the gross proceeds, then the tribe received nothing. In a net proceeds contract, a tribe faced more risk but could often negotiate a higher percentage than in a gross proceeds contract.

Under the third type of contract—production sharing—the tribes could sell the minerals themselves. After the company received enough of the product to cover its costs, the rest belonged to the tribe. Several tribes wrote provisions into their contracts for acquiring and marketing oil and gas themselves, often receiving their royalty interest "in kind," meaning as oil and gas rather than cash. None of the tribes in 1988 had true production sharing agreements because such contracts depended upon a relatively stable and predictable relationship between costs and prices. The rapidly rising and then falling energy prices of the 1970s and 1980s, as well as the tariffs that transportation companies were charging, made production sharing unattractive.[16]

About half of the sixty-seven negotiated agreements fell into the fourth category—leases—with only minor adaptations. With a negotiated lease the tribe held only a royalty interest and no ownership interest, but it could add clauses providing for escalating royalties, tribal employment and education, and tribal acquisition of equipment after production was completed. In some unusual cases tribes found they could make more money with leases than with negotiated joint ventures. For example, in 1976 the Navajos declined to enter a coal joint venture with El Paso Natural Gas Company and Consolidation Coal because they decided their profit share would not equal the potential interest on their money plus the standard royalty, which recently had been raised to 12.5 percent. Instead of an ownership interest the tribe chose to accept a lease with a bonus of $5.6 million and the new standard royalty. That decision was fortuitous because by 1988 the Conpaso mine never had produced coal commercially. The tribe got its bonus, and the companies paid the development costs. Although for several years experts predicted that the larger energy tribes would eventually abandon leasing and develop all their minerals in partnership with industry, in 1988 it was impossible to predict whether that trend would continue.[17]

AVOIDING RISK WHILE MAXIMIZING CONTROL
AND RETURN—RULES AND CASE STUDIES

To some extent the fears of the critics were realized after passage of the Indian Mineral Development Act. Dubious proposals from "black box" salesmen and Indian-owned companies fronting for other companies inundated tribal offices. Rumors of bribery by promoters persisted. As Sonosky suggested might happen, the Interior Department occasionally bowed to tribal self-determination and industry pressure against its better judgment. Sometimes tribal politics interfered with sound, businesslike decision making. Yet in most cases the tribes succeeded in smelling out the questionable deals and listened to either their own experts or to outside consultants, BIA, or other tribes. Between 1982 and mid-1988 BIA recommended disapproving few agreements that tribes had approved, although behind the scenes the agency sometimes recommended that a tribe change or reject a certain agreement. Although each agreement had its weak points, most deals that were consummated provided more return—sometimes considerably more—than leases would have.[18]

From the tribes' experiences, seven principal suggestions for avoiding problems emerged:

1. Avoid risk of exploration in unproven areas while preserving options.
2. Clearly establish accounting procedures.
3. Limit and identify acreage.
4. Choose the other party carefully.
5. Choose consultants carefully.
6. Protect tribal governmental rights.
7. Decide from the beginning who is in charge.

Each is discussed below with sample agreements to illustrate it. This is not an attempt to rank the agreements themselves because each has good and bad points and because in most cases it was too early at the time of this writing to determine whether each would be to the ultimate benefit or detriment of the tribe. I show how Indians and industry learned to accommodate one another's interests outside the confines of standard leases. I also show how much control the tribes in some cases have gained since the rubber-stamp era.

Avoid Risk of Exploration in Unproven Areas
while Preserving Options

Tribes could participate in the cost of exploration in many ways with varying degrees of risk. The Northern Cheyenne Tribe leased its lands and took a large bonus rather than a share of the profits, thus eliminating all risk but re-ducing their share of potential profits. The Blackfeet took some risk by not requiring a bonus prior to exploration in exchange for a greater share of the profits. The Fort Peck and Jicarilla Apache Tribes took the total risk and drilled their own wells but only in proven areas. In other cases these tribes backed into a working interest, acquiring their interest after exploration was completed. Each contract offered certain advantages to companies as well.

In 1981 an article in the *Billings Gazette* said the Northern Cheyenne Tribe was the laughingstock of the oil industry because it had leased its en-tire reservation to Atlantic Richfield Company (ARCO) instead of develop-ing the oil and gas itself. The Northern Cheyenne Tribal Council had chosen a conservative approach, opting for only a royalty interest and not an own-ership interest in the oil and gas development for the first thirty-three years. The tribe had the last laugh, however, when after spending $28 million, ARCO ended up three years later with only dry holes while the tribal mem-bers had $6 million in bonus monies in their pockets.[19]

The ARCO controversy involved questions about the tribal govern-ment's responsiveness to its members, conflicts with traditional values, and surface owner rights (see Chapters 3 and 6). The principal contract issue was whether the tribe would develop the oil and gas itself and reap all the benefits—and all the risk. Although tribal leaders had expressed interest in developing the tribal coal as far back as 1973, much less was known about the potential for oil and gas on the reservation. Gambling that there would not be much oil, the tribal council in 1980 chose a high bonus and a rela-tively low share of the potential profits (25 percent royalty). The tribe pre-served its long-term interests, specifying that it would get 75 percent owner-ship of any wells still producing after the 33-year term of the agreement. Except for the large amount of land involved and the high royalty, the Northern Cheyenne–ARCO agreement was no different from a standard lease.[20]

When ARCO was unsuccessful and found no oil and gas, the tribe lost nothing except the political costs of the prolonged disagreement. The tribe won its gamble but due only to blind luck; BIA did not have enough geo-logic information for an informed decision. If oil and gas had been discov-ered, other problems might have become more important, especially the tax holiday provided to ARCO. The problems were serious enough that when a staff person at the White House called Assistant Interior Secretary Tom Fre-dericks (Mandan-Hidatsa) and asked why he had not signed the Northern Cheyenne–ARCO agreement, Fredericks told him he would sign it only when he got an indemnification clause in it that said the department would not be liable.[21]

The situations of the two tribal governments—Fort Peck and Jicarilla Apache—that had successfully drilled their own wells differed from the Northern Cheyenne in two significant ways: Both had their own capital from existing development, and both could predict success for these wells with relative certainty. The Fort Peck Assiniboine and Sioux Tribes drilled their "Wenona" well in 1984 in the dead center of a known geologic area where production had been established years earlier. The Jicarilla Apache Tribe also drilled their eight tribally financed wells in already producing areas—four in 1983 and four in 1985.[22]

The Fort Peck and the Apache staffs expressed caution about tribes drilling their own wells. After spending $500,000 to drill their well the Fort Peck tribal leaders initially felt the pride of parents at its success, naming it Wenona, which means "first-born daughter" in Sioux. The well produced a slight profit for the tribes. Yet Fort Peck Minerals Director Larry Wetsit (Assiniboine) said he would not advocate tribally financed exploration again because drilling can be unsuccessful even in an established area and because tribes cannot write off the cost of dry holes on their taxes as companies can. "Unless a tribe has a bunch of money they want to dump in a hole, they should drill with someone else's money—someone who needs tax credits," he said. The Fort Peck tribes were much happier with their joint venture with U.S. Energy. A year after that contract was signed in 1983, five of the seven wells drilled were producing, with the tribes receiving 25 percent royalties before drilling costs were paid and 55 percent of net proceeds thereafter.[23]

Before ever drilling their own wells the Jicarilla Apaches gained considerable experience through joint ventures. That experience helped Thurman Velarde, the tribe's oil and gas administrator, choose trustworthy consultants to help him decide which drilling contractor and which tubing contractor to hire, what type of tanks to use, and whether to buy a used or a new pump jack. When drilling wells in wildcat areas, the Apaches always chose joint ventures, letting the companies take the risk rather than using tribal monies. In addition to its eight tribally drilled wells, the tribe had three joint venture agreements (Palmer Oil, Odessa Natural Gas, and ARCO). As the projects became more profitable, the three contracts provided the tribe with increasing participation, control, and revenue. The Jicarillas received a royalty (16.67 percent of gross proceeds) until the costs for each well were paid; then they received either 50 or 60 percent of the net proceeds (50 percent in the Palmer and Odessa contracts and 60 percent in ARCO). Such a provision normally protected a tribe against the cost of exploration, but the Jicarillas agreed to allow part of their profits from successful wells to repay the companies for some costs of unsuccessful wells, up to a certain limit.[24]

If a tribe did not want to pay costs of exploration but was willing to gamble a little, it could agree to require no bonus. If no oil was found, the tribe received little or nothing, but it gained a higher share of the return. For example, after requiring no bonus, Fort Peck got 55 percent of net proceeds from its U.S. Energy contract, and the Blackfeet got 50 percent from a 1975 agreement with Damson Oil Corporation. Damson Oil President Barrie

Thurman Velarde, an Apache, in 1987 shows off one of the many wells owned totally by the Jicarilla Apache Tribe. Although the tribe had over two thousand lease wells, it received 25 percent of its income from the thirty-four wells in which it had an ownership interest. (Photo by Sara Hunter-Wiles.)

Damson said independent companies often prefer to devote their limited cash to exploration rather than bonuses. Tribes did not always have to make the choice between bonuses and a high share of future proceeds, however. The 1977 Jicarilla Apache–Odessa agreement provided for a $500,000 bonus for the 1,920 acres as well as 50 percent of net proceeds.[25]

Clearly Establish Accounting Procedures

As the Blackfeet and Spokane Tribes learned from their experiences, contracts that looked good on paper did not always produce the expected revenues if they included vague or overly complicated accounting provisions. Once signed, the agreements were not self-enforcing. The Minerals Management Service did not begin taking any responsibility for accounting for alternative contracts until 1988 so tribes initially needed to devote their own resources to assuring compliance. At the minimum they needed a natural resources department staff. The Crow-Raven contract was so complicated that BIA said a full-time administrator would have been required to properly implement it. Tribes learned to write contracts that provided complete tribal access to records, clear definitions of overhead costs, funds for independent accounting and auditing, retention of all records for six or more years, and a minimum share of true profits at least equivalent to standard royalties.[26]

When the Blackfeet-Damson agreement was first signed in 1975, it was widely acclaimed as a model for the new era of tribal energy development. Although Damson did not share the benefits of its federal tax credits and allowances with the tribe, the contract included other apparently good provisions. The tribe was promised a royalty (16.67 percent) until exploration and development costs had been paid ("payout") and then after payout both a royalty and half of net proceeds. This guaranteed a certain minimum income to the tribe, even if there were no profits. Definitions soon became an issue in any contracts that mentioned "net proceeds" and "payout," however, leading to tribal dissatisfaction with Damson's and similar contracts.[27]

As tribes had learned through their royalty accounting work, it was difficult even with straight leases to certify how much of the minerals had been produced, whether the proper value had been used for calculating royalties, and whether royalties had been paid. Because royalties were just a percentage of gross proceeds, they could be calculated relatively easily compared with contracts that required calculation of costs. Companies in some cases siphoned off profits through management fees, sales commissions, technical assistance fees, overhead costs, and interest payments to affiliates. Higher costs meant lower profits to share with tribes. As consultant Charles Lipton said, tribes did not want to help pay for the company president's Rolls Royce. The Spokane Tribe discovered the problems with defining costs in its uranium contract with Western Nuclear. Although otherwise a relatively standard lease, the 1976 contract provided for the tribe to receive a percentage of net proceeds instead of a straight royalty. In 1985 MMS audited the lease and said the company had deprived the tribe of $250,000 because it had subtracted nonallowable costs, including meetings, conferences, and seminars, travel expenses, and office overhead. Western Nuclear lost its lawsuit in 1988.[28]

Payout could be defined by well or by contract area. Under the Damson and many other agreements, a company could continue drilling wells until all wells in the contract area were drilled and paid for before reaching payout. Until that time the tribe would continue receiving only a royalty and none of the expected share in profits. To solve this problem the Jicarilla Apache, the Ute Mountain Ute, and the Wind River–Chuska contracts provided for well-by-well payout. Under that provision the tribe would begin receiving its share of profits from each well after that well had paid out, which could be years earlier than under other contracts.[29]

A few contracts became accountants' nightmares—so complicated and full of loopholes that it was impossible to determine whether a tribe would make money. Some agreements factored apparently arbitrary multipliers into the definition of costs. A proposed contract between the Wind River tribes and Chuska Energy Company said the tribes' percentage would not increase until the operator received 250 percent of the costs of the well. Allowable costs were defined with a long, detailed list plus the sum of $100,000, thus providing Chuska a profit of at least $250,000 on each well before the tribes' share would increase. The Western Nuclear contract defined costs as 137 percent of actual costs.[30]

To allow monitoring of contracts, Velarde of the Jicarilla Apache Tribe recommended that contracts not only require access to raw data but also to data interpretations. The Jicarillas' Odessa contract specified that the tribe could have a representative on the derrick drilling floor during drilling and could appoint a liaison officer, hired by the tribe and paid for by the company as an overhead expense. Thus the tribal representatives could not only monitor record keeping but also learn more about the oil and gas business. In fact, Velarde—the tribe's oil and gas administrator and a tribal member— acquired most of his early experience working as the liaison officer under the Palmer contract. Prior to accepting the position in 1977 he had completed a bachelor's degree and some graduate work. The expertise he gained under the Palmer contract was reflected in the subsequent contract negotiated with Odessa.[31]

Limit and Identify Acreage

Under the old regulations of the 1938 leasing act, each lease was limited to 2,560 acres, and production had to be established within ten years. Although designed to encourage diligent development and discourage speculative holding of Indian leases, the requirements did not succeed.[32] Under the 1982 act industry and tribes frequently used the new freedom to include more acreage in a single contract, with potentially disastrous results. Unless special precautions were taken, a company could tie up large blocks of land for years without ever producing any significant revenues for the tribes, especially if the company did not have the money to develop the land. If oil and gas were found within such large blocks, the tribe could not—without special contract provisions—benefit from the increased competitive value of proven versus unproven fields. A company could "high grade," developing the lands with the most potential while tainting the prospects for other companies looking at the remaining, passed-over acreage. Thus tribes found they benefited in most cases from leasing smaller, clearly identified acreages. In special situations they leased large acreages but retained portions for themselves and imposed stiff diligence requirements so that lands would be quickly evaluated and either developed or relinquished.[33]

Three contracts were especially notorious because of the large acreages involved: the Wind River–Chuska contract, which included 800,000 acres, the Northern Cheyenne–ARCO, with 437,000 acres, and the Crow-Raven, with 200,000 acres. In contrast, the Ute Mountain Ute–Wintershall Corporation oil and gas deal involved 250,000 acres, but it provided significant safeguards. In 1985 former Interior Secretary James Watt approached the Arapahoe Tribe with an oil and gas proposal on behalf of Chuska Energy Company for exploring 800,000 acres of undeveloped land and for taking over some existing development on the Wind River Reservation in Wyoming. Chuska, a Texas-based corporation half owned by a Navajo, had developed oil and gas on the Navajo Reservation; Watt served Chuska as a consultant. The proposed contract for the undeveloped acreage was supported by most Arapahoe members of the joint business council because it would have provided the tribes with 27 percent royalties, escalating to 30 to 50 percent. The Shoshones opposed it, saying the Arapahoes were "listening to the whispers of a pot of gold." The contract contained many disincentives to diligence, the largest of which was the amount of land involved. Only a major oil and gas company could afford to expeditiously explore 800,000 acres of land. Chuska's only penalty if it did not meet minimal drilling requirements would have been losing acreage; when production was established, a single well could have held 5,120 acres. Rentals and bonuses were minimal and would not have been required during the early years.[34]

At first it looked as if Watt's political connections and his friendship with Interior Secretary Donald Hodel might force through the dubious proposal. An assistant Interior secretary suspended the area BIA director's authority to approve agreements, and tribal attorneys accused Watt of using his political influence to fight competing oil and gas firms' contracts for the same lands. After an economic analysis by the Interior Department endorsed a competing contract, however, the tribes rejected the Chuska proposal and turned Watt away.[35]

Unfortunately for the Crow Tribe the Interior Department approved a similar contract in 1984 between the Crows and an affiliate of O'Hare Energy Company of Denver over the objections of some field BIA personnel. In the contract's defense the Crow lands were much less attractive to industry than the Wind River lands, which had produced billions of dollars' worth of oil and gas since the turn of the century. In contrast, only thirty wells were producing on the Crow Reservation when the O'Hare partnership was struck, and companies' bonus bids had averaged only $11 an acre over the six years prior to 1982. Even considering the low potential of Crow oil and gas, the Crow-O'Hare contract was infamous for its poor terms, especially the large amount of acreage. In 1983 a newly formed O'Hare affiliate, Buffalo Exploration, joined with the Crow Tribe to form Raven Oil Company. Buffalo was the managing partner; the Crow Tribe owned 51 percent of Raven. The agreement covered 200,000 undesignated acres; worse yet, it provided Raven the right of first refusal for all Crow lands. Buffalo Exploration readily admitted that it would need to find outside investors. BIA's Energy and Minerals Division evaluation determined that if there was production, the tribe would make more money than it would from a standard lease. But agency and area BIA officials feared there would never be any production so they had recommended not approving the contract.[36]

By 1988 there still was no production, and the Raven contract was dissolved in June at the request of both Buffalo and the tribe. Tribal officials claimed the contract had stifled interest in the reservation, whereas O'Hare Energy said the company was the victim of tribal politics and the depressed energy market. The company had drilled only three wells, one of which was completed but none of which resulted in commercial production.[37]

In contrast, ARCO, a major mineral company, had the capital and the incentives to relatively quickly assess the Northern Cheyenne acreage and then leave, rather than tying up the reservation lands indefinitely. ARCO completed 950 miles of seismic testing and drilled seven dry holes. The contract specified that ARCO spend a minimum of $2 million during each of the first two years on seismic exploration and also specified how many miles of seismic testing would be conducted and how many exploratory wells would be drilled each year. ARCO was required to relinquish substantial unproductive portions of the reservation by certain dates. Adding to ARCO's incentive to move quickly were requirements that ARCO provide $1.4 million a year to the tribe during exploration for advance rentals, scholarships, and the tribal oil and gas office. Still, ARCO's quick appraisal reduced the likelihood that another company would look at any of the Cheyenne acreage.[38]

In none of the three, large agreements discussed above—ARCO, Chuska, and Raven—did the tribe have an opportunity to raise its price if oil and gas were discovered. The entire acreage was committed at the same price, based upon the lower value of unproven fields. To avoid this dilemma, the Ute Mountain Ute Tribe retained acreage in its 250,000-acre contract with Wintershall Corporation. Under the agreement with the German corporation, the tribe retained rights to one quarter of every other section (160 acres out of 1,280 acres). Because it had retained the acreage it could charge more for that acreage after Wintershall established production nearby, or it could drill its own wells. On some of its retained acreage, for example, the tribe decided to take a 20 percent working interest after payout, in addition to its 25 percent royalty.[39]

Choose the Other Party Carefully

The attitude and the financial capacity of the company were as important as the specific contract provisions. Tribes learned to scrutinize the company's financial records, past performance, and terms offered in similar agreements; to check with other tribes if a company claimed experience on other reservations; to prohibit assignments without written consent; and to refuse to guarantee debts. Some, notably the Northern Ute, found it too risky and expensive to experiment with a company that lacked credentials.

A Ute roughneck (center) works on a drilling rig on the Ute Mountain Ute Reservation. (Photo taken in 1983; courtesy of Wintershall Corporation.)

Despite conventional wisdom the Ute Mountain Ute Tribe decided to contract 250,000 acres to one company, primarily because of the company's attitude. Tribal Vice Chairman Judy Knight explained that the tribe had good experiences with Wintershall Corporation on an earlier, 1982 lease and poor experiences with other companies, which did not even comply with federal laws, much less agree to comply with tribal demands. The tribe wanted a company that would train its people and provide certain benefits. Wintershall wanted to have access to the large piece of land in the promising Paradox Basin, which the tribe had held off the market for many years. Wintershall agreed to use the tribally owned construction company; to restrictions on activity in the tribal archaeological park; to a bonus of $5 million; to escalating rentals; to an annual work commitment of $1.5 million; to royalties of 20 to 25 percent; and to $500,000 a year for special tribal projects. The company also spent $200,000 to straighten out lease status information for the land.[40]

Other tribes also recognized the importance of choosing a partner responsive to their special priorities. They began to find some industry representatives who were personally interested in negotiating innovative ways to encourage education and stimulate free enterprise on reservations. Many companies provided scholarship funds. The Northern Cheyenne–ARCO agreement provided $50,000 annually for the tribal oil and gas office. The Crow-Shell coal contract contained several pages of stipulations regarding employment quotas and said that Shell would pay $25,000 to the tribal employment rights office.[41]

The Northern Ute Tribe learned the hard way that a company should be judged by its financial capacity as well as its responsiveness to tribal concerns. The Northern Ute Tribe signed a joint venture contract in June 1986 with the Charging Ute Corporation (CUC) to build an oil refinery on the Uintah and Ouray Reservation in Utah, which would have refined the tribe's oil and sold fuel to the U.S. Defense Department under an Indian preference contract. According to BIA's analysis, the tribe could not have lost money on the proposal if the refinery had been built. The attitude of the company seemed exemplary. CUC immediately hired twelve tribal members, paying them twice what some of them earned in the tribal energy and minerals department. CUC set up welding and metering training for tribal members and sent one member to Houston to learn refinery management. CUC officers spent their investors' money freely, sharing it with everyone.[42] Suddenly CUC could not cover its paychecks, contractors' checks, or the $75,000 reclamation bond it had presented to BIA. Because the contract had been negotiated with the understanding that CUC would pay all costs, the tribal council had voted down a request for a $20 million loan guarantee. When the contractual deadline for presenting the financial package arrived, the company could not demonstrate any progress toward raising the necessary money or getting the defense contract. Consequently, in April 1987 the tribe canceled its joint venture contract with CUC, and the company forfeited the $50,000 it had supplied toward its bond.[43]

In the end, because of the precautions it and BIA had taken, the tribe lost little financially, but leaders suffered from the members' dashed dreams for refinery jobs. Outside investors lost close to a million dollars. "If there was ever a good course in business, those guys taught it to us," said tribal Chairman Lester Chapoose (Uintah-Ouray). He and BIA Agency Superintendent Perry Baker (Mandan-Hidatsa) recommended that other tribes set deadlines, as the Northern Ute Tribe did, require a performance bond, and conduct a more thorough financial check on a company's officers. Baker believed the company had good intentions but just failed to raise the necessary money.[44]

Chapoose said a recently formed company with no corporate history should be viewed with caution. Tribes often received ambitious offers from underfinanced companies, often formed just for the Indian proposal. Some intended to broker the Indian project, some to save their own failing operations with BIA's extremely limited economic development funds. Many companies underestimated how difficult it can be to finance operations on Indian lands. The Navajo Tribe learned to ask promoters first about their financial plans rather than wasting time on companies that planned to get cash from the tribe or a loan guarantee from BIA. Along with attitude and financial capacity, technical and legal experience were also important. The Southern Ute Tribe's staff developed a set of questions that covered environmental, geologic, and accounting subjects. They determined whether the company had experience with the geology specific to the region, federal regulations, and tribal regulations. The staff compared the company's geologic data with the tribe's, recognizing that an inexperienced company could miss a deposit, maligning the property for future development.[45]

Choose Consultants Carefully

Just as important as the choice of partners was the choice of consultants and of how best to integrate their advice into tribal decision making. Only a few tribes had enough energy development to justify hiring a full array of advisers on their own staffs. Tribes needed not only attorneys with experience in the minerals field but also technical experts to help assess resources and potential partners, to negotiate, and to implement agreements. Tribes had learned not to rely totally upon BIA because the bureau conducted very limited background checks on proposed industry partners, determining only if the company was incorporated where it claimed to be and not whether any of the parties had records with the Federal Bureau of Investigation (FBI) or a history of business failures. In most cases BIA did not have specific information about the value of tribal resources, forcing tribes to rely upon "blind luck" or others' data. The bureau's capacity to help negotiate and implement agreements was restricted by lack of training and lack of money. Sometimes critics said BIA could not even understand the proposals they evaluated.[46]

During the 1980s hundreds of consulting firms, many owned by Indians, made their fortunes conducting studies for tribes. Tribes soon found they could lean too heavily upon outside consultants and attorneys, limiting their own control and potential employment for tribal members. Consultants' reports frequently lay on shelves, unread by tribal officials or staffs. CERT economist Ahmed Kooros advised tribes that whenever possible, tribal staffs should work alongside consultants in performing research and evaluating the results. Then tribes can use their goals in the evaluation, better understand the reliability of the data, and reevaluate data themselves if conditions changed.[47]

The Jicarilla Apache Tribe and the Southern Ute Tribe effectively integrated consultant and staff expertise. The Southern Ute staff included both a geologist and an accountant. The Jicarilla Apache oil and gas staff included oil and gas administrator Velarde, an accountant, and ten other tribal members. Before putting its acreage up for a joint venture, the Jicarilla Apache Tribe utilized a geologist familiar with the area who told the tribe how much production and what royalties and bonuses to expect, as well as university consultants for economic projections. The tribal committee that considered project proposals included two council members, three tribal members, and two ex-officio members (the tribal oil and gas administrator and the BIA realty officer).[48]

During negotiations some tribes were represented primarily by their attorneys and some by their staffs, who consulted periodically during the bargaining process with council members on possible contract terms. Most effective seemed to be the tribes that sent elected tribal officials as part of the negotiations team instead of relying entirely upon attorneys or consultants. This helped assure that the finished product was consistent with tribal goals, that the contract would have an informed advocate when it came before the council for the final vote, and that some of the knowledge acquired during the negotiation process would be transferred into the "knowledge bank" of the tribe itself. Including BIA early in the negotiations helped to assure the terms would be acceptable under federal regulations and to prevent unnecessary delays in approval of the final product.[49]

Consultants were expensive. The Navajo Tribal Council, for example, was considering a $78.6 million budget in 1987 that included $3.5 million in consultants' fees, primarily for economic development proposals. Because Congress had failed to appropriate funds for hiring experts under the Indian Mineral Development Act, some tribes, such as the Hopi, had to turn to Congress for special appropriations to pay for specific negotiations, which was often time-consuming. Many company representatives recognized that an informed adversary was easier to negotiate with, compared with a tribe that lacked confidence in its information. When the Navajo Nation decided to participate in building a coal-fired power plant in New Mexico, the tribe hired a major corporate-utility law firm, a national engineering firm, and a financial banking house as consultants. The three industry partners agreed to help pay for the consultants. By the late 1980s many tribes had improved their ability to choose and pay for consultants who could match industry's expertise.[50]

Protect Tribal Governmental Rights

When coming onto reservations, companies understandably wanted to protect themselves from unforeseen costs of tribal taxes and regulations. Not wanting to enter contracts that seemed to be binding only upon one party, they distrusted tribal courts and resented the fact that in most cases they could not sue the tribal government unless the tribe agreed to waive its sovereign immunity. Some tribes, such as the Northern Cheyenne and the Blackfeet, resorted to granting blanket tax holidays to companies in exchange for other concessions. Usually, however, tribes and their attorneys found ways to protect tribal sovereignty while reassuring their industrial partners that they did not have a competitive disadvantage on Indian lands. In some cases a tribe agreed to set a tax ceiling or to provide for tribal taxation only if a state tax was voided; to set a ceiling on regulatory costs; to provide for arbitration; or to waive its immunity from suit to some extent. If a company would not agree to a blanket acceptance of future tribal regulations, it would sometimes agree to specific contract terms providing for such things as training and employing tribal members and protecting the environment.[51]

The most ambitious industry-Indian energy project on the drawing boards in 1988 was the 2,000-megawatt coal-fired power plant planned by the Navajo Nation, New Mexico Public Service Company, Combustion Engineering, and Bechtel Power Corporation. With billions of dollars at stake, the partners paid close attention to tribal sovereignty issues. In 1984 they agreed upon ground rules for negotiating the Dineh Power Project contract, which were later incorporated into a formal agreement. They acknowledged tribal sovereignty, including taxing powers, while agreeing that the tribe would exercise its powers in a way that would not threaten the viability of the project. Even though it in effect would be taxing itself, the tribe planned to tax the power plant because it wanted to use that revenue for alleviating local impacts while its share of the power plant profits would go to other purposes. In exchange for these concessions, the tribe made some of its own. The tribe in 1985 had created a new tribal enterprise, the Dineh Power Authority, that had the power to conduct business, sue and be sued, and borrow money. Because of low power demand, the project had been deferred until after the year 2000.[52]

By 1988 all of the troubling contract issues had been decided for the Dineh plant except the forum for resolving disputes. Although several other tribe-industry agreements provided for arbitration, the Dineh project planners said that that would not suffice because the arbitrators' decisions still had to be enforced. Tribal negotiators were considering creating a new forum, separate from tribal courts, but said they would reluctantly accept the jurisdiction of outside courts if no other solution could be agreed upon.[53] The method of settling disputes—which included both the nature of the forum and the extent of the tribe's liability—was the most difficult issue in many tribes' negotiations. Some tribes agreed to waive their immunity for specific enforcement actions or to itemize tribal assets that could be used to make settlements. The Blackfeet-Damson agreement said controversies involving more than $10,000 would be tried in federal court. The Crow-Shell coal contract and the Wind River–Chuska oil and gas contract provided for arbitration. Other agreements said appeals of tribal court decisions could be made to a federal district court or to the Interior Department.[54]

Although the Crow-Shell coal contract provided for tribal regulation and taxation, it set ceilings upon the costs. If the 30 percent state severance tax was found invalid, the company agreed to pay the tribe the same amount. The contract said,

Shell specifically acknowledges the authority of the tribe to enact tribal legislation relating to mining activities, including the tribal employment rights ordinance, the health and sanitation code, and the land use zoning code. The tribe may adopt laws that impose reasonable additional costs so long as the total of all externally imposed costs associated with mining activities . . . whether imposed by the tribe, the United States, the state, or local government will not exceed the governmental costs associated with the mining of Montana coal.[55]

Several companies agreed to make their projects subject to tribal regulation; two such projects were the Northern Cheyenne–ARCO and Wind River–Chuska. Wintershall Corporation agreed to pay the Ute Mountain Ute Tribe's severance tax, which the contract specified was 3 percent. Some contracts said certain new ordinances were applicable only after the two parties agreed to them through arbitration.[56]

Decide from the Beginning Who Is in Charge

Industry and tribes sometimes had trouble knowing who was responsible for negotiating and implementing contracts. Who had the authority to represent the tribe and the company in negotiations and afterward? Is a tribal council decision final or must it first be approved by the Interior Department? How does a tribe assure meaningful involvement in management decisions? Was a separate tribal business entity necessary? A notorious example of confusion occurred in the 1970s when two different Crow factions working independently tried to renegotiate coal contracts. Assistant Interior Secretary Forrest Gerard sent a memo to the local BIA superintendent saying the situation had to be resolved because the coal companies did not know who was authorized to negotiate on behalf of the tribe.[57]

Usually on other reservations it was clear with whom to negotiate and that only the tribal council had the final authority to approve contracts. In some cases, however, industry approached the wrong parties, either because of ignorance or because of deliberate attempts to subvert the process. On the Northern Cheyenne Reservation, Global Petro-Chemical Services made a billion dollar offer directly to the people at a tribal assembly in 1979 rather than going through official channels. When James Watt approached the Arapahoe Tribe on behalf of Chuska Energy in 1985, he apparently did not realize that the Arapahoe and Shoshone Tribes made decisions affecting the Wind River Reservation through their joint business council. Watt worked only with the Arapahoes, leading inevitably to conflict. Despite his reputation for fighting tribal sovereignty and defending industry interests on behalf of the Mountain State Legal Foundation, Watt wrote to Arapahoe Business Council Chairman Chester Armajo, saying, "I want to stand with you in fighting the major oil companies, those who might try to take your water,

Three Ute workers lay out seismic cable on the Ute Mountain Ute Reservation. (Photo taken in 1983; courtesy of Wintershall Corporation.)

or anyone else that would deny the rights of the Indian people of Wyoming." The Arapahoe Business Council chairman threatened to petition the Interior secretary to divide the minerals of the reservation between the tribes before the tribes finally agreed to reject Watt's proposal.[58]

Industry learned from an experience on the Blackfeet Reservation that despite the increased self-determination provided by the Indian Mineral Development Act, agreements were not legally binding until they had been approved by the Interior Department. Quantum Exploration Company and the Blackfeet Tribal Council negotiated an agreement that could have provided the tribe and allottees with an ownership interest in wildcat wells. The tribal council approved the proposed agreement in March 1984, but before the Interior Department signed it, a newly elected tribal council rescinded it in August, based partly upon BIA advice. Quantum sued the department, saying Interior had interfered with its contractual relations with the tribe. Quantum believed the department did not understand the agreement and had deliberately delayed approval until after the tribal election. The Ninth Circuit Court of Appeals ruled against Quantum, saying that Congress intended to give tribes the opportunity to disentangle themselves from agreements within certain time limitations, using advice from Interior on the potential risks and benefits. Thus contracts were not final until signed by Interior. Quantum's situation was not common. Although the 1982 Indian Mineral Development Act provided a "cooling-off period" for tribes to reconsider mineral contracts before final approval by Interior, in most cases tribal councils reaffirmed their decisions.[59]

If the lines of authority were not clear beforehand, serious problems sometimes occurred after the contract was signed and approved. The Northern Ute Tribe never knew from one day to the next which Charging Ute Corporation officer was president because they kept changing titles. At the same time, the corporation did not know whom to listen to because the appointed tribal liaison person was making decisions with which the entire council did not necessarily agree. Many joint venture contracts established management committees, but sometimes tribes only won the illusion of control, such as in the Blackfeet-Damson and Crow-Raven contracts. Holding the majority of seats on the management committee did not necessarily mean a tribe could influence crucial decisions. To have real power the tribal representatives needed knowledge, experience, and control over certain key decisions, such as closure, choice of general manager, production volume, major investments, sales plans, and employment plans. The Dineh structure offered a good model. For the Dineh power plant, the Navajo Nation was considered a full equity partner, contributing its resources (coal, water, and transmission line rights of way) rather than money. On that basis the four parties—the Navajo Nation, New Mexico Public Service Company, Combustion Engineering, and Bechtel Power Corporation—agreed to joint participation in planning, construction, and operation of the power plant, with each having veto authority.[60]

Because of concerns about conflicts between tribes' governmental roles and their roles as business partners, some tribes experimented with setting up separate energy development authorities to depoliticize energy decisions. The idea came from Third World countries that had set up state mineral corporations. The Indian corporations were scrapped, however, partly because tribal members distrusted the corporations' power and autonomy from electoral politics. The Crow Tribe created the Crow Development Corporation when it was planning a synthetic fuels plant in 1982. The non-Indian board could participate in exploration, development, and conversion of minerals as well as deal in water rights. Elected officials served only as advisers. The powers of the Navajo Energy Development Authority (NEDA), which was created in 1980, were not so broad. NEDA could sue and be sued, borrow money, pledge assets, and enter agreements with the tribe for developing minerals. One elected official was to be appointed to the board. Although the Dineh Power Authority had some autonomy, it had less than NEDA because it was overseen by a tribal council committee. Most tribes did not separate their role as energy developers from the tribal government structure. Thurman Velarde, for example, served as the Jicarilla Apache oil and gas administrator (an appointed government position) and also as head of the Jicarilla Energy Company, the tribe's energy development arm. Velarde retained his government position for more than ten years despite personnel changes in tribal administrations.[61]

RETURN TO LEASING?

Looking toward the future some tribal attorneys predicted the pendulum might swing back toward leasing rather than negotiated agreements with tribal ownership interests. After getting a taste of the volatile oil and gas market, some tribal officials wanted more stable, dependable income. Even with good contracts it was not easy being producers during poor markets. Although some elected officials preferred the higher return and higher risk of being producers, their successors sometimes got cold feet, as on the Blackfeet Reservation in 1984 when the new council rejected all of the earlier council's bold initiatives, including the Quantum agreement. Negotiations consumed months of time, and good contracts depended upon good geologic information. Thus some tribes, such as Fort Peck's Assiniboine and Sioux and the Ute Mountain Ute, chose a middle approach—they continued holding lease sales to test the market. The bid process showed the interest in the area at the companies' expense. If interest was high in a certain prospect, the tribe could exercise its option to reject the bid and open negotiations. Fort Peck Councilman Caleb Shields (Sioux) said the tribes still were interested in more joint ventures and taking over leases.[62] Leasing was made more attractive during the 1980s not only by the energy market's fall but also by the courts and Congress's endorsement of the exercise of tribal governmental authority. By adopting environmental and cultural regulations and codes, the tribes could get more control over energy development without sitting on management committees.[63] On the other hand, the Supreme Court's 1989 *Cotton* decision recognizing state taxing authority on reservation minerals made alternative contracts attractive again. Through such contracts tribes and companies in some cases could avoid state taxation.

CONCLUSION

Energy tribes face the turn of the century in a much better position than ever before as the result of tremendous changes in their own capacities and in the legislative and judicial atmosphere in which they operate. No longer must tribes devote their attention to defending their resource ownership and their power of consent. These questions are resolved in almost all cases, with recognition of legitimate water rights the most recent. Settling such basic questions freed tribes in the 1980s to focus upon improving the monetary return from old and new contracts, exercising their governmental powers, and developing minerals themselves. In a string of major court and legislative victories, the federal government has recognized tribes' authority to tax mineral producers, enforce royalty regulations, participate in off-reservation energy siting decisions, impose air and water quality standards, issue tax-free bonds, and negotiate with industry to develop their own minerals. Tribes' position of strength makes some state governments more willing to negotiate to find equitable solutions to taxing, water rights, and civil jurisdiction disputes. Beyond the rhetoric of the Red Power era, many tribes now see the advantages of increased cooperation with industry and other governments. Other tribes and states still feel they must insist upon litigation in some instances.

No longer the nation's energy colonies, several tribes have become energy developers, and others have gained significant control, making their own decisions with the help of consultants and attorneys, many of them Indian. Smaller tribes still must struggle to retain even minimal staffs and hired expertise. Although intertribal jealousy still exists, when tribes need advice they now often look to one another, either informally or through tribal organizations such as CERT. Allottees also often utilize consultants' and one another's expertise to their advantage, acquiring contracts that easily outmatch those of non-Indian mineral owners. The energy tribes have moved from a time when they were totally dependent upon the federal government to a time when a few pay most of the cost of providing services and economic development on the reservations.

Many of the remaining problems faced by Indian energy tribes are caused by the nation's vacillating policies toward Indians through the years that have fractionated ownership and confused jurisdiction, making it more difficult to regain economic self-sufficiency. During two periods (1871–1928 and the 1950s), the federal government tried to terminate tribal governments and integrate Indian people—and their resources—into the social mainstream. These periods were both followed by reversals when tribal governments and thus the need for economic development were once again recognized. Even when it adopted policies designed to encourage economic development on reservations, however, Congress time after time failed to provide the funds necessary. The most consistent policy has been treating Indian lands and resources similarly to federal lands and resources, which has resulted in persistent inequities toward and exploitation of Indians. Today the federal policy of self-determination has opened the door to the victories mentioned above. Yet the limits of the federal government's trust and financial responsibilities have not been determined. Indians want both freedom and protection from the avarice that nearly destroyed them in the past. Many federal agents distrust the tribes' and the allottees' capacities and want to protect their own jobs; the dedicated federal employees' advocacy of Indian interests is undermined by the Interior Department's conflicting responsibilities to protect non-Indian water users and to encourage domestic energy exploitation.

Energy tribes are not rich, but by developing their nonrenewable resources—under tight, tribal controls—they might be able to underwrite a broader economic development effort on their reservations. Beyond saying that American Indian tribes want control over their own resources, it is impossible to generalize about tribes and their attitudes toward energy development. Attitudes vary from tribe to tribe and often from one administration to another within tribes. Some leaders will consider oil and gas wells but not the more socially disruptive power plants or the environmentally degrading uranium mining.

Reservation economic development is necessary because most Indian people would like to continue to live on their reservations without remaining dependent upon federal subsidies. With a population growing faster than any other ethnic group and an unchanging land base, it will not be easy for tribes to provide for their people. Whether they will succeed will depend upon accommodating the concerns of their own members, of industry, and of state governments without compromising important tribal standards. It will also depend upon the American public's understanding certain concepts, including the need for reservations, the desirability of adaptability, tribal sovereignty, and self-determination. The public must reconcile its impulses to decry Indians' dependency while trying to defeat their efforts toward independence. Indian tribes must break the iron bonds linking them to dependency, paternalism, and exploitation.

Indian Land
and Mineral Timeline

1862	Homestead Act for public lands
1871	Congress ends treaty making
1887	General Allotment Act provides for dividing reservations into allotments
1891	Indian mineral leasing act; requires tribal consent and includes provision for leasing allotted lands
1902	peak in number of public lands homestead entries
1903	*Lone Wolf* decision says Congress can unilaterally change treaties
1906	Burke Act eliminates trust protection for "competent" Indians
1907	Congress authorizes selling noncompetents' allotments
1909	Congress passes mineral leasing act for Indian allotted lands; Enlarged Homestead Act leads to second highest peak in homestead entries on public lands; Congress splits minerals from surface of public lands for first time with Coal Lands Act of 1909
1910	Coal Lands Act of 1910 authorizes homesteading and desert land entries on public coal lands for first time
1912–1920	Dramatic escalation in sales of Indian lands and leasing, partly caused by the war's increasing of commodity prices; termination sentiment in Congress
1916	Stock Raising Homestead Act provides for 640-acre homesteads on public lands, reserving minerals to the United States
1917	Congress opens reservation coal lands to homesteading, with coal reserved to the United States
1919	Metalliferous mineral leasing act opens executive-order reservations in certain states to leasing of precious metals; no provision for tribal consent
1920	Public lands mineral leasing act institutes leasing system for energy minerals; federal government begins curtailing allotment of Indian lands
1923–1930	Overproduction of oil nationwide; precipitous drop in prices in March 1927
1924	Congress makes Indians "citizens"; Indian oil leasing act opens treaty reservations for oil leasing
1927	Indian oil act of 1927 provides for leasing on executive-order reservations; Indian office discontinues leasing on reservations in July

因 because of decline in oil prices and poor prospects for future development

1929   Mineral prospecting permits for all public domain and Indian lands are suspended

1933   Moratorium on allotment patents; low point in U.S. land prices

1934   Indian Reorganization Act ends allotment and provides for tribal self-government; Taylor Grazing Act ends settlement entry on public lands

1938   Indian minerals leasing act requires tribal consent and standardizes leasing policies for Indian lands

1944   National Congress of American Indians (NCAI) founded

1948   Average crude oil price at the wellhead climbs above $2 a barrel

1953   Congress adopts termination policy

1955–   First uranium boom
1965

1960   Both Democratic and Republican party platforms denounce termination

1972   Navajo Tribe and Exxon negotiate first Indian mineral contract providing for ownership interest

1973   Arab oil embargo causes dramatic rise in oil, gas, and coal prices; Interior Department declares moratorium on further public coal leasing because of environmental concerns

1975   Indian Self-Determination Act is passed; tribes form Council of Energy Resource Tribes (CERT); Blackfeet tribe and Damson Oil negotiate joint venture oil contract

1975–   U.S. uranium production peaks during nation's second uranium
1979   boom

1977   Interior Department resumes coal leasing; Congress passes law to control environmental effects of coal strip mining and provides for tribal control of strip mining of Indian coal

1979   Congress passes Energy Security Act to put billions of dollars into developing synthetic gas and oil; mobs seize fifty-two American hostages at U.S. Embassy in Iran

1980   Uranium prices start to fall following 1979 accidents at Three Mile Island nuclear power plant and at Church Rock uranium mill; Jicarilla Apache Tribe takes over ownership of Palmer Oil wells on reservation

1981   Oil prices climb to over $40 a barrel

1982   U.S. Supreme Court confirms Jicarilla Apache Tribe's right to impose a severance tax on mineral producers; coal, oil, and gas prices start dropping due to oversupply; Congress passes Indian Mineral Development Act (IMDA) and Indian Tribal Government Tax Status Act and includes tribes in Federal Oil and Gas Royalty Management Act (FOGRMA) and Nuclear Waste Policy Act

1983   Jicarilla Apache Tribe drills oil and gas wells using tribal funds; Fort Peck tribes and U.S. Energy produce oil from joint venture well

1984   Fort Peck tribes produce oil from tribally owned well; Environmental Protection Agency (EPA) adopts Indian policy supporting primary role of tribal governments in reservation environmental matters

1985   U.S. Supreme Court confirms Navaho Tribe's right to impose taxes without approval of secretary of Interior

1986   Congress adds amendments to Clean Water Act, Safe Drinking Water Act, and Superfund

Council of Energy Resource Tribes (CERT) Reservations, 1988

*Acoma Pueblo
*Blackfeet
Chemehuevi
Cherokee
Cheyenne Arapahoe
*Cheyenne River Sioux
Coeur d'Alene
*Crow
Flathead
*Fort Belknap
*Fort Berthold
*Fort Hall
*Fort Peck
*Hopi
Hualapai
*Jemez Pueblo
*Jicarilla Apache
Kalispel
*Laguna Pueblo
Muckleshoot
*Navajo
*Nez Perce

*Northern Cheyenne
Oglala Sioux
Pawnee
Penobscot
Ponca
*Rocky Boys'
Rosebud Sioux
Saginaw Chippewa
*Santa Ana Pueblo
Seminole of Florida
*Southern Ute
*Spokane
Standing Rock Sioux
Tule River
Turtle Mountain Chippewa
*Uintah and Ouray
Umatilla
*Ute Mountain Ute
Walker River
*Yakima
*Zia Pueblo

*Members of the first board of directors (two members of the first board—
the Colville Confederated tribes and the Wind River Reservation—were no
longer members of CERT in 1988).

Charles Lipton's Eighteen Points

1. *Limited period.* An agreement should be for a fixed period, preferably 20 to 25 years, not for "so long as minerals can be profitably produced," as the 1938 Minerals Leasing Act provides.

2. *Exploration work program.* A work program should detail what work is to be done, including drilling, and when it will be done and should require spending a specified minimum amount.

3. *Data delivery.* A tribe should obtain all information resulting from exploration, including interpretation. A tribe should have the property rights to that information, subject to confidentiality for a limited period and subject to the operator's right to use the information for carrying out the agreement.

4. *Prospect size and relinquishment.* If the agreement covers a large area, it should include a provision requiring the operator to relinquish percentages of the exploration area over a certain period. The area can also be divided into a number of blocks and specific blocks reserved for the tribe (traditionally in a checkerboard pattern) for future development should a discovery be made, which would increase the value of the reserved blocks.

5. *Sharing real profits.* The tribe should share in the true profits, including tax credits and allowances and other direct and indirect subsidies. The tribe's share preferably should increase on a sliding scale, based upon revenues or profitability or after the operator has recovered its original costs. A tribe should be sure that an operator does not siphon off profits by overpricing or underpricing affiliated company transactions.

6. *Limitation of recovery rates.* Where a tribe shares in profits, the operator's recovery of its original costs should be computed within certain limits: Either the number of years should be specified, or each year's costs should be limited by a specified percentage of the value of production.

7. *Royalty.* A tribe should be assured of a royalty equal to a percentage of the fair market value of production each year, regardless of profitability, preferably on a sliding scale that increases with price or revenue.

8. *Minimum cash payment.* A tribe should also be assured of a minimum annual revenue, as a rental or minimum royalty.

9. *Payments for surface and water rights.* The tribe should charge fair market prices for water and for the use of surface rights. Where a fixed annual charge is agreed upon, provision should be made for automatic adjustments according to a price index for changes in the cost of living.

10. *Bonuses.* Cash bonuses should be considered on signature, on dis-

covery, and perhaps at different production levels, keyed to the value rather than the quantity of production. A tribe should focus on long-term benefits, however. It should not trade off a share of the profits or a higher royalty for such front-end money if it is not to its long-term economic advantage.

11. *Employment preferences.* Tribal members should be assured of genuine employment and promotion preference in all employment categories, including supervisory, administrative, technical, and managerial. This preference should be combined with commitments to provide educational opportunities and on-the-job training.

12. *Preference for tribal enterprises.* Enterprises owned by a tribe and tribal members should be preferred for providing goods and services through mechanisms such as prequalification, advance notice, and a 15 percent edge in competitive bidding. (This 15 percent edge is required by the World Bank for local companies.)

13. *Tribal concurrence in basic decisions affecting the reservation and its resources.* These decisions should include: location of drill holes or wells, plant, equipment, offices, and access routes; size, method, and rate of operations; impact on air, surface and subsurface water, and community facilities; conservation, reclamation, and restoration programs; marketing arrangements; and annual operating budgets if the tribe shares profits.

14. *Indemnification.* A tribe and its members, officers, employees, and agents should be fully indemnified against all liabilities arising out of operations.

15. *Record keeping and reporting.* All pertinent information on exploration, production, and sales should be recorded on a regular basis, with financial information recorded in accordance with agreed-upon accounting principles (not just those "generally accepted"), and reports should be rendered at least quarterly to tribal officials.

16. *Inspection and monitoring procedures.* Tribal officials and their advisers should be assured the right to inspect all operations and books and records at all times.

17. *Assignment.* Tribal consent should be necessary for any assignment of any interest in the agreement.

18. *Insurance, guarantees, and performance bonds.* Operators should be required to carry appropriate insurance in adequate amounts for all operations. Performance bonds should also be required unless the operator's performance is guaranteed by parent corporations with adequate assets, including cash reserves.

Suggestions for Industry

To gain access to the vast mineral and water resources controlled by the tribes, industry must make some concessions without jeopardizing its crucial interests. Several suggestions may help industry in this pursuit:[1]

1. Do not generalize about tribes based upon your own—or worse, someone else's—bad experiences on one reservation. Before ever going to the reservation, recognize that the days are over when companies could deal with Indian lands as if they were federal lands and tribal governments as if they were rubber stamps for BIA contracts.

2. Be prepared to adjust your attitude. Tribal sovereignty is a reality that has to be faced. Rather than launching litigation, companies should try to determine compatibility of interests between the tribe and the company. Do your homework before negotiations begin. Visit the reservation. Try to understand the problems there. Look at each tribe's treaties, constitution, tribal newspaper, and ordinances to see how it views its powers. Analyze tribal courts.

3. Contact the tribal energy department. On most reservations you do not need to "know somebody" to get in the door. Find out if the tribe still holds lease sales and, if not, the process for making a proposal, which usually entails getting on the agenda for a tribal council meeting. On other reservations, you must contact the tribal attorney.

4. Be prepared for delays. It may take time to build trust and understanding. Because many tribes try to reach a consensus among council members—and sometimes among all tribal members—their decisions may take longer. Consensus can protect enterprises from future delays, however.

5. Avoid unnecessary delays by involving BIA in negotiations from the beginning. Although BIA under law has six months to review proposed agreements, the contracts are often approved faster when BIA is already familiar with them.

6. Do not align with any faction or take advantage of factionalism.

7. Expect unusual requests for funding and accept reasonable requests as part of the corporation's social responsibilities. Remember that Indian people often do not benefit from the more generalized giving of mineral corporation foundations. Tribes will ask for assistance with education and job training and may also ask for help with supporting tribal infrastructures (such as tribal courts), paying for ceremonials, and clarifying lease records.

NOTES

CHAPTER ONE. OUT OF THE MAINSTREAM:
THE IMPORTANCE OF THE RESERVATION

1. Simon Ortiz (Acoma), "The Land and the People are Speaking," in *Economic Development in American Indian Reservations*, ed. Roxanne Dunbar Ortiz (Albuquerque: University of New Mexico Native American Studies Development Series, 1979), 16.

2. Definition paraphrased from that used by the Administration of Native Americans as quoted in Ahmed Kooros and Anne C. Seip, "Tribal Self-Government: A Self-Sufficiency Analysis," unpublished paper written for CERT, October 1983, 7.

3. James Watt, "Conservative Counterpoint," news release, Satellite Program Network, 19 January 1983.

4. Ibid.

5. Associated Press (AP), "Reagan: Indians Humored," *Casper Star-Tribune*, 1 June 1988.

6. Watt, "Conservative Counterpoint."

7. U.S. Department of Interior (DOI), Bureau of Indian Affairs (BIA), *Report of the Task Force on Indian Economic Development* (Washington, D.C.: GPO, July 1986), 5. These unemployment figures (both Indian and non-Indian) include all men of employable age (sixteen and over) who are not working, thus encompassing the "discouraged workers" who are no longer looking for work.

8. DOI, *Task Force*, 5–7.

9. DOI, *Task Force*, 38. There are about 300 federal Indian reservations in the United States. The Bureau of Indian Affairs (BIA) figure of 755,000 includes its service areas: reservations, lands near reservations, native villages in Alaska, and areas of Oklahoma that once were reservations.

10. George Hardeen, "The Other Side of the Mountain: A Mixed Bag Visits Big Mountain Survival Camp," *Navajo Times*, 10 July 1986; Nancy Wood, *When Buffalo Free the Mountains: The Survival of America's Ute Indians* (Garden City, N.Y.: Doubleday & Company, 1980), book jacket; Loretta Fowler, *Shared Symbols, Contested Meanings: Gros Ventre Culture and History, 1778–1984* (Ithaca, N.Y.: Cornell University Press, 1987), 5–9.

11. D'Arcy McNickle, *Native American Tribalism: Indian Survivals and Renewals* (New York: Oxford University Press, 1973), 7; Andrew Skeeter & Co., "Jicarilla Apache Demographic Survey," study completed for tribe, 1978, 9, 17 (photocopy); Tori Adams, "NCC Faculty Works to Preserve Written Navajo," *Navajo Times*, 31 December 1987; American Indian Policy Review Commission (AIPRC), *Final Report*, vol. 1 (Washington, D.C.: GPO, 1986), 81.

12. Charles F. Wilkinson, *American Indians, Time and the Law* (New Haven, Conn.: Yale University Press, 1987), 68–75.

13. Ted Rising Sun (great-grandson of Dull Knife), in "Indians and Environ-

mentalists Conference" (Helena, Mont.: Northern Rockies Action Group, 1975), 44; James P. Boggs, "The Challenge of Reservation Resource Development: A Northern Cheyenne Instance," in *Native Americans and Energy Development II*, ed. Joseph G. Jorgensen (Boston: Anthropology Resource Center, 1984), 214.

14. Wyoming Governor Lester C. Hunt to Senator Joseph C. O'Mahoney, 30 March 1945, O'Mahoney Papers, University of Wyoming.

15. Francis Paul Prucha, *The Great Father: The United States Government and the American Indians*, abridged ed. (Lincoln: University of Nebraska Press, 1986), 108–121.

16. Ibid., 164–165.

17. Executive orders created all or part of many of today's energy reserations— Hopi, Jicarilla Apache, Northern Cheyenne, Uintah and Ouray, parts of Fort Berthold, and parts of Navajo.

18. Paul W. Gates, *History of Public Land Law Development* (Washington, D.C.: GPO, November 1968); Robert W. Swenson, "Chapter XXIII: Legal Aspects of Mineral Resources Exploitation," ibid., 699–764.

19. Congress made the Homestead Act available to individual Indians in 1875 for homesteading public lands, but few utilized it.

20. Board of Indian Commissioners, *Annual Report*, 1902, 90–91. John Aloysius Farrell, "The New Indian Wars," *Denver Post* (special reprint), 20–27 November 1983, 28.

21. D. S. Otis, "History of the Allotment Policy," in *Federal Indian Law Cases and Materials*, ed. David H. Getches, Daniel M. Rosenfelt, and Charles F. Wilkinson (St. Paul, Minn.: West Publishing Co., 1979), 70 [hereafter *Federal Indian Law I*]; Angie Debo, *And Still the Waters Run: The Betrayal of the Five Civilized Tribes* (Norman: University of Oklahoma Press, 1984), 24.

22. Otis, "History of Allotment," 71.

23. Sec. 5 of General Allotment Act, 24 Stat. 388; preamble of Fort Berthold opening act, 26 Stat. 1032.

24. Frederick E. Hoxie, "Building a Future on the Past: Crow Indian Leadership in an Era of Division and Reunion," in *Indian Leadership*, ed. Walter L. Williams (Manhattan, Kans.: Sunflower University Press, 1984), 78.

25. Frederick E. Hoxie, *A Final Promise: The Campaign to Assimilate the Indians, 1880–1920* (Lincoln: University of Nebraska Press, 1984), 147–187; *Federal Indian Law I*, 77–78.

26. *Lone Wolf v. Hitchcock*, 187 U.S. 553 (1903); *Federal Indian Law I*, 185–188; Gilbert L. Hall, *The Federal-Indian Trust Relationship* (Vienna, Va.: Institute for the Development of Indian Law, 1979), 36.

27. In the Northwest most reservations were allotted, including that of the only energy tribe there—the Spokane. In the Southwest many—including most of the pueblo tribes, the Hopis, the Ute Mountain Utes, and the Navajo Reservation— escaped allotment (see table, p. 4). Peter Iverson, "History of the Wind River Reservation," unpublished paper, n.d., 16–17; Floyd A. O'Neil and Kathryn L. MacKay, *A History of the Uintah-Ouray Ute Lands* (Salt Lake City, Utah: American West Center, 1982), 30.

28. Allotment had started earlier on some reservations under the authority of the 1887 General Allotment Act or executive orders establishing the reservations. The reservations listed in the text were not opened until after 1903, however. The 1906 Osage allotment act did not provide for opening Osage lands to non-Indian settlers, but other Oklahoma Indian lands had been allotted and opened earlier. Terry P. Wilson, *The Underground Reservation: Osage Oil* (Lincoln: University of Nebraska Press, 1985), 94, 97; Hoxie, "Crow Indian Leadership," 79; Crow opening law, 33

Stat. 352; Uintah and Ouray opening law, 33 Stat. 1069; Wind River allotment act, 33 Stat. 1016; Blackfeet allotment act, 34 Stat. 1015 (the Blackfeet Reservation was never entered by non-Indian settlers); Fort Peck allotment act, 35 Stat. 558; Fort Berthold allotment act, 36 Stat. 455.

29. Northern Cheyenne allotment act, June 3, 1926, 44 Stat. 690; James P. Boggs, "The Challenge of Reservation Resource Development: A Northern Cheyenne Instance," in Jorgensen, *Native Americans and Energy Development II*, 214–218. J. P. Kinney, *A Continent Lost—A Civilization Won: Indian Land Tenure in America* (Baltimore: Johns Hopkins University Press, 1937; reprint, New York: Arno Press, 1975), 275–277.

30. Peter Iverson, *The Navajo Nation* (Albuquerque: University of New Mexico Press, 1981, 1983), 16.

31. Veronica E. Velarde Tiller, *The Jicarilla Apache Tribe: A History, 1846–1970* (Lincoln: University of Nebraska Press, 1983), 110–117.

32. Lawrence C. Kelly, *The Navajo Indians and Federal Indian Policy* (Tucson: University of Arizona Press, 1968), 18, 19, 34; "Judge Rules for Navajos in Land Dispute," *Coal Week*, 12 September 1988.

33. Act of 28 February 1891, 26 Stat. 795; act of 3 March 1909, 35 Stat. 781; Sandra Calwalader and Marge Emery, "The Dawes Act Revisited," *Indian Truth*, newsletter of the Indian Rights Association, Philadelphia, n. d.

34. Hoxie, *Final Promise*, 166–167.

35. Ibid., 179–183.

36. American Indian Policy Review Commission (AIPRC), *Task Force Seven: Reservation and Resource Development and Protection* (Washington, D.C.: GPO, 1976), 39, part 2.

37. Presidential Commission on Indian Reservation Economies, *Report and Recommendations to the President of the United States* (Washington, D.C.: GPO, November 1984), II-73.

38. AIPRC, *Reservation and Resource Development*, 39; DOI, *Task Force*, 115; Farrell, "New Indian Wars," 66.

39. Prucha, *Great Father*, 276–316.

40. Oil leasing was reopened in 1932. According to Wilson, *Underground Reservation*, 150–151, the price of oil had dropped to 15 cents a barrel by 1930; Russel Lawrence Barsh and James Youngblood Henderson, *The Road: Indian Tribes and Political Liberty* (Berkeley: University of California Press, 1980), 289; Kelly, *Navajo Indians and Federal Indian Policy*, 101; Swenson, "Mineral Resources Exploitation," 749–750; Jerome H. Simonds, "The Acquisition of Rights to Prospect for and Mine Coal from Tribal and Allotted Indian Lands," *Rocky Mountain Mineral Law Institute* 21 (1975): 132.

41. *Federal Indian Law I*, 74. This figure included losses through the sale of "surplus" lands, sales by allottees after the trust period ended, and sales of heirship lands by the government.

42. McNickle, *Native American Tribalism*, 88.

43. John Collier, *Indians of the Americas: The Long Hope* (New York: New American Library, 1947), 155.

44. Indian Reorganization Act (IRA), 18 June 1934, P.L. 383, 48 Stat. 984; *Federal Indian Law I*, 79–86.

45. AIPRC, *Reservation and Resource Development*, 25; Indian Commissioner John Collier to secretary of Interior, "Restoration of Lands formerly Indian to Tribal Ownership," 10 August 1934, 54, *Interior Decisions*, 559–604; Iverson, "History of Wind River Reservation," 56–57.

46. Uintah-Ouray Reservation: solicitor's memo of 27 January 1947, 59 *Inte-*

rior Decisions, 393–396; Southern Ute: solicitor's memo of 17 May 1948, 60, *Interior Decisions*, 174–176; Fort Berthold, Southern Ute, and Ute Mountain Ute: Department of Interior (DOI), Office of Surface Mining (OSM), "Jurisdiction Study," in *Report to the U.S. Congress: Proposed Legislation Designed to Allow Indian Tribes to Elect to Regulate Surface Mining of Coal on Indian Lands* (Washington, D.C.: DOI, 1984), 237–238.

47. DOI, "Jurisdiction Study," 241–242; the act of 19 May 1958 (P.L. 85-420, 72 Stat., 121) states "to provide for the restoration to tribal ownership of all vacant and undisposed of ceded lands on certain Indian reservations." The act of 14 August 1958 (72 Stat. 575) amended the earlier 1958 act for the Crow Tribe, reducing the restored surface acreage by 4,900 acres (because of a reclamation project on that land) while retaining the tribe's rights to the minerals.

48. *Federal Indian Law I*, 83; Hall, *Federal-Indian Trust Relationship*, 91–95 (chart of tribes).

49. Debo, *Still the Waters Run*, 31–60.

50. Loretta Fowler, *Arapahoe Politics, 1851–1978: Symbols in Crises of Authority* (Lincoln: University of Nebraska Press, 1982), 174.

51. Joe S. Sando, *The Pueblo Indians* (San Francisco: Indian Historian Press, 1976), 8–14; Fowler, *Arapahoe Politics*; Tiller, *Jicarilla Apache*, 169; James Jefferson, Robert W. Delaney, and Gregory C. Thompson, *The Southern Utes: A Tribal History*, ed. Floyd A. O'Neil (Ignacio, Colo.: Southern Ute Tribe, 1972), 54–55.

52. Collier, *Indians of the Americas*, 169; Tiller, *Jicarilla Apache*, 161–166; Tom Weist, *A History of the Cheyenne People* (Billings: Montana Council for Indian Education, 1977), 196–197.

53. AIPRC, *Reservation and Resource Development*, 26–32. Section 5 of the Indian Reorganization Act (IRA) provided $2 million a year for tribes purchasing allotments. Instead of $100 million that could have been spent in the first fifty years after the IRA, Congress spent $6 million. The policy review commission estimated that it would cost $1.5 billion in 1976 to acquire all the allotted lands on all Indian reservations excluding mineral values.

54. The Northern Cheyenne Tribe and its members, for example, own nearly all reservation lands as the result of a tribal policy adopted in the 1950s to buy up fee land (Weist, *History of the Cheyenne*, 196–197).

55. *Federal Indian Law I*, 80; Kenneth R. Philp, "Termination: A Legacy of the Indian New Deal," *Western Historical Quarterly* 14 (April 1983): 165–180.

56. Peter Iverson, "Building toward Self-Determination: Plains and Southwestern Indians in the 1940s and 1950s," *Western Historical Quarterly* 16 (April 1985): 165–166.

57. H.R. Cong. Res. 108, 83d Cong., 1st sess., 67 Stat. B132 (1953): Felix Cohen, *Felix Cohen's Handbook of Federal Indian Law*, 2d ed. (Charlottesville, Va.: Michie/Bobbs-Merrill, 1982), 171 [hereafter *Cohen II*].

58. McNickle, *Native American Tribalism*, 111; William A. Brophy and Sophie D. Aberle, *The Indian: America's Unfinished Business—Report of the Commission on the Rights, Liberties, and Responsibilities of the American Indian* (Norman: University of Oklahoma Press, 1966), 73.

59. Michael Lawson, *Dammed Indians: The Pick-Sloan Plan and the Missouri River Sioux, 1944–1980* (Norman: University of Oklahoma Press, 1982), 94–134; *Cohen II*, 165; Brophy and Aberle, *Unfinished Business*, 72–73.

60. Kenneth R. Philp, *Indian Self-Rule: First-Hand Accounts of Indian-White Relations from Roosevelt to Reagan* (Salt Lake City: Howe Brothers Press, 1986), 125; *Cohen II*, 174; Fowler, *Arapahoe Politics*, 207; Wilson, *Underground Reservation*, 183–184.

61. *Congressional Record*, 92d Cong., 2d sess., 8 June 1972, 20268; *Old Problems—Present Issues: Nine Essays on American Indian Law* (Vienna, Va.: Institute for the Development of Indian Law, 1979), 19; Kirke Kickingbird and Karen Ducheneaux, *One Hundred Million Acres* (New York: Macmillan Publishing Co., 1973), 69-82; *Congressional Record*, 92d Cong., 2d sess., 8 June 1972, 20272. Studies showed that, in fact, the department had profited in some cases while holding these lands. On the Fort Peck Reservation, for example, the Interior Department earned $2 million in oil and gas royalties from lands that cost the government less than $500,000. The lands were turned back to the tribes by 89 Stat. 577, 17 October 1975, P.L. 94-114.

62. Collier, *Indians of the Americas*, 145; Philp, "Termination," 166; *Federal Indian Law I*, 94.

63. Justice Black in *Federal Power Commission v. Tuscarora Indian Nation*, as quoted in *Federal Indian Law I*, 201-204. The Supreme Court majority in that case ruled in 1960 that the eminent domain power in the Federal Power Act allowed condemnation of Tuscarora lands, which were not protected by trust status.

64. Observation based upon conversations with Indian leaders, especially Wilma Mankiller, principal chief of the Cherokee Nation, November 1986; Gerald Wilkinson (Cherokee), executive director of National Indian Youth Council, August 1983; and Ernest Stevens (Oneida), August 1980; U.S. Congress, Senate Select Committee on Indian Affairs, "Congress Repeals Termination Resolution," news release, April 1988.

65. AIPRC, *Reservation and Resource Development*, 27; General Accounting Office (GAO), *Farmers Home Administration: Information on Agricultural Credit Provided to Indians on 14 Reservations* (Gaithersburg, Md.: GAO, 1987), 50-51. The statistics include both tribal and individual borrowers. Public Law 280 (67 Stat. 588), passed on 15 August 1953, gave six states criminal and civil jurisdiction over Indian lands.

66. William T. Hagan, "Tribalism Rejuvenated: The Native American since the Era of Termination," *Western Historical Quarterly* 12 (January 1981): 5-16.

67. Hagan, "Tribalism Rejuvenated," 9-10; Iverson, *Navajo Nation*, 89-91.

68. *Federal Indian Law I*, 107. Nixon's Indian policies resulted largely from the input of Indian historian and author Alvin M. Josephy, Jr. For the text of his Indian recommendations, see Alvin M. Josephy, Jr., "The American Indian and the Bureau of Indian Affairs: A Study with Recommendations for President-elect Richard Nixon," in *Red Power*, ed. Alvin M. Josephy, Jr. (New York: McGraw-Hill Book Co., 1971), 93-127.

69. Indian Financing Act, P.L. 93-262, 88 Stat. 77; Daniel H. Israel, "The Reemergence of Tribal Nationalism and Its Impact on Reservation Resource Development," *University of Colorado Law Review* 47, 4 (Summer 1976): 626.

70. Indian Self-Determination and Educational Assistance Act, P.L. 96-638, 88 Stat. 2203.

71. Several groups with euphemistic names promoted legislation to abrogate treaties and promote state jurisdiction over hunting and fishing. These included the Interstate Congress for Equal Rights and Responsibilities, All Citizens Equal, Montanans Opposed to Discrimination, and Wyoming Citizens for Equality in Government. From various tribal newspapers and from Alvin M. Josephy, Jr., *Now That the Buffalo's Gone* (New York: Alfred A. Knopf, 1982), 206-211.

72. Andrus testimony before House appropriation subcommittee on 21 February 1979 regarding H.R. 9054, "Native Americans Equal Opportunity Act," introduced 12 September 1977 by Congressman John Cunningham of Washington State (where tribes had recently won a major fishing rights battle); Marjane Ambler, "In-

dians Wrestling for Control over Their Minerals," *High Country News*, 30 December 1977.

73. Tribal Governmental Tax Status Act, 14 January 1983, P.L. 97-473, 96 Stat. 2607; Robert A. Williams, Jr., "Small Steps on the Long Road to Self-Sufficiency for Indian Nations: The Indian Tribal Governmental Tax Status Act of 1982," in *Federal Indian Law Cases and Materials*, 2d ed., ed. David H. Getches and Charles F. Wilkinson (St. Paul, Minn.: West Publishing Co., 1986), 639–646 [hereafter *Federal Indian Law II*]; Susan M. Williams, "Legislative Memorandum Re Amendments to the Indian Tribal Governmental Tax Status Act," 29 January 1988 (photocopy); Indian Gaming Act, 17 October 1988, P.L. 100-497, 102 Stat. 2467.

74. AIPRC, *Final Report*, vol. 1, 8; President Ronald Reagan, "Statement by the President: Indian Policy," 14 January 1983 press release from the White House; Tim Giago, "Notes from Indian Country," *Lakota Times*, 18 April 1984; U.S. Congress, Senate, Select Committee on Indian Affairs, "Testimony of the National Congress of American Indians before the Senate Select Committee on Indian Affairs, United States Senate on the FY 82 Indian Budgets, 23 March 1981"; Farrell, "New Indian Wars," 57–71; Hazel W. Hertzberg, "Reagonomics on the Reservation," *New Republic*, 22 November 1982, 15–18.

75. Mary Hargrove, Grant Williams, and Edward M. Eveld, "BIA Chief's Message Sparks Indian Debate," *Tulsa Tribune*, 18 November 1987; Steve Hinchman, "Tribes Seek Freedoms They Say They Helped the U.S. Achieve," *High Country News*, 26 October 1987.

76. President Ronald Reagan, "Statement by the President: Indian Policy," 24 January 1983 press release from the White House.

77. Kooros and Seip, "Self-Sufficiency Analysis"; Presidential Commission on Indian Reservation Economies, *Report and Recommendations to the President of the United States* (Washington, D.C.: GPO, November 1984); DOI, *Task Force*; Ortiz, *Economic Development*; AIPRC, *Final Report*, vol. 1.

78. Hagan, "Tribalism Rejuvenated," 10; for the extent of government dependency, see also Roxanne Dunbar Ortiz, "Sources of Underdevelopment," in Ortiz, *Economic Development* 71; DOI, *Task Force*, 9. Between 1981 and 1984 federal spending for Indians declined in real dollars by 4 percent. In 1985 it fell by $145 million as Congress tried to reduce the national deficit. AIPRC, *Final Report*, vol. 1, 305; Presidential Commission, *Recommendations to the President*, II-62.

79. Kooros and Seip, "Self-Sufficiency Analysis," 21. The study estimated income multipliers for six reservations. Although the national income multiplier in the 1970s was 2.82, the reservation income multiplier was much lower—ranging from 0.99 to 1.19. A multiplier of 1 means that an extra dollar of investment (from the government or minerals, for example) results in exactly one extra dollar of income. A multiplier of less than 1 means that an extra dollar of investment cannot generate even one extra dollar of income.

80. Presidential Commission, *Recommendations to the President*, II-62.

81. See ibid., II-23-29, for tribal obstacles. Several of these same problems were identified by Navajo economist Al Henderson in his article, "Tribal Enterprises: Will They Survive?" in Ortiz, *Economic Development*, 114–118.

82. Presidential Commission, *Recommendations to the President*, I-23-61; II-45-54; DOI, *Task Force*, 169–264.

83. DOI, *Task Force*, and Presidential Commission, *Recommendations to the President*.

84. Kooros and Seip, "Self-Sufficiency Analysis," 1–3, 12; Lorraine Turner Ruffing, "Dependence and Underdevelopment," in Ortiz, *Economic Development*, 104–105.

85. Presidential Commission, *Recommendations to the President*, II-36-37; Kooros and Seip, "Self-Sufficiency Analysis," 1.

86. Ruffing, "Dependence," 97; Roxanne Dunbar Ortiz, "Choices and Directions," in Ortiz, *Economic Development*, 151-155. Ortiz defines neocolonialism as "a special form of domination within a capitalist democracy that is both political and economic and which creates new and more devastating forms of dependency."

87. Bureau of Mines coal figures, as quoted in *Business Week*, "U.S. Indians Demand a Better Energy Deal," 19 December 1977; "DOE Figures Say CERT Exaggerates Extent of Indian Uranium Wealth," *Nuclear Fuel*, 25 June 1979: AIPRC, *Final Report*, vol. 1, 339.

88. DOI, *Task Force*, 128, said that as many as 85 percent of reservation Indians were living on reservations where natural resources development was not likely to provide any significant economic development. Its estimate was based, however, on past production instead of potential resources.

89. For example, the Navajo Tribe derived 93 percent of its tribal revenue from mineral leasing in 1958 (Ruffing, "Navajo Mineral Development," 9).

90. Ruffing, "Dependence," 107; see also Philip Reno, *Mother Earth, Father Sky, and Economic Development: Navajo Resources and Their Use* (Albuquerque: University of New Mexico Press, 1981), 151-161. Reno discusses the usually uncalculated costs of reclaiming land and restoring people displaced by strip mining to a productive role in life.

91. For example, with tribal funds the Jicarilla Apache Tribe built a service station and bought two ranches and several condominiums; the Navajo Tribe bought several ranch properties and planned to build a resort on Lake Powell; the Arapahoe Tribe built a service station and a shopping center; the Fort Peck Assiniboine and Sioux Tribes built a kidney dialysis treatment center; and the Blackfeet Tribe started a bank. Based upon interviews with tribal officials and articles in tribal newspapers.

CHAPTER TWO. THE RUBBER-STAMP ERA: EARLY HISTORY OF INDIAN MINERAL LEASING

1. Earl Old Person (Blackfeet tribal chairman), speech given at Council of Energy Resource Tribes (CERT) annual meeting, Reno, Nev., October 1984.

2. D. S. Otis, *The Dawes Act and the Allotment of Indian Lands*, ed. Francis Paul Prucha (Norman: University of Oklahoma Press, 1973), 13; Francis Paul Prucha, *The Great Father: The United States Government and the American Indians*, abridged edition (Lincoln: University of Nebraska Press, 1986), 64-77; "Treaty between the United States of America and the Eastern Band of Shoshones and the Bannack Tribe of Indians," Article 2, 3 July 1868, Fort Bridger, Utah Territory.

3. Peter Iverson, "History of the Wind River Reservation Water Rights" unpublished paper, n.d., 6.

4. Ibid., 2; Veronica E. Velarde Tiller, *The Jicarilla Apache Tribe: A History, 1846-1970* (Lincoln: University of Nebraska Press, 1983), 87, 96, 102.

5. Jack Bickel, "The Dragline Reigned as King in the Firesteel Coal Country for 30 Years," *Timber Lake and Area Historical Society Newsletter*, May 1987 (U.S. Office of Surface Mining files, Denver, Colo.); Robert W. Swenson, "Legal Aspects of Mineral Resources Exploitation," in Paul W. Gates, *History of Public Land Law Development* (Washington, D.C.: GPO, November 1968), 699-764.

6. Tiller, *Jicarilla Apache*, 110, 116; Floyd A. O'Neil and Kathryn L. MacKay, *A History of the Uintah-Ouray Ute Lands* (Salt Lake City: American West Center, n.d.), 6, 10-12; James Jefferson, Robert W. Delaney, and Gregory C. Thompson,

The Southern Utes: A Tribal History, ed. Floyd A. O'Neil (Ignacio, Colo.: Southern Ute Tribe, 1972), 16–43.

7. Jason Cuch (Ute; Northern Ute Tribal Resources Division director), telephone interview, July 1987; Council of Energy Resource Tribes (CERT), "List of CERT Member Tribes with Known and Potential Reserves," Denver, Colo., March 1986 (photocopy).

8. J. P. Kinney, *A Continent Lost—A Civilization Won: Indian Land Tenure in America* (Baltimore: Johns Hopkins Press, 1937; reprint, New York: Arno Press, 1975), 275–277.

9. Gates, *Public Land Law*, 369–371, 452–453; David H. Anderson, "Strip Mining on Reservation Lands: Protecting the Environment and the Rights of Indian Allotment Owners," *Montana Law Review* 35 (1974): 217; Crow Tribe, *Section 710 Crow Indian Lands Study Final Report*, Crow Agency, Mont., September 1979, 2–3.

10. Felix Cohen, *Handbook of Federal Indian Law* (Albuquerque: University of New Mexico, 1972; reprint of 1942 edition), 312–313 [hereafter *Cohen I*]; *United States v. Shoshone Tribe of Indians*, 304 U.S. 111 (1938); David H. Getches, Daniel M. Rosenfelt, and Charles F. Wilkinson, eds., *Federal Indian Law Cases and Materials* (St. Paul, Minn.: West Publishing Co., 1979), 539–542 [hereafter *Federal Indian Law I*].

11. Loretta Fowler, *Arapahoe Politics, 1851–1978: Symbols in Crises of Authority* (Lincoln: University of Nebraska Press, 1982), 91; *Cohen I*, 327, 331; Iverson, "Wind River Reservation."

12. Angie Debo, *And Still the Waters Run: The Betrayal of the Five Civilized Tribes* (Norman: University of Oklahoma Press, 1984), 65, 80–85, 261; H. Craig Miner, *The Corporation and the Indian: Tribal Sovereignty and Industrial Civilization in Indian Territory, 1865–1907* (Columbia: University of Missouri Press, 1976), 58–74, 144–192; Terry P. Wilson, *The Underground Reservation: Osage Oil* (Lincoln: University of Nebraska Press, 1985), 100; Maggie Fox, "An Historical, Statutory, and Regulatory Review of Oil and Gas Leasing on Indian Lands and the Proposed Changes," prepared for the Sierra Club, June 1982, 33 (photocopy); *Cohen I*, 327, 331.

13. Michael A. Massie, "The Defeat of Assimilation and the Rise of Colonialism on the Fort Belknap Reservation, 1873–1925," *American Indian Culture and Research Journal* 7: 4 (1984): 40–41.

14. O'Neil and MacKay, *History of the Uintah-Ouray Ute Lands*, 13; Frederick E. Hoxie, *A Final Promise: The Campaign to Assimilate the Indians, 1880–1920* (Lincoln: University of Nebraska Press, 1984), 168; Prucha, *Great Father*, 298.

15. Act of 28 February 1891, 26 Stat. 794, Sec. 3; Felix Cohen, *Felix Cohen's Handbook of Federal Indian Law*, 2d edition (Charlottesville, Va.: Michie/Bobbs-Merrill, 1982), 523, 533 [hereafter *Cohen II*]; Swenson, "Minerals Resources Exploitation," 724, 731, 736.

16. Act of 3 March 1909, 35 Stat. 781; *Cohen II*, 529, note 5; Solicitor Leo M. Krulitz's 7 November 1977 opinion, "The Tax Status of the Production Of Oil and Gas from Fort Peck Lands," in Thomas E. Luebben, *American Indian Natural Resources: Oil & Gas* (Vienna, Va.: Institute for the Development of Indian Law, 1980), 78–89.

17. Hoxie, *Final Promise*, 147–187.

18. Ibid., 185, 187, 284.

19. U.S. Congress, Senate, Report no. 712, "Oil and Gas Lands, Wind River Reservation," 64th Congress, 1st sess., 22 July 1916.

20. Senate Report no. 712; Iverson, "History of the Wind River Reservation,"

45-48; *Congressional Record*, 64th Congress, 1st sess., 7 August 1916, 12159; Kinney, *Continent Lost—Civilization Won*, 282.

21. Wind River Ceded Land Oil and Gas Leasing Act, 21 August 1916, 39 Stat. 519.

22. The General Mineral Lands Leasing Act covered coal, oil, natural gas, oil shale, and several nonenergy minerals, including phosphate, sodium, and potassium. It also covered all minerals on lands acquired by the federal government. Act of 25 February 1920, P.L. 146, 41 Stat. 437; Lawrence C. Kelly, *The Navajo Indians and Federal Indian Policy* (Tucson: University of Arizona Press, 1968), 43; Swenson, "Mineral Resources Exploitation," 726-745.

23. Kelly, *Navajo Indians*, 40-42; Indian Appropriation Act of 30 June 1919, 41 Stat. 3; *Cohen I*, 327-328. The law was amended in 1926 to apply also to nonmetalliferous metals other than oil and gas. The states listed in the act were Arizona, California, Idaho, Montana, Nevada, New Mexico, Oregon, Washington, and Wyoming.

24. Kelly, *Navajo Indians*, 55-59; John Collier, *Indians of the Americas* (New York: New American Library, 1947), 144-145.

25. Collier became the commissioner of Indian affairs in 1933. Kelly, *Navajo Indians*, 55 59, 80; Collier, *Indians of the Americas*, 145, 152; Swenson, "Mineral Resources Exploitation," 746-747.

26. Kelly, *Navajo Indians*, 55-59, 73-74, 80; *Cohen I*, 313.

27. The 1924 law (43 Stat. 244) did not apply to the Five Civilized Tribes or the Osage Tribe. The 1927 law (44 Stat. 1347) was signed 3 March 1927, four months before the Indian office discontinued leasing on Indian reservations because of the decline in oil prices. Kelly, *Navajo Indians*, 78-92; *Cohen II*, 534.

28. *Cohen II*, 534; Kelly, *Navajo Indians*, 73-74, 90.

29. U.S. Department of Interior (DOI), Office of Surface Mining (OSM), "Jurisdiction Study," in *Report to the U.S. Congress: Proposed Legislation Designed to Allow Indian Tribes to Elect to Regulate Surface Mining of Coal on Indian Lands* (Washington, D.C.: OSMRE, 1984), 240; Terrence J. Lamb, "Some Considerations regarding Blackfeet Indian Water and Land Use in the Early Twentieth Century," paper delivered at the Western History Association meeting, Billings, Mont., 17 October 1986, 11; Solicitor's Opinion D. 41504, 11 July 1917, Records of the Office of the Secretary of Interior, Record Group 48, National Archives, Indian Office, Blackfeet allotments; Swenson, "Mineral Resources Exploitation," 706-728; Fort Peck Allotment Act, 30 May 1908, 35 Stat. 558.

30. Swenson, "Mineral Resources Exploitation," 726-738; Debo, *And Still the Waters Run*, 35.

31. Act of 3 March 1909, 35 Stat. 844 (allowed homestead entries on lands subsequently classified as coal, with coal reserved to the United States); act of 22 June 1910, 36 Stat. 584 (allowed patents for coal lands, with reservation of coal to the United States); act of 17 July 1914, 38 Stat. 509 (allowed agricultural entries on lands with phosphate, nitrate, potash, oil, and gas, with minerals reserved to the United States); Wilson, *Underground Reservation*, 85, 97; Gates, *Public Land Law*, 504.

32. Act of 27 February 1917, 39 Stat. 944; General Land Office, Circular no. 547, 16 April 1917. The 1917 act applied to all reservations except those of the Five Civilized Tribes (Cherokee, Creek, Choctaw, Chickasaw, and Seminole) where "surplus" lands may have contained minerals (*Cohen I*, 334).

33. *Crow Tribe of Indians v. United States*, 657 Fed. Supp. 573, District Court of Montana, 1985 (re the 27 February 1917 coal act); 28 April 1917 proclamation, 40 Stat. 1660 (re Fort Peck homesteading of coal lands); DOI, "Jurisdiction Study" (re Uintah-Ouray); act of 3 August 1914, 38 Stat. 681 (re Fort Berthold homesteading of

coal lands); Iverson, "History of Wind River Reservation," 49; *Decisions relating to the Public Lands*, 48 L.D. 448, 16 February 1922, "Christ C. Prange and William C. Braasch" (re Fort Berthold); 47 L.D. 560, 564, "Frank A. Kemp" (re Ute).

34. "Joint Resolution to Authorize Allotments to Indians of the Fort Berthold Indian Reservation of Lands Valuable for Coal," 3 April 1912, 37 Stat. 631, for coal under allotments; DOI, "Jurisdiction Study," 243-246.

35. Debo, *And Still the Waters Run*, 79-81, 197-202.

36. Collier, *Indians of the Americas*, 145; Hoxie, *Final Promise*, 178-187; Louis R. Moore, "Title Examination of Indian Lands" (Denver: Rocky Mountain Mineral Law Foundation, 1976), 61-62; 1908 Indian Appropriations Act, 1 March 1907, 34 Stat. 1035, 1037, 1038.

37. Prucha, *Great Father*, 301; Frederick E. Hoxie, "Building a Future on the Past: Crow Indian Leadership in an Era of Division and Reunion," in *Indian Leadership*, ed. Walter L. Williams (Manhattan, Kans.: Sunflower University Press, 1984), 83; *Northern Cheyenne v. Hollowbreast*, 425 U.S. 649 (1976); Debo, *And Still the Waters Run*, 286-287.

38. John Sledd, director of litigation for DNA-People's Legal Services, to author, 18 May 1987; Coal Lands Acts, 3 March 1909, 35 Stat. 844, and 22 June 1910, 36 Stat. 583; Oil and Gas Lands Act, 17 July 1914, 38 Stat. 509.

39. Kelly, *Navajo Indians*, 23-35, 122-124; Sledd letter to author.

40. *Etcitty v. Watt*, Civil Action No. 83-1408C, U.S. District Court, Albuquerque, N.Mex., complaint filed 31 August 1983, 13-14; Paul Frye (attorney for the allottees), telephone interview, March 1989.

41. Rich Schilf (director of Three Affiliated Tribes Natural Resources Department at Fort Berthold), telephone interview, April 1987; Interior Field Solicitor Wallace G. Dunker to Area Director Wyman D. Babby, 30 November 1972 (Fort Berthold tribal files); another solicitor made the same ruling for the Fort Belknap Reservation (Solicitor J. Reuel Armstrong to secretary of Interior, 7 September 1955). This policy seemed to conflict with the policy for public lands expressed in the 1909 Coal Lands Act, which said that if coal was found on land that had previously been homesteaded, the coal still would be reserved to the United States.

42. For example, after the IRA-sanctioned land acquisitions the boundary for the Hill Creek extension to the Uintah-Ouray Reservation in Utah was "so convoluted—to avoid mineral lands and white owned land—as to require four pages of text to describe it" (O'Neil and MacKay, *History of the Uintah-Ouray Ute Lands*, 37).

43. Based upon interviews with tribal minerals officials and documents cited elsewhere.

44. Debo, *And Still the Waters Run*, 35, 81-85.

45. *Opinions of the Solicitor of the Department of Interior relating to Indian Affairs, 1917-1974* (Washington, D.C.: GPO, 1979), 655 (3 July 1936 memorandum from Acting Solicitor Frederic L. Kirgis to commissioner of Indian affairs), quoting an unnamed congressman identified only as "the Chairman of the House Committee reporting on the measure"; Tiller, *Jicarilla Apache*, 112-113; Gates, *Public Land Law*, 369-370; Miner, *Corporation and the Indian*, 20-76; Hoxie, *Final Promise*, 170-171; Kelly, *Navajo Indians*, 81-99.

46. Fowler, *Arapahoe Politics*, 87-91, 132-205; Hoxie, *Final Promise*, 176; *Cohen II*, 138.

47. Fowler, *Arapahoe Politics*, 180-181, 195-196.

48. Wilson, *Underground Reservation*, 75, 99, 159; Fowler (*Arapahoe Politics*, 87) said that during the same period at Wind River, lease money replaced rations as a major source of subsistence.

49. Wilson, *Underground Reservation*, 102–103.

50. Ibid., 74–98, 108–118. The importance of retaining tribal mineral rights is illustrated by the fact that by 1943 more than half of the allotted surface had passed into non-Indian hands (Wilson, *Underground Reservation*, 187).

51. Kelly, *Navajo Indians*, 50–53, 65–72, 190–194; Peter Iverson, *The Navajo Nation* (Albuquerque: University of New Mexico Press, 1981, 1983), 20.

52. Kelly, *Navajo Indians*, 65–72, 190–194.

53. Indian Reorganization Act, Secs. 16 and 18, 18 June 1934, 48 Stat. 984; *Cohen II*, 559–562.

54. Omnibus Tribal Leasing Act, 11 May 1938, P.L. 506, 52 Stat. 347; Acting Solicitor Frederic L. Kirgis to commissioner of Indian affairs, 1 July 1936, in *Opinions of the Solicitor*, 655.

55. Assistant Commissioner William Zimmerman, Jr., in memorandum to the secretary of Interior, 27 January 1938, Fort Berthold tribal files.

56. *Cohen I*, 328.

57. *Cohen II*, 534; *Montana v. Blackfeet*, 471 U.S. 759 (1985); David H. Getches and Charles F. Wilkinson, eds., *Federal Indian Law Cases and Materials*, 2d ed. (St. Paul, Minn.: West Publishing Co., 1986), 627 [hereafter *Federal Indian Law II*].

58. Marvin J. Sonosky, "Oil, Gas, and Other Minerals on Indian Reservations," *Federal Bar Journal* 20 (1960): 231–232; David Harrison, "Negotiations: A Paradigm Shift of Tribal Self-Determination," paper presented at CERT annual meeting, Denver, Colo., 1986, 2.

59. Act of 15 August 1953, 67 Stat. 592; act of 27 August 1958, 72 Stat. 935; Marvin J. Sonosky (former Shoshone tribal attorney), telephone interview, February 1988.

60. *Cohen II*, 168; act of 30 June 1954, 68 Stat. 358.

61. Loretta Fowler, *Shared Symbols, Contested Meanings: Gros Ventre Culture and History, 1778–1984* (Ithaca: Cornell University Press, 1987), 98–107; Solicitor J. Reuel Armstrong to secretary of Interior, 7 September 1955, in *Opinions of the Solicitor*, 1682–1683, 1692–1693.

62. Congress passed legislation providing for perpetual tribal rights for the Northern Cheyenne Tribe in 1968, the Crow in 1968, and the Osage in 1964, DOI, "Jurisdiction Study," 242–246; Wilson, *Underground Reservation*, 187.

63. James Cannon, "Leased and Lost," *Economic Priorities Report* 5, no. 2 (1974): 1–48; Federal Trade Commission (FTC), *Staff Report on Mineral Leasing on Indian Lands* (Washington, D.C.: GPO, October 1975).

64. The CEP study (Cannon, "Leased and Lost," 33) said that Indian land leases came up for readjustment every ten years, but in fact—as the FTC found—the DOI had completely removed the readjustment clause from Indian coal leases, and in others it had added a discretionary readjustment (FTC, *Staff Report*, 30).

65. Cannon, "Leased and Lost," 33; Council of Energy Resource Tribes (CERT), *The Control and Reclamation of Surface Mining on Indian Lands*, by Douglas Richardson, for the Office of Surface Mining (Denver: CERT, 1979), 8–5.

66. FTC, *Staff Report*, 5, 32. Some industry spokesmen argued that the acreage limitation was designed for eastern, underground mining and was inappropriate for western strip mines.

67. FTC, *Staff Report*, 24, 59, 95.

68. FTC, *Staff Report*, 20–22. A 1982 court decision confirmed the lack of competition at a tribal oil and gas sale [*Jicarilla Apache Tribe v. Andrus*, 687 F.2d 1324 (10th Cir.)]. *Federal Indian Law I* 1983 supplement, 155.

69. FTC, *Staff Report*, 89.

70. FTC, *Staff Report*, 30–31, 83–89. A 1975 study, however, found significant differences between Indian and federal uranium proceeds caused by different sales prices. [Richard Nafziger, "Indian Uranium: Profits and Perils," in *You Don't Have to Be Poor to Be Indian*, ed. Maggie Gover (Washington, D.C.: Americans for Indian Opportunity, 1975), 170].

71. General Accounting Office (GAO), *Administration of Regulations for Surface Explorations, Mining and Reclamation of Public and Indian Coal Lands* (Washington, D.C.: GAO, August 1972); FTC, *Staff Report*, 29, 90; Cannon, "Leased and Lost," 28, 29, 34.

72. FTC, *Staff Report*, 44.

73. FTC, *Staff Report*, 23, 26; Cannon, "Leased and Lost," 31, 32, 34; Consolidation Coal Company leases were never developed (see Chapter 3).

74. Cannon, "Leased and Lost," 32; based upon the 1974 price of coal, which was $15 (FTC, *Staff Report*, 72–73).

75. FTC, *Staff Report*, 26, 31.

76. U.S. Department of Interior (DOI), Bureau of Land Management (BLM), *Jackpile-Paguate Uranium Mine Reclamation Project Environmental Impact Statement* (Albuquerque, N.Mex.: BLM, October 1986), 2: A-62.

77. Richard O. Clemmer, "Effects of the Energy Economy on Pueblo Peoples," in *Native Americans and Energy Development II*, ed. Joseph G. Jorgensen (Boston: Anthropology Resource Center, 1984), 90.

78. Alvin M. Josephy, Jr., "The Murder of the Southwest," *Audubon*, July 1971, 53–67.

79. Iverson, *Navajo Nation*, 105; Lynn A. Robbins, "Energy Developments and the Navajo Nation," in *Native Americans and Energy Development I*, ed. Joseph G. Jorgenson (Boston: Anthropology Resource Center, 1978), 43; Josephy, "Murder of the Southwest," 64; Christopher McLeod, "Navajos Resist Relocation," *High Country News*, 12 May 1986.

80. Richard O. Clemmer, "Black Mesa and the Hopi," in Jorgensen, *Native Americans and Energy Development I*, 17–34; *Federal Indian Law I*, 84–85; Robert T. Coulter, *Report to the Hopi Kikmongwis* (Washington, D.C.: Indian Law Resource Center, 1979), 148–155.

81. GAO, "Administration of Public and Indian Coal Lands"; the case (*Susenkewa v. Kleppe*) is discussed in *Federal Indian Law I*, 84–85. The traditionalists were never able to reach the merits in court because the tribe was not listed as a defendant, and it was considered an indispensable party. The case illustrates the "catch 22" of sovereign immunity, which had blocked other such suits. Hopi traditionalists had sued the tribe previously to protest oil leasing and had been told that the tribe had sovereign immunity and thus could not be sued.

82. FTC, *Staff Report*, 98–102.

CHAPTER THREE. EARLY HORSE TRADING: TRIBES BEGIN SETTING THE TERMS

1. Peter MacDonald (Navajo tribal chairman), "The Navajo Ten Year Plan for Economic Development, 1979," in Philip Reno, *Mother Earth, Father Sky, and Economic Development: Navajo Resources and Their Use* (Albuquerque: University of New Mexico Press, 1981), 143.

2. James Cannon, "Leased and Lost," *Economic Priorities Report 5*, no. 2 (1974): 1–48; Federal Trade Commission (FTC) *Staff Report on Mineral Leasing on Indian Lands* (Washington, D.C.: GPO, October 1975).

3. K. Ross Toole, *The Rape of the Great Plains: Northwest America, Cattle and Coal* (Boston: Atlantic Monthly Press Book, Little, Brown and Company, 1976), 56–66; James P. Boggs, "The Northern Cheyenne Coal Sales, 1966–1973," 175–183, and Steven H. Chestnut, "Coal Development on the Northern Cheyenne Reservation," 159–167, both in *Energy Resource Development: Implications for Women and Minorities in the Intermountain West*, prepared for the U.S. Commission on Civil Rights (Washington, D.C.: GPO, 1978).

4. Toole, *Rape of the Great Plains*, 57–64; Chestnut, "Northern Cheyenne Reservation," 160.

5. Peter Galuszka, "1911–1986: Coal's Rise and Fall and Rise," *Coal Age*, June 1986, 47–55; Alvin M. Josephy, Jr., "The Agony of the Northern Plains," *Audubon*, July 1973, 71; John A. Folk-Williams and James S. Cannon, *Water for the Energy Market* (Santa Fe, N.Mex.: Western Network, 1983), 4; Cannon, "Leased and Lost," 1, 14.

6. Marjane Ambler, "Cheyenne Fight, Again, for Land," *High Country News*, 11 October 1974; Chestnut, "Northern Cheyenne Reservation," 163.

7. Cannon, "Leased and Lost," 35; Chestnut, "Northern Cheyenne Reservation," 162–163; Ambler, "Cheyenne Fight"; Toole, *Rape of the Great Plains*, 51, 52.

8. Consolidation Coal Company to Northern Cheyenne Tribe, 7 July 1972, letter in author's files.

9. Boggs, "Coal Sales," 175; Richard O. Clemmer, "Black Mesa and the Hopi," in *Native Americans and Energy Development I*, ed. Joseph G. Jorgensen (Boston: Anthropology Resource Center, 1978), 31; Josephy, "Agony of the Northern Plains," 94–98; Brent Ashabranner, *Morning Star, Black Sun: The Northern Cheyenne Indians and America's Energy Crisis* (New York: Dodd, Mead & Company, 1982), 93.

10. Toole, *Rape of the Great Plains*, 56–66.

11. Cannon, "Leased and Lost," 32, 36–48.

12. Anson Baker, interview in New Town, N.Dak., April 1977; Jim Remsen, "Fort Berthold Says 'No' to Coal Development," *United Tribes News*, 25 September 1975; Cannon, "Leased and Lost," 33; Terry O'Connor (Peabody Coal spokesman), telephone interview, January 1988; Chris Baker (Southern Ute tribal chairman), interview in Denver, Colo., November 1987.

13. FTC, *Staff Report*, 5; Cannon, "Leased and Lost," 34; Toole, *Rape of the Great Plains*, 56–61.

14. FTC, *Staff Report*, 30, 83–89; Josephy, "Agony of the Northern Plains."

15. General Accounting Office (GAO), *Indian Natural Resources, Opportunities for Improving Management: Part II, Coal, Oil and Gas*, prepared for the Senate Committee on Interior and Insular Affairs (Washington, D.C.: GAO, 31 March 1976), 13.

16. Based upon interviews with members of these tribes in the early 1970s and upon reports in tribal newspapers; Marjane Ambler, "Visit with Navajos Convinces N. Cheyenne to Oppose Mining," *High Country News*, 22 October 1976.

17. Cannon, "Leased and Lost," 3, 23.

18. Cannon, "Leased and Lost," 12–13; Alvin M. Josephy, Jr., "The Murder of the Southwest," *Audubon*, July 1971, 54; Josephy, "Agony of the Northern Plains"; U.S. Department of Interior (DOI), *North Central Power Study* (Billings, Mont.: Bureau of Reclamation, October 1971).

19. GAO, *Indian Coal, Oil and Gas*, 1.

20. Energy Security Act of 1979, 30 June 1980, P.L. 96–294, 94 Stat. 611; John D. Smillie, "Whatever Happened to the Energy Crisis?" *The Plains Truth*, April

1986, 4. The synfuels predictions were contained in *Energy Activity in the West*, prepared by Robert L. McMahan and Porter B. Bennett of Abt/West for the Western Governors Policy Office, in cooperation with Four Corners Regional Commission (Denver: Western Governors Policy Office, March 1981). That report also predicted for seven western states 129 new coal mines, 28 coal-fired power plants, and 7 new uranium mills.

21. GAO, *Indian Coal, Oil and Gas*, 40-41.

22. See Josephy, "Agony of the Northern Plains" and "Murder of the Southwest," for discussions of how little the general public was involved in or aware of development decisions in the Northern Plains and Southwest.

23. Chestnut, "Northern Cheyenne Reservation," 172.

24. Northern Cheyenne coal leases act, 9 October 1980, P.L. 96-401, 94 Stat. 1701; Marjane Ambler, "Northern Cheyenne Lease Questions," *Coal Week*, 22 February 1982; Steven Chestnut (tribal attorney), telephone interview, December 1987.

25. Cannon, "Leased and Lost," 35; James P. Boggs, "The Challenge of Reservation Resource Development: A Northern Cheyenne Instance," in *Native Americans and Energy Development II*, ed. Joseph G. Jorgensen (Boston: Anthropology Resource Center, 1984), 221-223; Marjane Ambler, "Northern Cheyenne Coal Referendum Not Required," *CERT Report*, 2 March 1982, 23.

26. William H. White (Westmoreland attorney), speech presented at the University of Colorado Natural Resources Law Center conference, Boulder, Colo., 9 June 1988; Daniel H. Israel, "The Reemergence of Tribal Nationalism and Its Impact on Reservation Resource Development," *Colorado Law Review* 47 (Summer 1976): 645-646.

27. Terry O'Connor (Peabody attorney), telephone interview, February 1988; *Crow Tribe v. Frizzell*, U.S. District Court, District of Columbia, filed 19 September 1975 (later became *Crow Tribe v. Andrus*, Civil Action No. 76-10-BLG, U.S. District Court, District of Montana, Billings).

28. Dede Feldman, "Regs Would Give Tribes More Control over Mining," *High Country News*, 20 April 1979.

29. Remsen, "Fort Berthold Says 'No' "; GAO, *Indian Coal, Oil and Gas*, 9.

30. George Vlassis (Navajo tribal attorney), telephone interview, May 1980; Navajo Tribal Council Resolution, CAP-34-80, 29 April 1980, "Establishment and Adoption of the Navajo Nation Energy Policy"; Marjane Ambler, "Oil on the Reservation," *Western Energy*, March 1982, 41-44; Marjane Ambler, "Indians Wrestling for Control over Their Minerals," *High Country News*, 30 December 1977.

31. GAO, *Indian Coal, Oil and Gas*, 7.

32. Based upon interviews with tribal officials and on FTC, *Staff Report*, 10, 20, 59, 75, 95.

33. FTC, *Staff Report*, 22; Reno, *Mother Earth, Father Sky*, 145-146.

34. Reno, *Mother Earth, Father Sky*, 145-146.

35. Interior's coal figures varied from 7 to 30 percent of the nation's reserves and from 36 billion to 200 billion tons, depending upon the agency. GAO, *Indian Coal, Oil and Gas*, 2; "U.S. Indians Demand a Better Energy Deal," *Business Week*, 19 December 1977; FTC, *Staff Report*, 9; "DOE Figures Say CERT Exaggerates Extent of Indian Uranium Wealth," *Nuclear Fuel*, 25 June 1979.

36. Marjane Ambler, "Tribes Probe Possibilities of Their Coal, Uranium," *High Country News*, 29 July 1977.

37. Peter Iverson, *The Navajo Nation* (Albuquerque: University of New Mexico Press, 1981, 1983), 160-161; 185-188; Reno, *Mother Earth, Father Sky*, 126; *Manygoats v. Kleppe*, 558 F.2d 556 (10th Cir. 1977); David H. Getches, Daniel M. Rosenfelt, and Charles F. Wilkinson, eds., *Federal Indian Law Cases and Materials*

(St. Paul, Minn.: West Publishing Co., 1979), 585 [hereafter *Federal Indian Law I*]; Lynn A. Robbins, "Energy Developments and the Navajo Nation," 45-46, and Joseph G. Jorgensen, "Energy, Agriculture, and Social Science in the American West," 13, both Jorgensen, *Native Americans and Energy Development I*; *Peshlakai v. Schlesinger*, Civil Action No. 78-2416, U.S. District Court, District of Columbia, filed 22 December 1978.

38. Robbins, "Energy Developments and the Navajo," 44-45; Winona LaDuke, "CERT—An Outsider's View In," paper prepared for Women of All Red Nations, Rapid City, S.Dak., July 1980 (photocopy); Iverson, *Navajo Nation*, 161-163; Robert Schryver (former Navajo director of mineral development), telephone interview, January 1988.

39. "Navajos Occupy Strip Mine," *Americans before Columbus* (Albuquerque, N.Mex.: National Indian Youth Council), 8-4.

40. Iverson, *Navajo Nation*, 183-184; Robert Rudzik (WESCO project manager and vice president of Pacific Lighting), telephone interview, 15 December 1981; "Development Briefs," *CERT Report*, April 1979; also based upon 1988 telephone interviews with three former tribal employees and consultants: Tom Glenn, Eric Natwig, and Robert Schryver.

41. Schryver interview; Daniel S. Whipple, "OSM Pressured to Okay Mine on Navajo Land," *High Country News*, 10 August 1979.

42. Lorraine Turner Ruffing, "Dependence and Underdevelopment," in *Economic Development in American Indian Reservations*, ed. Roxanne Dunbar Ortiz (Albuquerque: University of New Mexico Native American Studies Development Series, 1979), 93, 101; Harris Arthur, "Preface," in Jorgensen, *Native Americans and Energy Development I*, 2: Jennie Whitt, "40 Burnham Families Face Eviction," *Navajo Times*, 20 October 1982; Robbins, "Energy Developments and the Navajo," 36; Lorraine Turner Ruffing, "Navajo Mineral Development," *American Indian Journal*, September 1978, 9.

43. Robbins, "Energy Developments and the Navajo," 44-46; Lynn A. Robbins, "Energy Developments and the Navajo, an Update," in Jorgensen, *Native Americans and Energy Development II*, 127; Iverson, *Navajo Nation*, 98, 185; Marjane Ambler, "Navajo Nation Power Plant Postponed," *Lakota Times*, 28 June 1988.

44. "Suit to Test Reclamation Requirement," *High Country News*, 10 March 1978.

45. Cannon, "Leased and Lost," 9. For discussions of corporate connections among the companies involved on Indian lands, see Joseph G. Jorgensen, "The Political Economy of the Native American Energy Business," 34, and Winona LaDuke, "The Council of Energy Resource Tribes," 60-64, both in Jorgensen, *Native Americans and Energy Development II*.

46. "Unlocking the Indians' Riches," *Business Week*, 24 March 1980.

47. FTC, *Staff Report*, 30.

48. "Domenici Heartbroken over MacDonald, CERT," *Navajo Times*, 16 August 1979; Kent Ware (Kiowa), interview in Denver, Colo., October 1980; Maggie Gover, ed., *You Don't Have to Be Poor to Be Indian* (Washington, D.C.: Americans for Indian Opportunity, n.d.), 165.

49. Based upon articles in the *Navajo Times* from 1981 to 1988 regarding contracts with Peabody, Utah International, and Pittsburg Midway and upon 1988 interviews with Robert Schryver and with tribal attorney Don Wharton.

50. Marjane Ambler, "In the Indian Camp, a New York Consultant," *High Country News*, 30 December 1977; "Charles Lipton: Indian Mineral Leases," *American Indian Journal*, May 1976, 9-10; Charles Lipton, "The Pros and Cons of Petroleum Agreements," *American Indian Journal*, February 1980.

51. Tom Luebben, "Mining Agreements with Indian Tribes," *American Indian Journal*, May 1976; Stephen Zorn, "Getting a Fair Deal in Mining Projects," Americans for Indian Opportunity Red Paper, August 1976 (photocopy), in Gover, *You Don't Have to Be Poor*, 53–63; Richard Nafziger, "Transnational Energy Corporations and American Indian Development," University of New Mexico Native American Studies Center, Albuquerque, 1979 (photocopy).

52. See Appendix 3 for a complete description of Lipton's eighteen points.

53. Based upon interviews with industry representatives and upon Dolores Proubasta, "Turning Energy into Power on Indian Lands," *Geophysics: The Leading Edge of Exploration*, December 1983, 32–39; Daniel H. Israel, "New Opportunities for Energy Development on Indian Reservations," *Mining Engineering*, June 1980; Russ Rountree, "Energy Company/Tribal Communication Lacking," *Western Oil Reporter*, August 1980, 78; Sam Maddox, "Can 17th Street Make Peace with the Energy Tribes?" *Denver Magazine*, May 1980, 32–37.

54. Reno, *Mother Earth, Father Sky*, 129.

55. Marjane Ambler, "Energy Companies Seek Peace and Resources with Indians," *High Country News*, 12 December 1980; Marjane Ambler, "CERT, Indians Cooperate on Energy," *Denver Post*, 28 November 1982.

56. Telephone interview in May 1980 with Ernest Stevens, Navajo economic development director; Ambler, "Energy Companies Seek Peace."

57. "Zia, a Child, Weans Away from Mother," *Pueblo News* (Albuquerque, N.Mex.: All Indian Pueblo Council), June 1980; James Cook, "New Hope on the Reservations," *Forbes*, 9 November 1981, 114.

58. Interviews with Claude Neely (formerly regional land manager for Amoco Production Company), in Denver, Colo., November 1987; Bob Aitkin and Bill Manning (both formerly of the Southern Ute staff), in Durango, Colo., October 1987; Marvin Cook (director of the Southern Ute Natural Resources Division), in Ignacio, Colo., October 1987; Chris Baker (chairman of the Southern Ute Tribe), by telephone, October 1987.

59. Interviews with Neely, Baker, and Cook.

60. Robert L. Rudzik (project director for the Crow Synfuels Project and vice president of Pacific Lighting Corporation), telephone interview, December 1981; telephone interviews, October 1980, with J. W. Bragg (spokesman for Exxon) and Marcus Wiley (Burnham mine superintendent for Consolidation Coal).

61. Reno, *Mother Earth, Father Sky*, 108, 150; Robbins, "Energy Developments and the Navajo," 44–45; Whipple, "OSM Pressured"; Iverson, *Navajo Nation*, 186.

62. Ambler, "Energy Companies Seek Peace."

63. Lynn A. Robbins, "Doing Business with Indian Tribes," in Jorgensen, *Native Americans and Energy Development II*, 52–57; Geoffrey O'Gara, "Canny CERT Gets Respect, Money, Problems," *High Country News*, 14 December 1979.

64. Kirk Blackard, "Developments in the Negotiation of Indian Resource Agreements," speech presented at Federal Bar Association's Indian Law Conference, 17 April 1980, Phoenix, Ariz.

65. William L. Bryan, Jr., *Montana's Indians Yesterday and Today* (Helena: Montana Magazine, 1985), 91; Patrick Dawson, "Reach Coal Accord, U.S. Warns Crows," *Billings Gazette,* 29 October 1977; Crow Tribe, "Section 710 Crow Indian Lands Study Final Report," September 1979, for CERT and DOI, author's files.

66. Bill Richards, "Crow Indians Find Huge Coal Reserves Help Them Very Little," *Wall Street Journal*, 31 January 1984; Rudzik interview.

67. Marjane Ambler, "Study Finds Crow Coal Gas Project To Be Uneconomi-

cal," *Coal Week*, 8 November 1982; Fluor Engineers and Constructors, "Crow Syn-fuels Feasibility Study," for DOE, August 1982, author's files.

68. Marjane Ambler, "Crow Tribe in Montana Packages 1,000 MW Plant Deal," *Electrical Week*, 23 August 1982; Council of Energy Resource Tribes, (CERT), *Crow Power Plant Feasibility Study*, for DOE, (Denver: CERT, 1982).

69. "Bylaws of the Crow Development Corporation," no date, author's files; author's notes from 9 April 1983 Crow Tribal Council meeting, Crow Agency, Mont.; interviews with tribal members.

70. Richards, "Crow Indians Find Coal"; Ellis Knows Gun (former director of the Crow Coal Reclamation Office and a tribal employee for thirteen years), telephone interview, January 1988.

71. John Irvin (Shell Oil Company spokesman), telephone interview, January 1986.

72. Based in part upon an August 1986 interview with David Harrison, an Osage tribal member who worked for BIA as special assistant to the commissioner and then as the director of trust responsibility until 1980. At the time of the interview he worked for CERT.

73. The Blackfeet Tribe had some oil and gas leases signed in the 1940s that provided a royalty interest as well as a net profits interest, but these were rare exceptions [Jeanne Whiteing (Blackfeet tribal attorney), telephone interview, June 1988].

74. Details of these and similar contracts are discussed at length in Chapter 9.

75. *Opinions of the Solicitor of the Department of Interior relating to Indian Affairs, 1917–1974* (Washington, D.C.: GPO, 1979), M-36007, p. 1514, 7 July 1949, for Northern Cheyenne Tribe; M-36040, p. 1529, 5 July 1950, for Laguna Pueblo (oil and gas). These two opinions by Solicitor Mastin G. White agreed that competitive sales were necessary and disagreed with an earlier, 1941 opinion by Acting Solicitor Felix S. Cohen that said competitive sales were not necessary under the Blackfeet tribal charter (*Opinions of the Solicitor*, p. 1055). In 1955 the Interior Department turned down a proposal for the Navajo Tribe to join in a partnership with an oil company for oil and gas development, according to Iverson, *Navajo Nation*, 76. The 1938 act did not apply to several reservations, including the Osage, where negotiations were allowed without competition (Omnibus Indian Mineral Leasing Act of 1938, 11 May 1938, P.L. 506, 52 Stat. 347).

76. *Federal Register* 42, no. 65 (5 April 1977): 18083–18088; *Federal Register* 45, no. 156 (11 August 1980): 53164–53181.

77. FTC, *Staff Report*, 19–20; Ruffing, "Navajo Mineral Development," 7.

78. David Baldwin, "Oklahoma Osage Top Oil Production, Systems Management," *CERT Report*, 13 September 1982, 9; Curtis Canard, "A Petroleum Geologist Branches Out in a Burgeoning Field," *American Indian Journal*, February 1980, 21–24; Ambler, "Oil on the Reservation"; Arthur Hill, "CERT Blasts DOI, BIA on Funding Commitments," *Western Oil Reporter*, October 1980.

79. Marjane Ambler, "CERT Stresses Production, Not Rebellion," *High Country News*, 19 September 1980, 6–7.

80. Telephone interviews with Ken Fredericks and Dick Wilson (both of the BIA Central Office) and Clyde O. Martz and Dave Jones (both of the Interior solicitor's office), September 1980.

81. Louis R. Moore, "Mineral Development on Indian Lands—Cooperation and Conflict," *Rocky Mountain Mineral Law Institute*, 1983, 65; Ambler, "Energy Companies Seek Peace"; Barrie Damson (founder and president of Damson Oil Company), telephone interview, November 1981.

82. Based upon interviews with tribal and BIA officials and with Charles Williams (officer of Tricentrol and later of Wintershall Corporation). Wintershall, a

U.S. subsidiary of a West German oil and gas firm, acquired Tricentrol in 1984. Jim Hendon, "Indians, Oil Firm OK $7.7 Million Deal," *Rocky Mountain News*, 19 November 1982.

83. Harrison interview.

84. Martz interview; Richard B. Collins, "Memo to Council of Energy Resource Tribes re the Secretary of the Interior's Authority to Approve Joint Venture Mineral Development Agreements between Tribes and Mining Companies," 13 October 1981, National Indian Law Library, Boulder, Colo.

85. "Alternative Agreements Bill Passes Both Houses, Awaits Final Actions," *CERT Report*, 13 September 1982; Indian Mineral Development Act, 22 December 1982, P.L. 97-382, 96 Stat. 1938.

86. "GRI Projects 1% Annual Growth Rate in U.S. Energy Use through 2010," *Inside Energy*, 29 August 1988; John D. Smillie, "Whatever Happened to the Energy Crisis?" *The Plains Truth* (Billings, Mont.: Northern Plains Resource Council), April 1986, 4.

87. "Conferees Spread $525 Million in Clean Coal Funding over 3 Years," *Inside Energy*, 15 August 1988; Peter Carrels, "Syngas Plant Survives the '80s," *High Country News*, 13 March 1989, 1.

88. The six states were Colorado, Montana, New Mexico, North Dakota, Utah, and Wyoming (Folk-Williams and Cannon, *Water for Energy*, 6–10). The Indian figures include all minerals, not just energy minerals, and reflect price increases as well as increases in production [U.S. Department of Interior (DOI), *American Indians,* by Vince Lovett and Larry Rummel (Washington, D.C.: BIA, 1984), 37].

89. Williams interview; "Utilities Continue to Seek Approaches for Keeping Long Term Coal Supply Costs Down," *Coal Week*, 18 July 1988; Council of Energy Resource Tribes (CERT), "Proceedings of the Energy Forum—Strategic Planning for Marketing Indian Gas," Denver, Colo., April 1986, section 12.

90. Peter G. Chronis, "Indian OPEC Spurs Tribal Development," *Denver Post*, 15 December 1986.

CHAPTER FOUR. INDIAN OPEC?
THE COUNCIL OF ENERGY RESOURCE TRIBES

1. A. David Lester, then commissioner of the Administration for Native Americans, speech at CERT annual meeting in Washington, D.C., 4 September 1980.

2. LaDonna Harris, "Congratulations to CERT," *Red Alert* 13 (1978).

3. Native American Natural Resources Development Federation (NANRDF) of the Northern Great Plains, "Declaration of Indian Rights to the Natural Resources in the Northern Great Plains States," prepared by member tribes, in conjunction with Native American Rights Fund, Bureau of Indian Affairs, and private consultants, Denver, Colo., June 1974 (photocopy). Document can be obtained from the National Indian Law Library, Boulder, Colo. (hereafter NILL).

4. Dale Vigil (vice president of Jicarilla Apache Tribe), interview in Dulce, N.Mex., October 1987; Council of Energy Resource Tribes (CERT), "CERT Board of Directors," Denver, Colo., 1976, CERT files (photocopy).

5. CERT, "Recommendations Submitted to FEA regarding Energy Resource Development on Indian Reservations," Washington, D.C., 18 September 1975, photocopy, NILL.

6. Ibid.

7. LaDonna Harris, interview in Denver, Colo., November 1986.

8. Harris interview. The Interior Department said Indian reservations held 3

percent of the nation's known reserves of oil and gas and 30 percent of the coal west of the Mississippi. Interior's estimates for Indian coal ranged from 36 billion tons in 1968 to 200 billion tons in 1975 [U.S. Department of Interior (DOI), *Indian Natural Resources—Part II: Coal, Oil, and Gas. Better Management Can Improve Development and Increase Indian Income and Employment*, prepared by the comptroller general of the United States for the Senate Committee on Interior and Insular Affairs (Washington, D.C.: GPO, 1976), 2; "U.S. Indians Demand a Better Energy Deal," *Business Week*, 19 December 1977]. The uranium figures also varied, from 16 percent in 1975 (recoverable at $8 a pound) to 37 percent (total potential resources) [Federal Trade Commission (FTC), *Staff Report on Mineral Leasing on Indian Lands* (Washington, D.C.: GPO, October 1975)].

9. "DOE Figures Say CERT Exaggerates Extent of Indian Uranium Wealth," *Nuclear Fuel*, 25 June 1979; Harris interview.

10. Allen Rowland, telephone interview, December 1980; Tom Fredericks, telephone interview, February 1987.

11. Based upon the author's observations of CERT annual meetings and discussions with tribal leaders.

12. Bill Strabala, "Zarb Meets 23 Indian Tribal Leaders in Resources Powwow," *Denver Post*, 26 September 1976; Marjane Ambler, "Tribes Probe Possibilities of Their Coal, Uranium," *High Country News*, 7 July 1977, p. 1; FTC, *Staff Report*, 9.

13. Ken Fredericks (BIA national director of realty), telephone interview, December 1980.

14. Bill Strabala, "Indian Tribes Seek to Form OPEC-Style Energy Cartel," *Denver Post*, 10 July 1977.

15. Ibid.

16. Ahmed Kooros, telephone interview, January 1987; Roger Cohn, "Oil Tips: Teheran to Tepees," *Philadelphia Inquirer*, 17 September 1980; Ann Crittenden, "Fuel-Rich U.S. Tribes Call on Iranian," *New York Times*, 7 August 1979.

17. William Endicott, "Indians Seek Help from OPEC," *Los Angeles Times*, 16 October 1977; Cohn, "Oil Tips: Teheran to Tepees."

18. "Indians in OPEC?" *Denver Post*, 13 August 1979.

19. Associated Press (AP), "Reagan: Indians Humored," *Casper Star-Tribune*, 1 June 1988.

20. DOI, *Report of the Task Force on Indian Economic Development* (Washington, D.C.: GPO, July 1986), 46, 126–128.

21. "Indian Energy Message," *Wotanin Wowapi*, 27 August 1979; John D. Smillie, "Whatever Happened to the Energy Crisis?" *The Plains Truth* (Billings, Mont.: Northern Plains Resource Council), April 1986.

22. Peter MacDonald to President Carter, 20 July 1979, author's files; "Indian Energy Message," *Wotanin Wowapi*.

23. Richard Stone (assistant to Energy Secretary Charles Duncan), telephone interview, December 1980.

24. Geoffrey O'Gara, "Canny CERT Gets Respect, Money, Problems," *High Country News*, 14 December 1979, 1.

25. O'Gara, "Canny CERT."

26. Winona LaDuke (Chippewa-Ojibway), "CERT, an Outsider's View In," *Akwesasne Notes*, Summer 1980, 20–22; Ken Peres and Fran Swan, "The New Indian Elite: Bureaucratic Entrepreneurs," *Akwesasne Notes*, Late Spring 1980, 18; Mike Meyers, "Ahmed Kooros: A Discussion," *Akwesasne Notes*, Late Spring 1980.

27. Marjane Ambler, "CERT, Indians Cooperate on Energy," *Denver Post*, 28

November 1982; Ed Gabriel (former executive director of CERT), interview in Denver, Colo., November 1982; O'Gara, "Canny CERT."

28. Marjane Ambler, "Uncertainty in CERT," in *Native Americans and Energy Development II*, ed. Joseph G. Jorgensen (Boston: Anthropology Resource Center, 1984), 71–78.

29. Ibid.

30. "Letters" column, *New Age*, October 1980, 5–7.

31. CERT, "Fiscal Year 1980 Activities Report," CERT files, 22 (photocopy).

32. Stone interview; Bruce Hamilton, "Carter Energy Mobilization Demobilized," *High Country News*, 10 August 1979; Daniel H. Israel, "New Opportunities for Energy Development on Indian Reservations," *Mining Engineering*, June 1980.

33. The project was abandoned in 1982, but in 1984 when BIA awarded the Crows a grant to update it, the Crow Tribe named Ed Gabriel (then a private consultant) as the project manager. The project is discussed at more length in Chapter 3.

34. Stone interview.

35. CERT, *The Control and Reclamation of Surface Mining on Indian Lands*, by Douglas Richardson, for the Office of Surface Mining (Denver: CERT, 29 September 1979).

36. Ambler, "Uncertainty in CERT"; CERT, "Prefeasibility Analysis of Coal Development on the Southern Ute Reservation: Six Options," Denver, Colo., June 1981, 75, 135. Bob Aitken, interview in Durango, Colo., October 1987; Bill Manning, interview in Ignacio, Colo., November 1980.

37. Ambler, "Uncertainty in CERT."

38. Carol Connor, telephone interview, January 1981.

39. Wilfred Scott, telephone interview, November 1980; Chris Baker (Southern Ute tribal chairman), telephone interview, November 1986.

40. The comparative figures had to be viewed with caution, however, because the government changed the parameters of its statistics frequently. Governmental figures did not include some of the largest producers, such as the Osages and, in later years, the Jicarilla Apaches. [DOI, Minerals Management Service (MMS), "Royalties: A Report on Federal and Indian Mineral Revenues for 1981" (Washington, D.C.: GPO, 1982), 37–41; same report for 1982]; FTC, *Staff Report*, 43. Wilfred Scott, speech at CERT annual meeting, Washington, D.C., 5 September 1980; David Harrison, "Negotiations: A Paradigm Shift of Tribal Self-Determination," paper presented at CERT annual meeting, Denver, Colo., 18 November 1986, 4.

41. Ambler, "Uncertainty in CERT"; interviews with representatives of CERT member tribes.

42. U.S. Congress, Senate, Select Committee on Indian Affairs, *Hearing on the Nomination of Thomas W. Fredericks to be an Assistant Secretary of the Interior*, 96th Cong., 2d sess. (Washington, D.C.: GPO, 1980); Ambler, "Uncertainty in CERT"; Tom Fredericks, telephone interview, January 1987; Ed Gabriel, telephone interview, January 1981.

43. Interviews with ex-staff members Paul Epley, Barbara Nagel, Bill Roberts, Bill Nagle, and Mahmood Rana in Denver, Colo., November 1980; Touche Ross & Co., Washington, D.C., 11 January 1980 letter to CERT board of directors following examination, author's files; CERT budgets, 1979–1981, author's files.

44. "Joint Budget Committee Review of the BIA Budget for Fiscal '81," 42.

45. Bobby Moore (GAO investigator), telephone interview, November 1981; James DeFrancis, DOE's deputy assistant secretary for external affairs, to GAO, 2 July 1981; Kenneth L. Payton, DOI's acting deputy assistant secretary for Indian affairs, to GAO, 24 June 1981; Touche & Ross Co. to CERT board of directors, 11 January 1980. All correspondence is located in author's files.

46. Peter MacDonald, "Starship CERT," speech delivered at the CERT annual meeting in Denver, Colo., 18 November 1982.

47. Bill Donovan, "Council Takes Matters into Own Hands," *Navajo Times*, 2 March 1989, p. 1.

48. Ambler, "Uncertainty in CERT."

49. A. David Lester, interview in Spokane, Wash., October 1983; John Aloysius Farrell, "CERT Plagued by Financial Problems," *Denver Post*, 9 October 1983.

50. Farrell, "CERT Plagued by Financial Problems."

51. Senator Peter Domenici, "Tribal Independence under a Reagan Administration," speech at CERT annual meeting in Washington, D.C., 5 September 1980.

52. Lester interview.

53. Wilfred Scott, speech at CERT annual board meeting in Spokane, Wash., 27 October 1983; Lester interview.

54. CERT, *First Decade* (Denver: CERT, November 1986); interviews with CERT and other tribal organizations' staffs.

55. Marjane Ambler, "Tribes Refining Goals for Resources," *Western Business* (Billings, Mont.), February 1985, 23, 28, 29; Mark Trahant, "CERT Wants to Change Its Image," *Navajo Times*, 1 June 1983; Robert Wilson (Southern Cheyenne allottee), interview in Concho, Oklahoma, April 1985.

56. CERT membership list, April 1988, CERT files; see also Appendix 2.

57. A. David Lester, speech presented at CERT annual meeting in Washington, D.C., September 1980, and telephone interview, November 1983; Ambler, "Tribes Refining Goals for Resources"; Peter G. Chronis, "Indian OPEC Spurs Tribal Development," *Denver Post*, 15 December 1986.

58. Lester speech.

59. "Council to Convene July 19," *Navajo Times*, 13 July 1983.

60. "CERT Membership Considered," *Navajo Times*, 25 May 1983; CERT, *First Decade* (Denver: CERT, November 1986).

61. Peterson Zah, speech given at CERT annual meeting in Spokane, Wash., 27 October 1983.

62. Ross Swimmer, speech given at CERT annual meeting in Spokane, Wash., 27 October 1983; Ross Swimmer, speech given at CERT annual meeting in Billings, Mont., 24 September 1985.

63. Marjane Ambler, "Swimmer Slaps and Praises CERT Simultaneously," *Lakota Times*, 3 December 1986; "Indian Leaders Step Up Efforts to Remove BIA Head," *NCAI Sentinel*, 15 June 1987.

64. Ambler, "Tribes Refining Goals for Resources."

65. Marjane Ambler, "Despite Controversy, CERT Hangs Tough," *Lakota Times*, 18 September 1985, 1.

66. Norman Hollow, speech given at CERT annual meeting in Billings, Mont., 23 September 1985.

67. Mark Tatge, "Montana Bank First to Be Owned by Indians," *Denver Post*, 13 August 1987.

68. Martha Wetmore (Chemeheuvi Tribe's vice chairman), speech given at CERT annual meeting in Billings, Mont., September 1985; telephone interviews with CERT staff members Glen Lane and David Lester and with CERT consultant Eric Natwig.

69. Marjane Ambler, "Nuke Waste Sites Border Indian Lands," *Navajo Times*, 8 October 1984; Environmental Protection Agency (EPA), *EPA Policy for the Administration of Environmental Programs on Indian Reservations* (Washington, D.C.: EPA, 8 November 1984).

70. Marie Monsen (local and Indian affairs chief for DOE), telephone inter-

view, December 1986; Tanna Chattin (CERT communications coordinator) to author 7 February 1987.

71. Monsen interview, December 1986; EPA, *Inventory of Hazardous Waste Generators and Sites on Selected Indian Reservations*, by Warner Reeser (Denver: CERT, July 1985).

72. Monsen interview.

73. CERT, *First Decade*, 12; CERT Board Chairman Judy Knight, interview in Denver, Colo., November 1986.

CHAPTER FIVE. WHO'S MINDING THE STORE? INDIAN ROYALTY MANAGEMENT

1. Richard LaCourse and John Butler, "45 Indian Tribes in a Dozen States Form Heartland of Indian Oil Production," *CERT Report*, 13 September 1982.

2. Ibid.

3. The allottees owned oil and gas individually as a result of receiving land allotments, usually by authority of the General Allotment Act of 1887. The heirs of the original allottees are also referred to as allottees.

4. The Minerals Management Service (MMS) figures included income to both tribes and individual Indian mineral owners. None of the MMS annual reports includes the Osage royalties, which represent 26 percent of the total 1982 royalties cited in this text. The annual reports also exclude royalties that are received "in kind," meaning in oil rather than cash, which in some years were quite substantial. U.S. Department of Interior (DOI), Minerals Management Service (MMS), *Mineral Revenues: The 1982 Report on Receipts from Federal and Indian Leases* (Washington, D.C.: GPO, 1983), 38; Peter G. Chronis, "Indian OPEC Spurs Tribal Development," *Denver Post*, 15 December 1986; Osage income figures taken from Grant Williams, Mary Hargrove, and Edward M. Eveld, "Indian Destinies Shaped by Oil," *Tulsa Tribune*, 17 November 1987.

5. Jicarilla Apache Tribal President Leonard Atole and Jicarilla Apache BIA Agency Superintendent Sherryl Vigil, interviews in Dulce, N.Mex., October 1987; Chronis, "Indian OPEC"; William L. Bryan, Jr., *Montana's Indians Yesterday and Today* (Helena: *Montana Magazine*, 1985), 63.

6. Sherry Keene-Osborn and Jeff Rundles, "Wyoming Oil Thefts before Grand Jury," *Rocky Mountain Journal*, 29 October 1980; U.S. Congress, Senate, Select Committee on Indian Affairs, *Hearing on Federal Supervision of Oil and Gas Leases on Indian Lands* (Washington, D.C.: GPO, 1981), part 1 (Billings, Mont., 27 February 1981), 5–11; and part 2 (Washington, D.C., 6 April 1981), 316–317.

7. Keene-Osborn and Rundles, "Wyoming Oil Thefts"; also based upon articles in *Riverton Ranger* and *Casper Star-Tribune* during 1980 and 1981 and conversations with tribal members.

8. Senate Select Committee, *Oil and Gas Leases on Indian Lands*, part 1, 70–75; "FBI's Energy Fraud Unit Investigating Oil Thefts in Wyoming and Colorado," *CERT Report*, 21 October 1981; DOI, *Report of the Commission on Fiscal Accountability of the Nation's Energy Resources*, by David F. Linowes et al. (Washington, D.C.: DOI, January 1982), 27, 30 [hereafter Linowes Commission Report]; "Industry Groups Protest Tentative Royalty Reforms," *Platt's Oilgram News*, 10 December 1981; Richard LaCourse, "Oil Companies Support Status Quo on USGS Royalty Accounting," *CERT Report*, 20 November 1981.

9. General Accounting Office (GAO), *Indian Natural Resources, Opportunities for Improving Management: Part II, Coal, Oil and Gas*, prepared for the Senate

Committee on Interior and Insular Affairs (Washington, D.C.: GAO, 31 March 1976), 29–30, 35; GAO, *Oil and Gas Royalty Collections—Serious Financial Management Problems Need Congressional Attention* (Washington, D.C.: GAO, April 1979).

10. Linowes Commission Report, 4–5.

11. Mike Quinn, "Senate Oil Report Calls USGS Derelict," *Riverton Ranger*, 27 August 1981; Senate Select Committee, *Oil and Gas Leases on Indian Lands*, part I, 12, 72.

12. Senate Select Committee, *Oil and Gas Leases on Indian Lands*, part 1, 60–70.

13. Senate Select Committee, *Oil and Gas Leases on Indian Lands*, part 3, 144; Linowes Commission Report, 117.

14. Richard LaCourse, "Oil Tribes Voice Fears over Theft, Underreporting of Resource Royalties," *CERT Report*, 7 October 1981; Senate Select Committee, *Oil and Gas Leases on Indian Lands*, Part 1, 31, 100; part 3, 128–129, 149–162; Peter Iverson, *The Navajo Nation* (Albuquerque: University of New Mexico Press, 1981, 1983), 79; Susan Williams and David C. Cole, "Resource, Revenue, and Rights Reclamation: The Tax Program for the Navajo Nation," 1 August 1977, 5, author's files (photocopy).

15. Linowes Commission Report, 121; David L. Baldwin, "Oklahoma Osage Top Oil Production, Systems Management," *CERT Report*, 13 September 1982, 9; Williams, Hargrove, and Eveld, "Indian Destinies."

16. Senate Select Committee, *Oil and Gas Leases on Indian Lands*, part 3, 265; "BIA Will Launch Oil Lease Compliance Training for Tribes," *CERT Report*, 7 October 1981.   •

17. Howard Kohn, "The Lawman Who Corralled the Oil Rustlers," *Reader's Digest*, August 1984, 43–48.

18. Marjane Ambler, "Who Controls the Oil?" *Nations, the Native American Magazine*, 20–23; Richard LaCourse, "Interstate Oil Theft Network in Existence, Witnesses Tell Commission," *CERT Report*, 21 Sept 1981.

19. William R. Ritz, "Security Slipshod at Blackfeet Oil Fields," *Denver Post*, 31 May 1981.

20. Ambler, "Who Controls the Oil?"; Richard LaCourse, "BIA Will Launch Oil Lease Compliance Training," *CERT Report*, 7 October 1981; Garrett Big Leggins, "Tribes Began Oil and Gas Monitoring," *Wotanin Wowapi*, 12 June 1981.

21. Based upon conversations with tribal members; a series of articles by Dan Neal in the *Riverton Ranger* during the spring of 1981; and "Indians' Oil Theft Probe Deserves Wide Support," *Casper Star-Tribune*, 29 May 1981, editorial page.

22. *Riverton Ranger* and *Casper Star-Tribune* articles.

23. Keith Haugland, "Fired Blackfeet Minerals Investigator Vows to Expose Misconduct on Council," *Great Falls Tribune*, 6 December 1981; Dan Neal, "Blackfeet May Ax Thomas," *Casper Star-Tribune*, 6 December 1981; interviews with members of the Blackfeet tribal Natural Resources Department staff in Browning, Mont., September 1981.

24. Kathy Sawyer, "A Modern Brand of Wild West Outlaw: The Big-Time Oil Rustler," *Washington Post*, 11 January 1981; Ambler, "Who Controls the Oil?"; Richard LaCourse, "Amoco, Sohio Seek Show Cause Hearing on Wind River Cancellations," *CERT Report*, 9 July 1982; Jim Hendon, "Two Oil Firms May Lose Leases on Indian Lands," *Rocky Mountain News*, 7 January 1982. Amoco operated the two leases for itself and Sohio, which held a 13.5 percent interest in the production.

25. Linowes Commission Report, 38. BIA had canceled Osage leases for noncompliance on grounds of sloppy operations and violating lease terms. LaCourse,

"Oil Tribes Voice Fears." As discussed in Chapter 3 the secretary of Interior refused to cancel the Northern Cheyenne and the Crow Tribes coal permits and leases in the early 1970s, which the tribes wanted voided because the terms violated federal regulations. Although the secretary recognized problems in the leases, he left those cancellation decisions up to Congress and the courts.

26. Penelope Purdy, "Amoco Responds to Tribes' Hidden Pipeline Charges," *Casper Star-Tribune*, 19 February 1982; James P. Morris (USGS) to Saul Goodman (Shoshone Tribe's attorney), 9 November 1981, Shoshone Tribe's files; Ruth B. Johnson (Amoco Production Company attorney) to author, 22 December 1987.

27. "Allegation Company Mismanaging Reservation Fields Surprises Amoco," *Casper Star-Tribune*, 17 September 1981; LaCourse, "Amoco, Sohio Seek Hearing."

28. *Amoco v. Department of Interior*, C82-0398-B1, U.S. District Court, Cheyenne, Wyo., filed 4 October 1982 (Amoco exhibits L and M).

29. Linowes Commission Report, 107; National Congress of American Indians (NCAI) resolution, 12 February 1986, NCAI files (photocopy); Tom Acevedo (Arapahoe tribal attorney), telephone interview, August 1986; Ross Swimmer (assistant secretary for Interior), interview in Denver, Colo., November 1986.

30. Ruth Johnson (attorney for Amoco), telephone interview, February 1988; Orville St. Clair (Shoshone business councilman), telephone interview, January 1988.

31. Linowes Commission Report, 34-36.

32. Linowes Commission Report, 31; Senate Select Committee, *Oil and Gas Leases on Indian Lands*, part 2, 315.

33. "FBI's Energy Fraud Unit Investigating Oil Thefts in Wyoming and Colorado," *CERT Report*, 21 October 1981; Senate Select Committee, *Oil and Gas Leases on Indian Lands*, part 1, 9; Linowes Commission Report, 32-33.

34. Linowes Commission Report, 180.

35. Senate Select Committee, *Oil and Gas Leases on Indian Lands*, part 1, 70-71; Richard LaCourse, "Interstate Oil Theft Network in Existence, Witnesses Tell Commission," *CERT Report*, 21 September 1981; Linowes Commission Report, 29.

36. Linowes Commission Report, 25-29, 85; LaCourse, "45 Tribes Form Heartland."

37. "Oil Theft Investigator Sees Mismanagement," *Platt's Oilgram News*, 18 September 1981; "Interior Department Inspector Downplays Royalty Oil Thefts," *Platt's Oilgram News*, 31 August 1981.

38. Linowes Commission Report, 11, 13; Dan Neal, "Tribes Try for Tighter Oil Leases," *Riverton Ranger*, 16 April 1981.

39. U.S. Congress, House, Committee on Government Operations, *Hearing on Problems Associated with the Department of Interior's Distribution of Oil and Gas Royalty Payments to Indians*, 99th Cong., 1st sess. (Washington, D.C.: GPO: 1985), 48; Linowes Commission Report, 15.

40. Baldwin, "Oklahoma Osage"; GAO, *Coal, Oil and Gas*, 32; Linowes Commission Report, 33, 37, 106.

41. Linowes Commission Report, 14.

42. Ibid., 143-144.

43. Ibid., 62-65, 168.

44. Federal Oil and Gas Royalty Management Act (FOGRMA), 28 December 1982, P.L. 97-451, 96 Stat. 2447.

45. U.S. Congress, House, Committee on Interior and Insular Affairs, with the assistance of the General Accounting Office (GAO), *Federal Minerals Royalty Man-*

agement, 98th Cong., 2d sess., December 1984 (Washington, D.C.: GPO, 1985), 86–116.

46. William Bettenberg (director of Minerals Management Service), interview in Denver, Colo., November 1986; Marjane Ambler, "Minerals Management Service Puts Royalty Management Problems in Perspective," *Lakota Times*, 24 December 1986; House Committee on Interior, *Federal Minerals Royalty Management*, 19–20; U.S. Department of Interior (DOI), *Office of Inspector General Audit Report: Review of Bureau of Land Management's Inspection and Enforcement Program* (Washington, D.C.: DOI, June 1986).

47. House Committee on Interior, *Federal Mineral Royalty Management*, 18, 29.

48. U.S. Congress, House, Committee on Government Operations, *Hearing on Problems Associated with the Department of Interior's Distribution of Oil and Gas Royalty Payments to Indians*, 99th Cong., 1st sess. (Washington, D.C.: GPO, 8 April 1985), 100–149; General Accounting Office (GAO), *Indian Royalties: Interior Has Not Solved Indian Oil and Gas Royalty Payment Problems* (Washington, D.C.: GAO, March 1986), 35–37; interviews with tribal royalty personnel in 1987 and 1988.

49. FOGRMA, P.L. 97–451; "Congress Approves Royalty Management Bill without Royalty Increase," *Federal Lands*, 27 December 1982.

50. U.S. Department of Interior (DOI), *Implementation Status Report on the Management Action Plan for the Royalty Management Program*, prepared by the Minerals Management Service for Congress (Washington, D.C.: DOI, October 1987), 18–19; Associated Press (AP), "Probe Reveals Company Stole Oil from Indians," *Casper Star-Tribune*, 10 May 1989.

51. Joe Chesser (program analyst with BLM's fluid minerals division in Washington, D.C.), telephone interview, February 1988; Grant Williams, Mary Hargrove, and Edward M. Eveld, "Billion-Dollar Headache," *Tulsa Tribune*, 3 November 1987; U.S. Department of Interior (DOI), Office of Inspector General, *Biennial Report on the Royalty Management System for the Period Oct. 1, 1985, through Sept. 30, 1987* (Washington, D.C.: DOI, April 1988), i.

52. Linowes Commission Report, 210–224; "Interior Tabs Royalty Group to Oversee Department Procedures, Practices," *Inside Energy*, 23 December 1985; Thurman Velarde (Jicarilla Apache oil and gas administrator and advisory committee member), interview in Dulce, N.Mex., October 1987; Vern Ingraham (MMS chief of office of external affairs), interview in Lakewood, Colo., August 1986.

53. Based upon MMS press releases and interviews with tribal personnel and with Floyd Gonzales of MMS in Lakewood, Colo., August 1986.

54. Senate Select Committee, *Oil and Gas Leases on Indian Lands*, parts 1 and 3.

55. Linowes Commission Report, 117–118. A spot survey by CERT in 1982 found twenty-five tribes with energy and natural resources departments (LaCourse, "45 Tribes Form Heartland"). The Blackfeet, Northern Ute, and Southern Ute Tribes created their natural resources departments in 1980.

56. U.S. Department of Interior (DOI), *Report of the Task Force on Indian Economic Development* (Washington, D.C.: GPO, July 1986), 104.

57. Rich Schilf, Fort Berthold natural resources office, speech at CERT meeting in Denver, Colo., 17 November 1986. The Oklahoma allotted lands listed in the BIA task force chart as three reservations were the Kiowa, Comanche, and Apache (98 percent allotted); the Wichita-Caddo-Delaware (96 percent allotted); and the Five Civilized Tribes (93 to 100 percent allotted). The allotment percentages are from Gil-

bert L. Hall, *The Federal-Indian Trust Relationship* (Vienna, Va: Institute for the Development of Indian Law, 1979), 116–118.

58. Williams, Hargrove, and Eveld, "Indian Destinies"; Bryan, *Montana's Indians*, 88; Loretta Fowler, *Arapahoe Politics, 1851–1978: Symbols in Crises of Authority* (Lincoln: University of Nebraska Press, 1982), 201–209.

59. Marjane Ambler, "Royalty Training for Tribal Staff," *Lakota Times*, 31 December 1986; telephone interviews with Northern Ute Terry Cuch, November 1986; Navajo Perry Shirley, August 1986; and MMS IPA (Interagency Personnel Act) director Todd McCutcheon, November 1986.

60. Interviews with Floyd Gonzales (MMS staff member), Perry Shirley (Navajo auditor general), and Richard Ortiz (director of the Wind River Tax Commission), June 1988; Thurman Velarde (Jicarilla Apache oil and gas administrator), "Testimony before the U.S. Senate Energy and Natural Resources Committee, 12 July 1988," *Jicarilla Chieftain*, 15 August 1988.

61. Vern Ingraham (MMS chief of office of external affairs), telephone interview, February 1988; "Southern Utes and Federal Officials Discuss Oil and Gas Production from Tribal Lands," *Southern Ute Drum*, 4 March 1988.

62. Bill Donovan, "Tribe Gets Authority to Audit Collection of Mineral Royalties," *Navajo Times Today*, 10 June 1985; Todd McCutcheon (MMS state-tribal office), telephone interview, January 1988.

63. U.S. Department of Interior (DOI), *Office of Inspector General Audit Report: Review of Bureau of Land Management's Inspection and Enforcement Program* (Washington, D.C.: DOI, 1986), 23–25. In March 1988 BLM still had no cooperative agreements with tribes. Interviews with BLM officials in February 1988, including Dick Forester, Gil Lockwood, and Frank Salwerowicz; interviews with tribal staff on the Southern Ute, Ute Mountain Ute, Uintah and Ouray, and Blackfeet reservations.

64. William Bettenberg, MMS director, report to CERT in Denver, Colo., 19 November 1986 (unpublished); Bill Trujillo (MMS ombudsman), telephone interview, July 1988.

65. Ambler, "Who Controls the Oil?"; Wes Pettingil (director of Northern Ute Energy and Minerals Resource Department), telephone interview, February 1988; Gary Comes at Night (Blackfeet Revenue Department inspector), telephone interview, March 1988; Kee Ike Yazzie (director of the Navajo Tax Commission), telephone interview, May 1988.

66. CERT staff report to board on 1986 accomplishments (photocopy) and 1985 workshop in Billings; interviews with tribal staffs in 1987 and 1988, including Darwin Whiteman (Ute Mountain Ute), October 1987; George Adams (Northern Ute royalty auditor), February 1988; Wes Pettingil (director of Northern Ute Energy and Minerals Resource Department), February 1988; Geraldine Oscar (Blackfeet Revenue Department director) and Gary Comes at Night (Blackfeet Revenue Department inspector), March 1988; and Richard Ortiz (director of the Wind River Tax Commission), June 1988.

67. CERT workshops in Billings, Mont., 24 September 1985, and in Denver, Colo., 16 November 1986; MMS and tribal interviews.

68. Richard TeCube, Jicarilla Apache vice president, speech at June 1984 CERT meeting, Denver, Colo.; Senate Select Committee, *Oil and Gas Leases on Indian Lands*, part 3, 144; Perry Baker, "Office of the Superintendent Report," *Ute Bulletin*, 30 March 1988.

69. David Wong, interview in Dulce, N.Mex., October 1987; Bill Donovan, "Tribe Gets Authority to Audit Collection of Mineral Royalties," *Navajo Times Today*, 10 June 1985; McCutcheon, telephone interview; Marvin Cook, Southern Ute

Natural Resources Division director, speech at a CERT workshop in 1985, Denver, Colo.

70. Following the Jicarilla Apaches and the Navajos, in descending order, the Wind River Shoshone and Arapahoe Tribes had 236 producing tribal wells, the Blackfeet had 390, and the Northern Utes had 236, according to testimony presented to Senator Melcher in 1981 (Senate Select Committee, *Oil and Gas Leases on Indian Lands*, parts 1 and 3). Neither the Southern Ute Tribe nor the Ute Mountain Ute Tribe testified at that time. In 1985 the Southern Ute energy division director, Marvin Cook, said at a CERT meeting that his tribe had 1,700 producing tribal wells. In 1987 the Ute Mountain Utes had 94 producing tribal wells. Gordon Hammond (Ute Mountain Ute staff), interview in Towaoc, Colo., October 1987.

71. Production accounting was then being used only for offshore leases and solid minerals onshore [U.S. Department of Interior (DOI), Minerals Management Service (MMS), *Management Action Plan for the Royalty Management Program,* (Washington, D.C.: DOI, April 1985), 25].

72. U.S. Department of Interior (DOI), *Implementation Status Report on the Management Action Plan for the Royalty Management Program*, prepared by the Minerals Management Service for Congress (Washington, D.C.: DOI, October 1987), 20.

73. DOI, *Inspector General Audit Report: BLM's Inspection and Enforcement Program*, 3. The report scoring BLM's drainage protection program was dated 13 November 1986.

74. Sec. 101 (b) (1) of FOGRMA; telephone interviews with Joe Chesser, Gil Lockwood, and George Brown of BLM, February 1988. The inspection statistics do not include inspections by anyone other than its own employees, and BLM does not have such information [Dave Allison (acting chief of BIA Energy and Minerals Division), interview in Boulder, Colo., June 1988].

75. Bill Haltom, Jicarilla Apache tribal attorney, to author 13 July 1988; Patricia L. Brown, "Federal Valuation of Indian Oil and Gas for Royalty Purposes," presentation at the Rocky Mountain Mineral Law Institute in Denver, Colo., 17 March 1988.

76. Of eleven major natural gas royalty audits conducted by the inspector general from 1977 through 1981, ten showed underpayments, nine of which were due to undervaluation (Linowes Commission Report, 64–65).

77. David Harrison (CERT staff), interview in Denver, Colo., November 1986; "Oil Product Valuation Revision Opposed in Indian Country," *NCAI Sentinel*, 15 June 1987; Matt Winters, "Griles Says U.S. Energy Policy One of Minimum Interference," *Casper Star-Tribune*, 18 May 1988.

78. *Jicarilla Apache Tribe v. Supron Energy Co.*, 728 F.2d 1567 (1975) (Judge Seymour dissenting). The other two lawsuits (filed in federal district court in Albuquerque, N.Mex.) were *Jicarilla Apache v. El Paso Natural Gas* (1974) and *Jicarilla Apache Tribe v. Continental Oil Co.* (1976).

79. DOI, Inspector General, *Biennial Report on the Royalty Management System*, ii; Chuck Cook, Mike Masterson, and Mark N. Trahant, "Fraud in Indian Country," *Arizona Republic*, 4–11 October 1987 (reprint), 5; Pat Paquette, "Royalty Rule Will Stand, despite Protests," *Inside Energy*, 27 February 1989.

80. Terry P. Wilson, *The Underground Reservation: Osage Oil* (Lincoln: University of Nebraska Press, 1985), 100-105; Fowler, *Arapahoe Politics*, 194–196; Federal Trade Commission (FTC), *Staff Report on Mineral Leasing on Indian Lands* (Washington, D.C.: GPO, October 1975), 31–32, 44; "Federal Trust Responsibility at Issue in Oil and Gas Cases," *NARF Legal Review*, Summer 1988; interviews with tribal officials.

81. Marjane Ambler, "Downturn in Oil Market, BIA Develops Stripper Well Policy," *Lakota Times*, 15 October 1986; Ross Swimmer, assistant secretary for Indian affairs, memo to area directors, "Stripper and Marginal Well Guidelines for Suspension of Production Requirements," 4 November 1986, author's files; Dave Allison (Wind River Agency BIA superintendent), interview in Boulder, Colo., June 1988.

82. Carey Vicenti, "Trial Ends on Jicarilla Apache Lease Challenge, *CERT Report*, 1 September 1981; "Settlement Implemented with Benson-Montin-Greer," *Southern Ute Drum*, 26 June 1987. Albert R. Greer to author, 3 May 1988.

83. Charles D. Williams (senior vice president of exploration, Wintershall Corporation), interviews in Denver, Colo., November 1987 and February 1988; David Harrison, "Considerations for Determining Whether or Not to Grant Permission to Suspend Operations on Stripper Oil Wells," memorandum from CERT for member tribes, Summer 1986, author's files (photocopy); Larry Wetsit (Fort Peck minerals director), telephone interview, January 1987.

CHAPTER SIX. THE FORGOTTEN PEOPLE: INDIAN ALLOTTEES

1. U.S. Congress, Senate, Select Committee on Indian Affairs, *Hearings on S. 1894* [Indian Mineral Development Act] *to Permit Indian Tribes to Enter into Certain Agreements for the Disposition of Tribal Mineral Resources*, 97th Cong., 2d sess. (Washington, D.C.: GAO, 1982), 56–59 [hereafter *Hearings on S. 1894 (IMDA)*].

2. The term "allottees" refers to both those who originally received the allotments and their heirs [U.S. Department of Interior (DOI), *Annual Report of Indian Land and Income from Surface and Subsurface Leases* (Washington, D.C.: DOI, 30 September 1984)].

3. U.S. Congress, House, Committee on Government Operations, *Hearing on Problems Associated with the Department of Interior's Distribution of Oil and Gas Royalty Payments to Indians*, 99th Cong., 1st sess. (Washington, D.C.: GPO, 8 April 1985), 74; Marjane Ambler, "Oil Rich and Penny Poor," *Denver Post Empire Magazine*, 13 October 1985; Don Aubertin (BIA Energy and Minerals Division staff), telephone interview, March 1988.

4. DOI, Bureau of Indian Affairs (BIA), *American Indians: U.S. Indian Policy, Tribes and Reservations, BIA—Past and Present Economic Development* (Washington, D.C.: BIA, 1984), 37; Carol Connor (attorney), telephone interview, December 1979. Connor said close to 40 percent of 1979 CERT members' lands, excluding Navajo lands, were allotted.

5. David H. Getches, Daniel M. Rosenfelt, and Charles F. Wilkinson, eds., *Federal Indian Law Cases and Materials* (St. Paul, Minn.: West Publishing Co., 1979), 74 [hereafter *Federal Indian Law I*].

6. Ethel J. Williams, "Too Little Land, Too Many Heirs—The Indian Heirship Land Problem," *Washington Law Review* 46 (1971): 718; South Dakota figures from *Hodel v. Irving*, 481 U.S. 704 (1987).

7. Williams, "Too Little Land," 724–741.

8. Dorothy Vail (BIA Billings, Mont., area realty specialist), telephone interview, April 1987; Land Consolidation Act, 12 January 1983, 96 Stat. 2515; *Hodel v. Irving*.

9. Marjane Ambler, "The Three Affiliated Tribes at Fort Berthold—Mandan, Hidatsa, Arikara—Seek to Control Their Energy Resources," in *Native Americans*

and *Energy Development II*, ed. Joseph G. Jorgensen (Boston: Anthropology Resource Center, 1984), 194-199.

10. General Accounting Office (GAO), *Farmers Home Administration Information on Agricultural Credit Provided to Indians on 14 Reservations* (Gaithersburg, Md.: GAO, March 1987), 19-23; John Fredericks III, "Financing Indian Agriculture: Mortgaged Indian Lands and the Federal Trust Responsibility," *American Indian Law Review* (1989), 14: 1; Vail interview; Patrick Dawson, "The U.S. Has Spent a Century Chiseling Away Crow Land," *High Country News*, 6 June 1988.

11. U.S. Congress, Senate, Select Committee on Indian Affairs, *Federal Supervision of Oil and Gas Leases on Indian Lands*, part 1, 97th Cong., 1st sess. (Washington, D.C.: GPO, 27 February 1981); David Allison (former realty officer at Fort Peck and acting chief of energy and minerals for BIA), interview in Boulder, Colo., June 1988; "Mineral Rights Sales Halted," *Wotanin Wowapi*, 13 August 1982; Williams, "Too Little Land."

12. Jerome H. Simonds, "The Acquisition of Rights to Prospect for and Mine Coal from Tribal and Allotted Indian Lands," *Rocky Mountain Mineral Law Institute* 21 (1975): 143-146.

13. Terry P. Wilson, *The Underground Reservation: Osage Oil* (Lincoln: University of Nebraska Press, 1985), 110-120.

14. The Coal Lands Act of 1910, 36 Stat. 583, provided full authority to mine coal but required payment of damages. U.S. Department of Interior (DOI), Office of Surface Mining (OSM), "Jurisdiction Study," in *Report to the U.S. Congress: Proposed Legislation Designed to Allow Indian Tribes to Elect to Regulate Surface Mining of Coal on Indian Lands* (Washington, D.C.: OSMRE, 1984), 230, 237; David Baldwin, "Oklahoma Osage Top Oil Production, Systems Management," *CERT Report*, 13 September 1982, 9.

15. David H. Anderson, "Strip Mining on Reservation Lands: Protecting the Environment and the Rights of Indian Allotment Owners," *Montana Law Review* 35 (1974): 217-218; Jim Remsen, "Fort Berthold Says No To Coal Development," *United Tribes News*, 25 September 1975.

16. The tribal actions in the coal conflicts are discussed in Chapter 3; Northern Cheyenne Allotment Act, 3 June 1926, 44 Stat. 690; act of July 24, 1968, P.L. 90-424, 82 Stat. 424; *Northern Cheyenne Tribe v. Hollowbreast*, 425 U.S. 649 (1976); *Adams v. Osage*, 59 F.2d 653 (10th Cir., 1932), cert. denied, 287 U.S. 652 (1932); *Federal Indian Law I*, 563-565; Frank H. Alverez and J. Kevin Poorman, "Real Property: Congressional Control of Allotted Mineral Interests," *American Indian Law Review* 3 (1975): 165.

17. Based upon conversations with tribal members at Lame Deer, Mont., in 1981.

18. Act of 17 May 1968, P.L. 90-308, 82 Stat. 123.

19. Surface Mining Control and Reclamation Act, P.L. 95-97, 30 USC 1201, Sec. 710 and Sec. 714; *Congressional Record*, 95th Cong., 1st sess., S8043-44.

20. U.S. Department of Interior (DOI), Bureau of Land Management (BLM), *Uranium Development in the San Juan Basin Region, Final Report* (Albuquerque, N.Mex.: BLM, Fall 1980), 11-1; Lawrence C. Kelly, *The Navajo Indians and Federal Indian Policy* (Tucson: University of Arizona Press, 1968), 23-35.

21. *Etcitty v. Watt*, Civil Action No. 83-1408C, U.S. District Court, Albuquerque, N.Mex., filed 31 August 1983. As of March 1989 the lawsuit still had not been heard [John Sledd (DNA director of litigation), telephone interview, April 1986; Paul Frye (attorney for Etcitty), telephone interview, March 1989].

22. Kelly, *Navajo Indians*, 23-35; Navajo Tribal Chairman Peter MacDonald to BLM State Director Charles Luscher, 2 November 1981, author's files; Bureau of

Land Management (BLM), Albuquerque, N.Mex., District Office, "Know Your Rights over Federal Coal" (mimeographed poster, 1981); Ed Plummer (Crownpoint BIA superintendent), interview in Denver, Colo., May 1982.

23. BLM required compensation for "authorized occupants," defined as those living on individual allotments, tribal trust lands, and tribal fee lands. BLM suggested that coal companies also reach agreement on compensation for certain other squatters. U.S. Department of Interior (DOI), Bureau of Land Management (BLM), *Farmington Resource Management Plan, Final Environmental Impact Statement* (Farmington, N.Mex.: BLM, September 1987), 1-8, 1-9, 1-10; DOI, *Record of Decision, Final San Juan River Regional Coal Environmental Impact Statement* (Farmington, N.Mex.: BLM, October 1987), 24–25.

24. Ellis Knows Gun (Crow reclamation office), telephone interview, January 1988; DOI, *Proposed Legislation to Regulate Surface Mining*, 26. The proposed Indian strip mining legislation is discussed in Chapter 7.

25. *Peshlakai v. Schlesinger*, Civil Action No. 78-2416, U.S. District Court, District of Columbia, filed 22 December 1978 (later became *Peshlakai v. Duncan*); Hansley Hadley, "Between Sacred Mountains," *Navajo Times*, 7 December 1983.

26. Tom Barry, "Navajo Legal Services and Friends of the Earth Sue Six Federal Agencies over Alleged Careless Uranium Mining Policies," *American Indian Journal*, February 1979, 5–7; *Peshlakai v. Schlesinger*, 28.

27. *Peshlakai v. Duncan*, 476 F.Supp. 1247, 1261 (1978); Alan Taradash, telephone interview, April 1986; DOI, *Final San Juan River Impact Statement*, 21.

28. Wayne Littlewhiteman (Northern Cheyenne elder), interview in Lame Deer, Mont., February 1981; Marjane Ambler, "Modern Law v. Ancient Indian Tradition," *Denver Post*, 1 November 1981.

29. *Gilbert Red Neck v. Northern Cheyenne Tribe*, Tribal Court Civil No. 81-117, Northern Cheyenne Appellate Court, Lame Deer, Mont. (4 September 1981); Ambler, "Modern Law"; Allen Rowland (Northern Cheyenne tribal president), interview in Lame Deer, Mont., February 1981.

30. Senate Select Committee, *Hearings on S. 1894 (IMDA)*, 86–103; Associated Press (AP), "ARCO Exploration on Reservation Ends," *Char-Koosta*, 21 August 1984; specific terms of the Northern Cheyenne–ARCO contract are discussed in Chapter 9.

31. Lawrin "Hugh" Baker (founder and landman for Baker Services), interview in Denver, Colo., November 1987; "Company Withdraws Bid for Exploration," *Great Falls Tribune*, 2 October 1981.

32. Native American Natural Resources Development Federation to Assistant Interior Secretary Forrest Gerard, 26 April 1978, author's files; Carol A. Dentz Connor, "Oil and Gas Leasing on Allotted Indian Lands," unpublished report prepared for the American Indian Law Center, Albuquerque, N.Mex., 24 April 1978; Federal Trade Commission (FTC), *Staff Report on Mineral Leasing on Indian Lands* (Washington, D.C.: GPO, October 1975), 23, 88, 95, 96, 102.

33. John Aloysius Farrell, "The New Indian Wars," *Denver Post* (special reprint of articles appearing 20–27 November 1983), 28; U.S. Congress, Senate, Select Committee on Indian Affairs, *Hearing on Federal Supervision of Oil and Gas Leases on Indian Lands*, part 3, 97th Cong., 1st sess. (Washington, D.C.: GPO, 1 June 1981), 333–343; Richard LaCourse, "Individual Indian Owners of Oil and Gas Face Uncaring Bureaucracy," *CERT Report*, 13 September 1982, 14.

34. Connor, "Oil and Gas Leasing on Allotted Indian Lands," 28–29.

35. Senate Select Committee, *Hearings on S. 1894 (IMDA)*, 56–59; U.S. Senate, Select Committee on Indian Affairs, *Hearing on Federal Supervision of Oil and Gas*

Leases on Indian Lands, part 1, 97th Cong., 1st sess. (Washington, D.C.: GPO, 27 February 1981), 26–29.

36. Minerals Management Service Director William Bettenberg to Robert J. Nordhaus, 11 October, 1984, author's files; Marjane Ambler, "Oil Rich and Penny Poor," *Denver Post Empire Magazine*, 13 October 1985, 12–27.

37. *Frank P. Poafpybitty v. Skelly Oil Co.*, 390 U.S. 365 (1968).

38. Senate Select Committee, *Hearing on Federal Supervision*, part 3, 333–343; FTC, *Staff Report*.

39. Senate Select Committee, *Hearings on S. 1894 (IMDA)*, 56–59; Senate Select Committee, *Hearing on Federal Supervision*, part 1, 18–69.

40. Senate Select Committee, *Hearing on S. 1894 (IMDA)*, 146–151.

41. The *Sanguine Ltd. v. Department of Interior* case is discussed in Senate Select Committee, *Hearing on S. 1894 (IMDA)*, 146.

42. Wilson, *Underground Reservation*, 121–147; Senate Select Committee, *Hearing on S. 1894 (IMDA)*, 146–148.

43. "Notice of Intent to Develop Guidelines for Approval of Oil and Gas Leases on Allotted Land" (dated 27 May 1982), *CERT Report*, 23 July 1982; *Federal Register*, 12 July 1983, 31984 and 31991; *Federal Register*, 24 August 1987, 31935.

44. Paul E. Frye to Jennie Boylan, 11 May 1982, in Senate Select Committee, *Hearing on S. 1894 (IMDA)*, 184; Edward M. Eveld, Mary Hargrove, and Grant Williams, "Dubious Legal Theory Sparks Judge's Ire," *Tulsa Tribune*, 10 November 1987.

45. Ed Plummer, telephone interview, April 1986; Farrell, "New Indian Wars," 28; Chuck Cook, Mike Masterson, and M. N. Trahant, "Uranium, Coal Cases Reveal U.S. Duplicity," *Arizona Republic*, 5 October 1987; David Allison (acting chief of BIA Energy and Minerals division), interview in Boulder, Colo., June 1988.

46. Ambler, "Oil Rich and Penny Poor"; interviews with allottees in Oklahoma, April 1985.

47. House Committee on Government Operations, *Hearing on Distribution of Royalty Payments*, 454.

48. LaCourse, "Uncaring Bureaucracy," 13; Louis R. Moore, "Mineral Development on Indian Lands—Cooperation and Conflict," *Rocky Mountain Mineral Law Institute* (Denver: Rocky Mountain Mineral Law Institute, 1983), 66–67; Senate Select Committee, *Hearing on S. 1894 (IMDA)*, 148.

49. Senate Select Committee, *Hearing on S. 1894 (IMDA)*, 126–142.

50. Acting Deputy Assistant Interior Secretary Kenneth L. Payton to Billings area director, "Waiver of 25 CFR 172.4," 17 August 1981, in Senate Select Committee, *Hearing on S. 1894 (IMDA)*, 140–141; ibid., 126, 140.

51. John J. McNamara to ARCO Exploration Company, 18 September 1981, and Gidley testimony, Senate Select Committee, *Hearing on S. 1894 (IMDA)*, 134–135; "Battle of Unique Waiver Grows," *Wotanin Wowapi*, 30 April 1982; "BIA Superintendent Takes Strong Stand," *Wotanin Wowapi*, 26 November 1982.

52. Senate Select Committee, *Hearing on S. 1894 (IMDA)*, 128, 136–138.

53. Mike Fairbanks (BIA Blackfeet Agency superintendent), interview in Browning, Mont., September 1981; interviews with several tribes' staffs.

54. FTC, *Staff Report*, 96; interviews with BIA personnel; "Allottees Oppose Tribal Interference," *Navajo Times*, 15 September 1988.

55. G. G. Kipp, interview in Browning, Mont., September 1981; Robert Wilson (Southern Cheyenne), interview in Concho, Okla., April 1985.

56. Based upon interviews with various tribal staff members and especially with Ron Chohamin (Northern Ute minerals staff member), in Denver, Colo., October 1981.

57. Marvin J. Sonosky (Fort Peck tribal attorney), telephone interview, April 1987.

58. Interviews in Browning, Mont., September 1981, with BIA Agency Superintendent Mike Fairbanks and G. G. Kipp, John Murray, and Gary Comes at Night of the tribal minerals staff; Tesoro landman John Brock, telephone interview, September 1980; Tesoro Senior Vice President Monte N. Swetnam to the Blackfeet Tribe, 25 June 1980, author's files.

59. Blackfeet Tribal Business Councilman Leland Ground to Senator John Melcher, 15 September 1980, author's files; BIA Billings Area Director Anson Baker (Mandan-Hidatsa), interview in Billings, Mont., February 1981; Marjane Ambler, "Energy Companies Seek Peace and Resources with Indians," *High Country News*, 12 December 1980, 1; Marjane Ambler, "Oil on the Reservation," *Western Energy*, March 1982.

60. Section 3(b) of the Indian Mineral Development Act, 96 Stat. 1938; Marjane Ambler, "Mineral Law Hamstrung by BIA Delay," *Lakota Times*, 14 March 1989.

61. Rich Schilf (tribal Natural Resources Department director), interview in Denver, Colo., November 1986; assistant solicitor for Indian affairs to chief of BIA Division of Energy and Mineral Resources, 2 August 1984, author's files.

62. Marjane Ambler, "Interior Disputes Fort Peck Claims," *Lakota Times*, 18 April 1984; minutes, Fort Peck Business Council Meeting, 27 February 1984, 16 June 1988, and 19 June 1988, author's files.

63. Joe Chesser (program analyst with fluid minerals division of BLM, Washington, D.C.), telephone interview, February 1988; Richard LaCourse, "Oil Tribes Voice Fears over Theft, Underreporting of Resource Royalties," *CERT Report*, 7 October 1981; interviews with various MMS, BLM, and tribal officials.

64. "Tribes Not Moved by Pressure," *Wotanin Wowapi*, 21 May 1981. Ambler, "Three Affiliated Tribes Seek Control"; Interior Associate Solicitor Fredericks to assistant secretary for Indian affairs, 19 January 1978, memorandum entitled "Authority of Indian Tribes to Regulate the Use of Lands Held in Trust for Individual Indians by the United States," National Indian Law Library, Boulder, Colo.

65. Based upon interviews with allottees in Wyoming, Montana, and Oklahoma; Marjane Ambler, "The Forgotten People," *Lakota Times*, 2 November 1983; Marjane Ambler, "Indian Landowners Go after Amoco," *High Country News*, 31 October 1983; Senate Select Committee, *Hearing on Federal Supervision*, part 3, 333–343.

66. Carol Connor, telephone interview, April 1986.

67. *Shii Shi Keyah Association v. Clark*, Civil Action No. 84-1622M, U.S. District Court, Albuquerque, N.Mex., filed 30 October 1984; Ambler, "Oil Rich and Penny Poor"; Chuck Cook, Mike Masterson, and Mark N. Trahant, "Agencies Let Oil-Rich Indians Wallow in Poverty," *Arizona Republic*, 4 October 1987.

68. *Kauley v. United States*, Civil Action No. 84-3306T, U.S. District Court, Western District of Oklahoma, filed 14 December 1984 (this suit was later taken over by the Native American Rights Fund).

69. Interviews with BIA officials and allottees in Oklahoma in April 1985; Ambler, "Oil Rich and Penny Poor."

70. Ambler, "Oil Rich and Penny Poor."

71. House Committee on Government Operations, *Hearing on Distribution of Royalty Payments*, 9, 17–64.

72. Ibid., 116.

73. Chuck Cook, Mike Masterson, and M. N. Trahant, "Indian's Oil, Identity Stolen," *Arizona Republic*, 5 October 1987.

74. MMS Director William Bettenberg, interview in Denver, Colo., November 1986; Marjane Ambler, "Minerals Management Services Puts Royalty Management Problems in Perspective," *Lakota Times*, 24 December 1986. General Accounting Office (GAO), *Indian Royalties: Interior Has Not Solved Indian Oil and Gas Royalty Payment Problems* (Washington, D.C.: GAO, March 1986), 33; U.S. Department of Interior (DOI), Minerals Management Service (MMS), "Royalty Management Advisory Committee Minutes for the Inaugural Meeting, January 9–10, 1986," MMS files.

75. Perry Baker, "Office of the Superintendent Report," *Ute Bulletin*, 30 March 1988; Alan Taradash, telephone interview, April 1988; John Martin, "Allottees Meeting," *Ute Bulletin*, 30 September 1987.

76. Aubertin interview.

77. Bernice Muskrat, member of the mineral-wealthy Jicarilla Apache Tribe, unsuccessfully tried to get tribal funds divided among the members in 1985 ["Solicitor Denies Petition for Pro Rata Distribution of Tribal Funds, Cites 1934 Indian Reorganization Act," Department of Interior news release, 8 May 1985; "Beulah Chapin Campaign Message," *Ahead of the Herd*, 11 January 1985].

CHAPTER SEVEN. AFTER THE CONTRACT IS SIGNED: THE TRIBE AS REGULATOR

1. Richard Nafziger, "Indian Uranium: Profits and Perils," in *You Don't Have to Be Poor to Be Indian*, ed. Maggie Gover (Washington, D.C.: Americans for Indian Opportunity, no date), 149; U.S. Department of Interior (DOI), *Indian Lands Map: Oil, Gas, and Minerals on Indian Reservations* (Washington, D.C.: GPO, 1978); "DOE Figures Say CERT Exaggerates Extent of Indian Uranium Wealth," *Nuclear Fuel*, 25 June 1979.

2. General Accounting Office (GAO), *Surface Mining: Issues Associated with Indian Assumption of Regulatory Authority* (Washington, D.C.: GAO, May 1986), 8; Keith Lamb (Office of Surface Mining abandoned mines lands coordinator), telephone interview, February 1988; John A. Folk-Williams and James S. Cannon, *Water for the Energy Market* (Santa Fe, N.Mex.: Western Network, 1983), 70; *1986 Keystone Coal Mining Directory* (New York: Mining Informational Services, McGraw-Hill, 1986), 847, 994, 1000, 1001.

3. Council of Energy Resource Tribes (CERT), *The Control and Reclamation of Surface Mining on Indian Lands*, by Douglas R. Richardson, for the Office of Surface Mining, Department of Interior (Denver: CERT, 29 September 1979), 3–1. One of the biggest coal strip mines in the country in the 1920s was partly on the Cheyenne River Sioux Reservation in South Dakota [Jack Bickel, "The Dragline Reigned as King in the Firesteel Coal Country for 30 Years," *Timber Lake and Area Historical Society Newsletter*, May 1987 (OSMRE files)].

4. Janet Siskind, "A Beautiful River That Turned Sour," *Mine Talk*, Summer/Fall 1982, 37–59; Michele Strutin, "Sacrificing Southwest Splendor for Energy's Sake," *Rocky Mountain News*, 27 January 1980; J. W. Schomisch, "EID Lifts Ban on Eating Church Rock Livestock," *Gallup Independent*, 22 May 1980; Steve Hinchman, "Rebottling the Nuclear Genie," *High Country News*, 19 January 1987.

5. New Mexico Governor Bruce King to Thomas E. Baca, 9 January 1981, author's files; Steve Hinchman, "Film Gets Lousy Reviews from New Mexico," *High Country News*, 15 February 1988; Chris Shuey, telephone interview, May 1988; Chris Shuey, "The Puerco River: Where Did the Water Go?" *The Workbook* 11: 1–10.

6. "EID Finds That Church Rock Dam Break Had Little or No Effect on Residents," *Nuclear Fuel*, 14 March 1983; Shuey interview; Lawrence Lano (Navajo water division administrator), telephone interview, May 1988.

7. Shuey, "Where Did the Water Go?"; Frank Pitman, "Navajos-UNC Settle Tailings Spill Lawsuits," *Nuclear Fuel*, 22 April 1985.

8. Siskind, "River Turned Sour"; Lano interview; Navajo Nation, "Navajo Nation Water Code," Sec. 101 of Title 22, Navajo Tribal Code, Chapter 7.

9. Siskind, "River Turned Sour"; Shuey, "Where Did the Water Go?"; Frank Pitman, "Mine Dewatering Operation in New Mexico Seen Violating Arizona Water Standards," *Nuclear Fuel*, 1 March 1982; Hinchman, "Nuclear Genie."

10. Safe Drinking Water Act, 16 December 1974, P.L. 93-523, 88 Stat. 1660; Clean Water Act, 18 October 1972, P.L. 92-500, 86 Stat. 816.

11. Jim LeBret (BIA geologist in Portland area office), telephone interview, August 1987.

12. The uranium 238 levels were four thousand times the natural levels in that area and forty times what EPA later required in its discharge permit. Paul B. Hahn, chief of evaluation branch, U.S. EPA Office of Radiation Programs, Las Vegas Facility, to Richard Parkin, chief of water compliance section, EPA, Region 10, 18 February 1987, author's files; Shuey interview; Grover Partee (environmental engineer with EPA Region 10), telephone interview, April 1988; Barry Stern, IHS sanitarian, to Jim LeBret, 2 March 1983, author's files.

13. Telephone interviews in April 1988 with Grover Partee (environmental engineer with EPA, Region 10); Marcel DeGuire (president of Dawn Mining Company); Glenn Ford (Spokane tribal councilman); and Gary Farrell (Spokane tribal attorney).

14. Leonard Robbins (director of Navajo Department of Surface Mining), telephone interview, April 1988; CERT, *Surface Mining on Indian Lands,* appendix G (legislative history); Surface Mining Control and Reclamation Act of 1977, Sec. 102, "Purposes," P.L. 95–97.

15. Environmental Protection Agency (EPA), *Radiological Quality of the Environment in the United States, 1977* (Washington, D.C.: GPO, 1977), 58–67.

16. Ibid., 62–67; U.S. Department of Energy (DOE), *Fact Sheet on Uranium Mill Tailings* (Washington, D.C.: DOI, September 1982); Justas Bavarskis, "Uranium: The West Mines, Mills and Worships Radioactive Fuel," *High Country News*, 10 March 1978, 4.

17. Navajo Nation, "Resolution of the Navajo Tribal Council Establishing a Navajo Tribal Environmental Protection Commission and Adopting the Plan of Operation Thereof," 10 August 1972, author's files; Harold Tso, speech given at CERT annual meeting in Denver, Colo., October 1981; EPA, *Radiological Quality*, 65; Ford, Bacon & Davis Utah, *A Summary of the Phase II–Title I Engineering Assessment of Inactive Uranium Mill Tailings, Shiprock Site, Shiprock, N.M.*, prepared for the U.S. Energy Research and Development Administration (Salt Lake City, Utah: Ford, Bacon & Davis Utah, March 1977), 65.

18. EPA, *Radiological Quality*, 62–66; Marjane Ambler, "Study of Radioactive Homes Lost for Eight Years," *High Country News*, 25 January 1980; Victor E. Archer, J. Dean Gillam, and Joseph K. Wagoner, "Respiratory Disease Mortality among Uranium Miners," *Annals of the New York Academy of Sciences* 271 (28 May 1976): 280–293; Ford, Bacon & Davis Utah, *Shiprock Site*, 1–6, 7.

19. Uranium Mill Tailings Radiation Control Act, 8 November 1978, P.L. 95-604, 92 Stat. 3021; Tracy Loughead (public relations for DOE), telephone interview, April 1988.

20. Marjane Ambler, "Wyoming to Study Tailings Issue," *Denver Post*, 5 Feb-

ruary 1984; U.S. Department of Energy (DOE), *Environmental Assessment on Remedial Action at the Riverton Uranium Mill Tailings Site, Riverton, Wyo.* (Albuquerque, N.Mex.: DOE, June 1987).

21. Marjane Ambler, "Man Exposed to U Ore in His Home Drops Suit in Return for Tests," *Nuclear Fuel*, 27 August 1984 (concerning man in Moab, Utah); Harry Tome (Navajo tribal councilman), telephone interview, August 1984; Joseph M. Hans, Jr. (chief of field studies branch for EPA, Las Vegas), telephone interview, August 1984.

22. U.S. Health and Human Services (HHS) Department, Public Health Service, *Health Hazards related to Nuclear Resource Development on Indian Land* (Washington, D.C.: HHS, November 1982), 10; Eric Eberhard, lobbyist for Navajo Nation, to Congressman Morris K. Udall, 30 April 1985, author's files; Environmental Protection Agency (EPA), *Potential Health and Environmental Hazards of Uranium Mine Wastes* (Washington, D.C.: EPA, June 1983), 1:23; telephone interviews in August 1984 with EPA personnel including Joseph M. Hans, Jr., and Wayne Bliss of EPA, Las Vegas, Dave Duncan of EPA, Region 9, and John Giedt of EPA, Region 8.

23. "Congress Unlocks $30 Million Mine Land Funds," *Navajo Times*, 10 July 1987; LeVon Benally (Navajo reclamation specialist), telephone interview, May 1988; Kim St. Aubin and Sue Massie, *The Abandoned Mine Land Program, 1977-1987* (Springfield, Ill.: Association of Abandoned Mined Land Programs, May 1987), 7.

24. Judge Justin L. Quackenbush ruled that the Interior Department had the right to raise the amount of the bond but had not followed the correct procedures [19 December 1984 decision, page 4, on *Dawn Mining Co. v. Clark* (Civil Action No. 82-974-JLQ, filed in 1982 in U.S. District Court, Eastern District of Washington)].

25. Interviews with Ford, Partee, LeBret, and DeGuire; Ross Swimmer, assistant Interior secretary, to Marcel F. DeGuire, 18 June 1987, demanding $9.7 million bond, author's files; Charles W. Luscher, Oregon state director for BLM, to Marcel F. DeGuire, 19 May 1987, demanding new operations plan, author's files.

26. Stan Dayton, "Washington's Sherwood Project: A Newcomer in an Orphan District," *Engineering & Mining Journal*, November 1978, 112ff.; U.S. Department of Interior (DOI), Bureau of Indian Affairs (BIA), *Sherwood Uranium Project, Spokane Indian Reservation, Final Environmental Statement* (Portland, Oreg.: BIA, 19 August 1976).

27. U.S. Department of Interior (DOI), Bureau of Indian Affairs (BIA), *Mineral Resource Facilities Maintenance Contract for the Sherwood Mine-Mill Complex* (Portland, Oreg.: BIA, 1989); Jim LeBret, telephone interview, March 1989.

28. Dan Jackson, "Mine Development on U.S. Indian Lands," *Engineering and Mining Journal*, January 1980; John Aloysius Farrell, "The New Indian Wars," *Denver Post*, 20-27 November 1983 (reprint), 31; interviews in Denver, Colo., November 1987, with Laguna Governor Chester Fernando and Ron Solimon, a tribal member on the negotiating team from 1977 to 1984; Gerald F. Seib, "Indians Awaken to Their Lands' Energy Riches and Seek to Wrest Development from Companies," *Wall Street Journal*, 20 September 1979.

29. U.S. Department of Interior (DOI), Bureau of Land Management (BLM), *Jackpile-Paquate Uranium Mine Reclamation Project Environmental Impact Statement* (Albuquerque, N.Mex.: BLM, October 1986), 1: 2-27, 2: letter 10, A-32.

30. Marjane Ambler, "Lagunas Face Fifth Delay of Uranium Cleanup," *Navajo Times Today*, 4 February 1986; DOI, *Jackpile EIS*, 1: 3-23, 2: L-11.

31. DOI, *Jackpile EIS*, A-40; "Agreement Signed for Reclamation of Jackpile Mine in New Mexico," Interior Department press release, 12 December 1986; Marc

Nelson (Jacobs Engineering staff), telephone interview, January 1987; Chester Fernando, interview in Denver, Colo., November 1986; Neal Kasper (manager of Laguna Construction Company), interview in Laguna, N.Mex., February 1989.

32. Tom Barry, "Bury My Lungs at Red Rock," *The Progressive*, February 1979, 25–28; Christopher McLeod, "Uranium Mines and Mills May Have Caused Birth Defects among Navajo Indians," *High Country News*, 4 February 1985; Jessica S. Pearson, *A Sociological Analysis of the Reduction of Hazardous Radiation in Uranium Mines,* prepared for the National Institute for Occupational Safety and Health, Public Health Service (Washington, D.C.: GPO, 1975).

33. *Nance v. EPA*, 645 F.2d 701 (9th Cir. 1981), cert. denied; Marjane Ambler, "Northern Cheyennes Want Class I Air," *High Country News*, 16 July 1976; Patrick Smith and Jerry D. Guenther, "Environmental Law—Protecting Clean Air: The Authority of Indian Governments to Regulate Reservation Airsheds," *American Indian Law Review* 9 (1981): 83–119; Edwin Dahl (former Northern Cheyenne tribal councilman and later MPC agreement administrator), telephone interview, April 1988.

34. Smith and Guenther, "Protecting Clean Air," 83, 90; Associated Press (AP), "Rancher Challenges Cheyennes' Air Rating," *Missoulian*, 12 September 1977; David H. Getches and Charles F. Wilkinson, eds., *Federal Indian Law Cases and Materials*, 2d ed. (St. Paul, Minn.: West Publishing Co., 1986), 536 [hereafter *Federal Indian Law II*].

35. Ken Peres and Fran Swan, "The New Indian Elite: Bureaucratic Entrepreneurs," *Akwesasne Notes*, Late Spring 1980, 18; Doug Richardson, "Northern Cheyenne," *CERT Report*, 28 April 1980; "Montana Power Says It and Cheyenne Indians Agree to End Lawsuits," *Wall Street Journal*, 1 May 1980; Dahl interview.

36. Marjane Ambler, "Clean Air Rules New Weapon in Indian Control of Coal Mining," *Coal Week*, 11 April 1983; Ford interview; J. Dallas C. Gudgell, "Fort Peck Project Environmental Study for the Fort Peck Reservation" (senior paper, Fort Peck Community College, 22 May 1985), 20.

37. CERT, *First Decade* (Denver: CERT, November 1986), presented at 1986 annual meeting. Jason Whiteman (Northern Cheyenne tribal member and director of air-quality staff), telephone interview, April 1988; Associated Press (AP), "Monitoring Acid Rain on Reservation Much-Needed Step, USFWS [U.S. Fish and Wildlife Service] Officials Say," *Casper Star-Tribune*, 4 July 1988.

38. Senator John Melcher, speech presented to CERT board of directors meeting in Washington, D.C., 4 September 1980; Doug Richardson, telephone interview, April 1988; Dahl interview; CERT, *Surface Mining on Indian Lands*, appendix G.

39. CERT, *Surface Mining on Indian Lands*. The six subcontracting tribes were the Crow, Three Affiliated Tribes of Fort Berthold, Navajo, Northern Cheyenne, Southern Ute, and Northern Ute. The Hopis participated as observers, and the other coal tribes were less involved.

40. Marjane Ambler, "$24 Million Not Drawing Interest," *Lakota Times*, 29 February 1984.

41. U.S. Department of Interior (DOI), Office of Surface Mining (OSM), *Report to the U.S. Congress: Proposed Legislation Designed to Allow Indian Tribes to Elect to Regulate Surface Mining of Coal on Indian Lands* (Washington, D.C.: OSM, 1984).

42. Peter Iverson, "Knowing the Land, Leaving the Land: Navajos, Hopis, and Relocation in the American West," *Montana: The Magazine of Western History*, 38, no. 1 (Winter 1988): 67–70; General Accounting Office (GAO), *Surface Mining: Issues Associated with Indian Assumption of Regulatory Authority* (Washington, D.C.: GAO, May 1986), 14–16; David W. Simpson, Westmoreland Resources, 26 Au-

gust 1982 letter to OSM, in DOI, OSM, *Report to Congress*, Appendix 5; "Policy Sought for Mining on Tribal Lands," *Phoenix Gazette*, April 1984.

43. United Press International (UPI), "Court Upholds Indian Mineral Claim," *Navajo Times Today*, 22 October 1985.

44. Christopher McLeod, "Navajos Resist Relocation," *High Country News*, 12 May 1986; Felix Cohen, *Felix Cohen's Handbook of Federal Indian Law*, 2d ed. (Charlottesville, Va.: Michie/Bobbs-Merrill, 1982), 480–481 [hereafter *Cohen II*]; Iverson, "Knowing the Land."

45. Loretta Fowler, *Shared Symbols, Contested Meanings: Gros Ventre Culture and History, 1778–1984* (Ithaca, N.Y.: Cornell University Press, 1987), 257, note 1: "Of the 13 reservations in the northern and central plains, only two are occupied by one tribal group—Lower Brule and Northern Cheyenne"; David H. Getches, Daniel M. Rosenfelt, and Charles F. Wilkinson, eds., *Federal Indian Law Cases and Materials* (St. Paul, Minn.: West Publishing Co., 1979), 539–542 [hereafter *Federal Indian Law I*]; Loretta Fowler, *Arapahoe Politics, 1851–1978: Symbols in Crises of Authority* (Lincoln: University of Nebraska Press, 1982), 142–144; *Shoshone Tribe v. United States*, 299 U.S. 476 (1937); *United States v. Shoshone Tribe*, 304 U.S. 111 (1938).

46. *Healing v. Jones*, 373 U.S. 758 (1963); Interior Department solicitor to assistant secretary for Indian affairs, 5 December 1983, "Memorandum re Hopi Coal Severance License Fee Ordinance No. 38," author's files; Iverson, "Knowing the Land," 69; Bill Walker, "Many Believe Greed for Coal Forcing Them from Homes," *Denver Post*, 22 June 1986.

47. Interior solicitor, "Hopi Coal Severance Fee," 2; DOI, *Report to Congress: Proposed Legislation*, Hopi comments section; Peterson Zah, speech given at CERT annual meeting in Spokane, Wash., 27 October 1983.

48. DOI, *Report to Congress: Proposed Legislation*, Navajo comments section; Interior Assistant Secretary Ken Smith to Hopi Tribal Chairman Ivan Sidney, 5 December 1983, author's files; Marjane Ambler, "Interior Blocks Hopi Proposal to Tax Peabody Operations," *Coal Week*, 9 January 1984.

49. DOI, *Report to Congress: Proposed Legislation*, 17.

50. CERT, *Surface Mining on Indian Lands*, 10-16; 12-15 through 12-24; DOI, *Report to Congress: Proposed Legislation*, 26.

51. CERT, *Surface Mining on Indian Lands*, 4-1 through 4-42, 7-17; "Tribe to Get $315,000 for Mine Inspector Visits," *Navajo Times*, 26 November 1987; "Sarpy Mine Inspection: Production Down, Seeding Continues," in *Crow Tribe Abandoned Mine Lands Report* (Crow Agency, Mont.: Crow Tribe, July 1983); Don Ami (Hopi), speech given at CERT meeting in Denver, Colo., 16 November 1986; Keith Lamb (OSMRE Indian AML coordinator), telephone interview, February 1988.

52. Arthur Abbs of OSMRE to Mark Boster of OSMRE, "Tribal Capability Study," 2 February 1987, OSMRE files.

53. Based upon conversations with tribal staff members and Tim Giago, "Too Much Politics Can Spoil the Stew of Reservation Life," *Lakota Times*, 2 September 1987.

54. Based upon conversations with tribal and OSMRE coal personnel.

55. CERT, *Surface Mining on Indian Lands*, 5-31; GAO, *Indian Assumption of Regulatory Authority*, 20–21; Marjane Ambler, "Strip Bill Unacceptable," *Lakota Times*, 8 August 1984; "Navajo Attorney General Upshaw Speaks at ABA Conference on Reservation Environment Laws," *Navajo Times*, 10 March 1988.

56. CERT, *Surface Mining on Indian Lands*, 11-2 through 11-12; *Federal Indian Law II*, 599; DOI, *Report to Congress: Proposed Legislation*, Shell Oil comments section.

57. GAO, *Indian Assumption of Regulatory Authority*, 22; *Jicarilla Apache v. Andrus*, 687 F.2d 1324 (10th Cir. 1982); *Federal Indian Law II*, 615.

58. Frank Pitman, "U.S. Appeals Court Decides U.S. Can't Be Sued for Deaths of Uranium Miners," *Nuclear Fuel*, 26 August 1985; Bill Donovan, "Tribe's Tax Deputy Fired for Acting without Authority," *Navajo Times Today*, 28 July 1986.

59. CERT, *Surface Mining on Indian Lands*, 11-8, 11-13, 12-25 through 12-27.

60. Based upon tribal comments upon the Interior Department's proposed legislation and interviews in 1984 with representatives of the Crow, Hopi, and Three Affiliated Tribes, as well as with Ed Grandis of the Environmental Policy Institute; Bill Donovan, "Environmentalists-OSM Squabble May Delay Millions of Dollars to Navajo Communities," *Navajo Times*, 21 October 1981.

61. CERT, *Surface Mining on Indian Lands*, 11-9 through 11-12; Lamb interview.

62. Americans for Indian Opportunity (AIO), *Survey of American Indian Environmental Protection Needs on Reservation Lands*, written for the Environmental Protection Agency (Washington, D.C.: GPO, September 1986), vi; Council of Energy Resource Tribes (CERT), *Inventory of Hazardous Waste Generators and Sites on Selected Indian Reservations*, by Warner Reeser, for the Environmental Protection Agency (Denver: CERT, July 1985).

63. Environmental Protection Agency (EPA), "EPA Policy for the Administration of Environmental Programs on Indian Reservations," 8 November 1984 (photocopy); B. Kevin Gover, "Indian Tribal Governments Look to Take Control of Reservation Environments," *The Workbook* 12 (July/September 1987): 89; Jack Lewis, "An Indian Policy at EPA," *EPA Journal*, January-February 1986, 23–26; EPA, *Report to Congress: Indian Wastewater Treatment and Assistance* (Washington, D.C.: GPO, 1989); EPA, *Indian Drinking Water Supply Study* (Washington, D.C.: GPO, 1988); Leigh Price, "Tribal Programs under the Clean Water Act," speech presented at the Rocky Mountain Mineral Law Institute, Mineral Development on Indian Lands, 17 February 1989, in Albuquerque, N.Mex.

64. Smith and Guenther, "Protecting Clean Air," 85; Federal Oil and Gas Royalty Management Act (FOGRMA), 22 December 1982, P.L. 97-451, 96 Stat. 2447.

65. *Washington Department of Ecology v. EPA*, 752 F.2d 1465 (9th Cir. 1985); Marjane Ambler, "Washington Tribes and EPA Win Court Battle to Regulate Hazardous Wastes," *Navajo Times Today*, 14 May 1985; *Federal Indian Law II*, 531ff.

66. Superfund Amendments and Reauthorization Act, P.L. 99-499; Safe Drinking Water Act amendments, P.L. 99-339, 100 Stat. 666; Water Quality Act of 1987, P.L. 100-4, 101 Stat. 7; Gover, "Reservation Environments"; Lawrence Lano (director of Navajo Water Division), telephone interview, May 1988; Roy Popkin, "Indians Act for a Cleaner Environment," *EPA Journal*, April 1987, 28–31; Anne Miller (director of special programs for EPA), telephone interview, May 1988.

67. Based upon interviews with tribal staffs and 19 July 1988 letter from Warner Reeser, private consultant, to author; Marjane Ambler, "The Lands the Feds Forgot," *Sierra Magazine*, May 1989.

68. Crow Tribe, "Section 710 Crow Indian Lands Study Final Report," for the Crow Tribe, CERT, and DOI, September 1979, E-8, author's files; Marjane Ambler, "Crow Tribe Protests OSM Move Giving States Lead on Coal Site," *Coal Week*, 30 September 1985.

69. GAO, *Indian Assumption of Regulatory Authority*, 24–26; Marjane Ambler, "CERT Tribes Hope for Hearing to Protest AML Proposal," *Inside Energy*, 23 December 1985; Ellis Knows Gun, "Nationwide Concerns over OSM Plan to Transfer Indian AML Funds to State Projects," *Crow Office of Reclamation News*, Win-

ter 1986; U.S. Department of Interior (DOI), "Crow, Hopi, Navajo Access $36.9 Million in Mining Funds," *Indian News*, 20 July 1987.

70. David H. Getches, "Common Concerns of States and Tribes in Natural Resources Development," speech presented at CERT annual meeting in Reno, Nev., 30 October 1984; *Federal Indian Law II*, 547; telephone interviews in April 1988 with Nat Nutongla (director of the Hopi Division of Mining and Reclamation Enforcement) and Leonard Robbins (director of the Navajo Department of Surface Mining).

71. Deborah Gates (EPA regional counsel), telephone interview, January 1989; Richard A. DuBey, Mervyn T. Tano, and Grant D. Parker, "Protection of the Reservation Environment: Hazardous Waste Management on Indian Lands," *Environmental Law* 18 (1988): 504.

72. Interviews with tribal officials and review of tribal natural resources departments' reports; Marjane Ambler, "The Three Affiliated Tribes at Fort Berthold Seek to Control Their Energy Resources," in *Native Americans and Energy Development II*, ed. Joseph G. Jorgensen (Boston: Anthropology Resource Center, 1984), 194–199.

73. Ken Peres and Fran Swan, "The New Indian Elite: Bureaucratic Entrepreneurs," *Akwesasne Notes*, Late Spring 1980, 18.

74. CERT, *Surface Mining on Indian Lands*, 4–41; U.S. Congress, Senate, Select Committee on Indian Affairs, *Hearing on S. 1039 to Review and Determine the Impact of Indian Tribal Taxation on Indian Reservations and Residents, November 12, 1987*, 100th Cong., 1st sess. (Washington, D.C.: GPO, 1988), 14, 34; Philip Reno, *Mother Earth, Father Sky and Economic Development: Navajo Resources and Their Use* (Albuquerque: University of New Mexico Press, 1981), 135, 141, 142.

75. Robert Nordhaus, "*Merrion v. Jicarilla Apache Tribe*: The Tribal Severance Tax Case," paper delivered at the CERT Tax Policy Forum, 14 July 1982, in Billings, Mont.; *Federal Indian Law I*, 1983 supplement, 108–120.

76. James G. Watt, "Brief of the Mountain States Legal Foundation as Amicus Curiae in Support of Petitioners, *Merrion v. Jicarilla Apache Tribe*," 20 November 1980 (submitted to U.S. Supreme Court).

77. *Merrion v. Jicarilla Apache*, 455 U.S. 130 (1982).

78. Jicarilla Apache Tribal President Leonard Atole, interview in Dulce, N.Mex., October 1987; Frank G. Long, "Energy Critics Jump the Gun on Tribal Mineral Tax," *Rocky Mountain News*, 11 May 1982; "Landmark Jicarilla Apache Severance Tax Ruling Establishes National Precedent," *CERT Report*, 8 February 1982.

79. *Kerr-McGee Corp. v. Navajo Tribe*, 471 U.S. 195 (1985); U.S. Department of Interior, *Guidelines for Review of Tribal Ordinances Imposing Taxes on Mineral Activities* (Washington, D.C.: DOI, 18 January 1983); "Interior Leaning against CERT Request for Revocation of Tax Guidelines," *Inside Energy*, 11 April 1983; John Aloysius Farrell, "The New Indian Wars," *Denver Post*, November 1983 (reprint), 18; Susan M. Williams and David C. Cole, "Resource, Revenue, and Rights Reclamation: The Tax Program for the Navajo Nation," 1 August 1977, 12, 13, author's files (photocopy).

80. Susan Williams, telephone interview, April 1988; Nancy Zeilig, "Face to Face: Interview with Susan M. Williams on Indian Water Rights," *Journal of the American Water Works Association*, March 1985, 110.

81. Williams and Cole, "Tax Program for Navajo Nation," 14; David Redhorse and Theodore Reynolds Smith, "American Indian Tribal Taxation of Energy Resources," *Natural Resources Journal* 22 (July 1982): 659–671; Susan Williams, tel-

ephone interview, May 1988; Alvin M. Josephy, Jr., "The Murder of the Southwest," *Audubon*, July 1971, 53.

82. Williams and Cole, "Tax Program for Navajo Nation," 4; Henry Adams (attorney with the Navajo Tax Commission), telephone interview, October 1988.

83. Susan Williams (chair of the Navajo Tax Commission), telephone interview, April 1985; Jeanne Whiteing (Blackfeet attorney), interview, June 1988.

84. Lorna Thackeray, "High Court Backs Crow: State May Lose $90 Million in Coal Taxes," *Billings Gazette*, 12 January 1988; *Crow Tribe of Indians v. Montana*, 819 F.2d 895 (9th Cir. 1987), affirmed, 108 U.S. 685 (1988).

85. The tax total is through 1982 and the royalty total through 1983. "Tribal Coal Grants Nixed," *Billings Gazette*, 8 April 1983.

86. John Irvin, telephone interview, January 1986; Ellis Knows Gun, "Shell Withdraws from Crow Coal Market; Tax Changes, Other Interests Cited," *Crow Office of Reclamation News*, Winter 1986; "NSP [Northern States Power] Sues Westmoreland on Absaloka Taxes; Seeks Reimbursement of Crow, Montana Levies," *Coal Week*, 18 July 1988.

87. Susan M. Williams, "Multiple Taxation of Mineral Extraction in Indian Country: State and Indian Tribal Jurisdiction," speech presented at Rocky Mountain Mineral Law Institute, Mineral Development on Indian Lands, 16 February 1989, in Albuquerque, N.Mex.; Anne MacKinnon, "Ruling May Affect Wyoming Pact with Tribes on Taxes," *Casper Star-Tribune*, 30 April 1989; *Cotton Petroleum v. New Mexico* (Civil Action No. 87-1327), 104 Lawyers' Edition 2d. 209.

88. Farrell, "New Indian Wars," 32; "Henderson Wants Reservation Taxes on Gasoline Used for Road Improvement," *Navajo Times*, 24 March 1988; Mike Chase, "Governor's Redi-Fund Not Ready to Aid South Dakota Reservations," *Lakota Times*, 10 February 1988.

89. Lorraine Turner Ruffing, "Navajo Mineral Development," *American Indian Journal*, September 1978, 78.

90. The Wind River study was conducted by National Economic Research Associates of White Plains, N.Y. Scott Farris, "Tribes: State Taxing Them Excessively," *Casper Star-Tribune*, 5 August 1988; Robert Nicholas, Wyoming assistant attorney general, "Confidential Memorandum Re: Severance Taxes on the Wind River Indian Reservation," 5 October 1987, 23, author's files.

91. Anne MacKinnon, "State, Tribe Announce Water, Tax Agreement," *Casper Star-Tribune*, 18 February 1989; Fremont County Association of Governments, "Impact of Wind River Reservation Tribes Supplanting State Authority to Levy Severance Tax on Reservation Oil/Gas and Other Minerals," 1987, author's files; based also upon interviews in Fort Washakie, Wyo., August 1987, with Wind River Tax Commission members Orville St. Clair, Wes Martel, and Gary Collins and in Boulder, Colo., June 1988, with Richard Ortiz, executive director for the commission.

92. Based upon comments from state and tribal representatives at a CERT tax workshop in Denver, Colo., 4 November 1987; Senate Select Committee, *Hearing on Tribal Taxation*, 17, 37, 44–45.

93. Marjane Ambler, "Few Respond at Tribal Hearing on Plans to Claim All Oil, Gas Revenue from Lands," *Casper Star-Tribune*, 29 August 1987; Senate Select Committee, *Hearing on Tribal Taxation*.

94. Angie Debo, *And Still the Waters Run: The Betrayal of the Five Civilized Tribes* (Princeton, N.J.: Princeton University Press, 1940; reprint, Norman: University of Oklahoma Press, 1984), 18; Williams, "Multiple Taxation."

95. Lawrence Wetsit, "From the Heart," *EPA Journal*, April 1977, 30.

96. Judy Knight, "Conference Summary," speech presented at CERT annual meeting, Denver, Colo., 4 November 1987.

CHAPTER EIGHT. NO RESERVATION
IS AN ISLAND: WATER AND
OFF-RESERVATION ENERGY DEVELOPMENT

1. Peterson Zah, "Water: Key to Tribal Economic Development," in *Indian Water 1985, Collected Essays*, ed. Christine L. Miklas and Steven J. Shupe (Oakland, Calif.: American Indian Resources Institute, 1986), 76.

2. Gary Paul Nabhan, *Enduring Seeds* (San Francisco: North Point Press, 1989), "Harvest Time: Northern Plains Agricultural Change."

3. Michael L. Lawson, *Dammed Indians* (Norman: University of Oklahoma Press, 1982), 59–62; U.S. Department of Interior, "Joint Tribal Advisory Committee Hearing," unpublished transcript of 10 January 1986 hearing at New Town, N.Dak., 130, 154; DOI, Bureau of Indian Affairs (BIA), *Indian Service Population and Labor Force Estimates* (Washington, D.C.: BIA, January 1983).

4. Some of the other dams that flooded Indian lands include Oahe, Big Bend, Fort Randall, Fort Peck, Yellowtail, Boysen, Grand Coulee, Roosevelt, and Glen Canyon. Marjane Ambler, "Nation Now Molding Its First Indian Water Policy," *High Country News*, 23 February 1979, 1; Michael Lawson, "Pick-Sloan and the Tribes," *Boundaries Carved in Water* (Missoula, Mont.: Northern Lights Institute), April 1988, 1.

5. Norman K. Johnson, "Indian Water Rights in the West," prepared for the Western States Water Council, Salt Lake City, Utah, December 1983, 4–5; James Cannon, "Leased and Lost," *Economic Priorities Report* 5, no. 2 (1974): 18.

6. Lawson, *Dammed Indians*, 45, 59–62.

7. Native American Natural Resources Development Federation (NANRDF) of the Northern Great Plains, "Declaration of Indian Rights to the Natural Resources in the Northern Great Plains States," June 1974, 25, 33 (photocopy); U.S. Congress, Senate, Committee on Interior and Insular Affairs, Subcommittee on Energy Resources and Water Resources, *Hearing on the Sale of Water from the Upper Missouri River Basin by the Federal Government for the Development of Energy*, 94th Congress, 1st sess. (Washington, D.C.: GPO, 18 July and 26 and 28 August 1975), 62.

8. John A. Folk-Williams and James S. Cannon, *Water for the Energy Market* (Santa Fe, N.Mex.: Western Network, 1983), 10, 44.

9. Ibid., 10; Western States Water Council, "Western States Water Requirements for Energy Development to 1990," Salt Lake City, Utah, November 1974, 31.

10. Alvin M. Josephy, Jr., "The Agony of the Northern Plains,"*Audubon*, July 1973, 87; Gary D. Weatherford and Gordon C. Jacoby, "Impact of Energy Development on the Law of the Colorado River," *Natural Resources Journal* 15 (January 1975), 182; Folk-Williams and Cannon, *Water for Energy*, 72.

11. Marjane Ambler, "Dialog Opens for Protecting Common Water Hole," *High Country News*, 2 January 1976, 6; Senate Committee on Interior, "Missouri Water Sale," 127–128, discussing the *United States v. Powers* case of 1939.

12. Monroe E Price and Gary D. Weatherford, "Indian Water Rights in Theory and Practice: Navajo Experience in the Colorado River Basin," in *American Indians and the Law*, ed. Lawrence Rosen (New Brunswick, N.J.: Transaction Books, 1976), 119ff.; Folk-Williams and Cannon, *Water for Energy*, 70–77.

13. Senate Committee on Interior, "Missouri Water Sale," 9, 10, 254–255.

14. Ibid., 8, 240.

15. Ibid., 470.

16. U.S. Department of Interior (DOI), memo from the associate solicitor, 20 September 1976, in Senate Committee on Interior, "Missouri Water Sale," 136–138. The Congressional Research Service reported in 1976 that Upper Missouri Basin tribes had potential diversion rights to 8.6 million acre-feet, according to Lawson, "Pick-Sloan and the Tribes," 10; Senate Committee on Interior, "Missouri Water Sale," 465, 472.

17. Ibid., 198–199 (letter from Assistant Interior Secretary Jack Horton to South Dakota Governor Richard Kneip offering the state first option for selling water to Energy Transportation Systems use in a coal slurry pipeline); Folk-Williams and Cannon, *Water for Energy*, 67–68; Reid Peyton Chambers, telephone interview, May 1988.

18. Senate Committee on Interior, "Missouri Water Sale," 464–473.

19. Reid P. Chambers, "Discharge of the Federal Trust Responsibility to Enforce Legal Claims of Indian Tribes: Case Studies of Bureaucratic Conflict of Interest" in *Federal Indian Law Cases and Materials*, ed. David H. Getches, Daniel M. Rosenfelt, and Charles F. Wilkinson (St. Paul, Minn.: West Publishing Co., 1979), 219–220 [hereafter *Federal Indian Law I*]; Senate Committee on Interior, "Missouri Water Sale," 114–142.

20. Daniel McCool, *Command of the Waters: Iron Triangles, Federal Water Development, and Indian Water* (Berkeley: University of California Press, 1987), 162–175; Chambers, "Bureaucratic Conflict of Interest"; Senate Committee on Interior, "Missouri Water Sale," 135, 467–470.

21. Marjane Ambler, "Bighorn Water Battle Goes to Court," *High Country News*, 22 April 1977; U.S. Department of Interior (DOI), Bureau of Reclamation (BuRec), *Water for Energy: Missouri River Reservoirs, Pick–Sloan Missouri Basin Program, Final Environmental Impact Statement* (Billings, Mont.: BuRec, December 1977), F-85, 1-15.

22. "Water Service Contract," *CERT Report*, 17 August 1979; "ANG Uncertain about Tribes," *Hazen* (N.Dak.) *Star*, 15 November 1979; Missouri River Basin Commission, "Draft Water Assessment of the Great Plains Gasification Associates Project," prepared for the U.S. Water Resources Council, Omaha, 4 April 1980; Three Affiliated Tribes Councilman Lawrin "Hugh" Baker, telephone interview, April 1988; Pat Paquette, "For $85 Million Down, Electric Coop Would Obtain Coal-Gas Plant," *Inside Energy*, 8 August 1988.

23. Folk-Williams and Cannon, *Water for Energy*, 61–62; Robert Madsen, Bureau of Reclamation Regional Planning Office, to author, 12 April 1988.

24. *ETSI Pipeline Project v. Missouri et al*, Nos. 86–939 and 86–941 (1988); U.S. Supreme Court decision, 23 February 1988; Mike Chase, "Sioux Tribes Seek ETSI Money," *Lakota Times*, 24 May 1988; Marjane Ambler, "The Missouri River Branch of the Hatfield-McCoy Feud," *High Country News*, 27 October 1986; "Judge: Railroads Conspired to Thwart ETSI Slurry Line," *Casper Star-Tribune*, 8 March 1989; Philip Shabecoff, "U.S. to Have Role in Sales of Water," *New York Times*, 1 January 1989, 17.

25. Les Taylor (attorney for the Jicarilla Apache Tribe), telephone interview, March 1989; *Jicarilla Apache Tribe v. United States, Utah International Inc., and Paragon Resources*, Civil Action No. 82-1327 (1982), U.S. District Court, Albuquerque, N.Mex.; Folk-Williams and Cannon, *Water for Energy*, 75–77; U.S. Department of Interior (DOI), Bureau of Land Management (BLM), *Draft Environmental Impact Statement on Public Service Company of New Mexico's Proposed New Mex-*

ico Generating Station and Possible New Town (Santa Fe, N.Mex.: BLM, November 1982), S-4.

26. John A. Folk-Williams, *What Indian Water Means to the West* (Santa Fe, N.Mex.: Western Network, 1982), 72; Taylor interview.

27. Wyoming, *In re: The General Adjudication of All Rights to Use Water in the Bighorn River System and All Other Sources*, No. 85-204, Wyoming State Supreme Court decision, 28 February 1988; Stanley Pollack (Navajo tribal attorney), telephone interview, March 1988; Scott McElroy (Southern Ute tribal attorney), telephone interview, October 1988; Ute Water Settlement Act, 3 November 1988, P.L. 100-585, 102 Stat. 2973; the case number for *Wyoming v. United States et al.* is 88-309 (Docket Nos. 85-203, 85-204, 85-205, 85-217, 85-218, 85-225, 85-226, 85-236). [Page numbers for *United States Reports* are not available yet because the decision was made on June 26, 1989.]

28. Norris Hundley, Jr., "The *Winters* Decision and Indian Water Rights: A Mystery Reexamined," in *The Plains Indians of the Twentieth Century*, ed. Peter Iverson (Norman: University of Oklahoma Press, 1985), 79–80.

29. *Winters v. U.S.*, 207 U.S. 564 (1908); Hundley, "Mystery Reexamined."

30. McCool, *Command of the Waters*, 65, 256; Bob Delk (chief of water services, BIA Billings area office), telephone interview, March 1988.

31. McCool, *Command of the Waters*, 169.

32. Ambler, "Nation Molding Indian Water Policy"; McCool, *Command of the Waters*, 4; Richard B. Collins, "The Future Course of the *Winters* Doctrine," in Miklas and Shupe, *Indian Water 1985, Collected Essays*, 89–99.

33. Charles Philip Corke, "Corke Testimony Exposing Interior's Failure to Protect Indian Water Rights, April 4, 1978," *Wassaja*, May 1979; Associate Interior Solicitor Alexander Good memorandum to under secretary of Interior, "Nature and Extent of the Secretary's Trust Responsibility when Indian Interests Conflict with Other National or Departmental Interests," 28 October 1981, National Indian Law Library, Boulder, Colo. [hereafter NILL].

34. David H. Getches and Charles F. Wilkinson, eds. *Federal Indian Law Cases and Materials*, 2d ed. (St. Paul, Minn.: West Publishing Co., 1986), 8 [hereafter *Federal Indian Law II*]; Zah, "Water: Key to Development"; Doris Giago, "Water Problems Plague Reservation: 60 Percent of Residents Haul Their Own," *Lakota Times*, 7 August 1985.

35. General Accounting Office (GAO), *Reserved Water Rights for Federal and Indian Reservations: A Growing Controversy in Need of Resolution* (Washington, D.C.: GAO, 16 November 1978), 18, 100; Charles Wilkinson, "Western Water Law in Transition," *University of Colorado Law Review* 56 (Spring 1985): 323; McCool, *Command of the Waters*, 162–175.

36. *Colville Confederated Tribes v. Walton*, 647 F.2d 42 (9th Cir. 1981); *Federal Indian Law II*, 664, 707.

37. Marjane Ambler, "Paper Rights and Wet Water: The Question of Indians," *Northern Lights*, November–December 1985; GAO, "Growing Controversy," 21, 23; Tom Fredericks (Mandan-Hidatsa), former assistant secretary for Interior and currently an attorney representing several tribes, telephone interview, September 1985; DOI Deputy Assistant Secretary for Indian Affairs John W. Fritz, "Contracting with Indian Tribes for Collection of Evidence to Support Water Rights Quantifications," memorandum to area directors, 22 February 1985, NILL (the memo said the department could not delegate its responsibility to conduct studies).

38. *Federal Indian Law II*, 656–675, 701; GAO, "Growing Controversy," 30–31; Johnson, "Indian Water Rights," 122–124; Presidential Commission on Indian

Reservation Economies, *Report and Recommendations to the President of the United States* (Washington, D.C.: GPO, November 1984), II-23, 48.

39. McCool, *Command of the Waters*, 49–60; telephone interviews in March 1989 with Shoshone Business Council Chairman John Washakie, Wyoming Assistant Attorney General Jane Caton, and BIA spokesman Eric Wilson.

40. David Harrison, "Negotiations: A Paradigm Shift of Tribal Self-Determination," speech given at CERT annual meeting, Denver, Colo., November 1986; Lloyd Burton, "The American Indian Water Rights Dilemma: Historical Perspective and Dispute-Settling Policy Recommendations," *UCLA Journal of Environmental Law and Policy* 7 (Spring 1988); GAO, "Growing Controversy," 24; William H. Veeder, "Indian Water Rights and the Energy Crisis," in *Energy Resource Development: Implications for Women and Minorities in the Intermountain West*, ed. U.S. Commission on Civil Rights (Washington, D.C.: GPO, 1978), 188–197; Ken Peres and Fran Swan, "The New Indian Elite: Bureaucratic Entrepreneurs," *Akwesasne Notes*, Late Spring 1980, 5.

41. *Federal Indian Law II*, 674–675, 685–705.

42. "Secretary Approves Southern Ute Tribe's Water Rights Settlement," *NARF Legal Review*, Fall 1986; McCool, *Command of the Waters*, 124, 236, 242–244; Ambler, "Paper Rights"; *Federal Indian Law II*, 702–705; "Water Settlement Enables Ak-Chin to Expand Profitable Farm Operations," *Indian News Notes*, 14 December 1984; Richard LaCourse, "Smith Goes to Bat for Ak-Chin Water Project," *CERT Report*, 7 October 1981; Ute Water Settlement Act.

43. Price and Weatherford, "Navajo Experience," 123; McCool, *Command of the Waters*, 241; Folk-Williams, *What Indian Water Means*, 13; Charles DuMars and Helen Ingram, "Congressional Quantification of Indian Reserved Water Rights: A Definitive Solution or a Mirage?" *Natural Resources Journal* 20, no. 17 (1980): 35–36.

44. Stanley Pollack (Navajo tribal attorney), telephone interview, March 1988; John Atkins (vice president of BHP-Utah International), telephone interview, March 1988.

45. John Echohawk (executive director of NARF), telephone interview, February 1988; Marjane Ambler, "Watt Tries to Reassure Indians," *High Country News*, 24 December 1982.

46. U.S. Department of Interior (DOI), "Western Governors State Policy on Indian Issues," *Indian News*, 20 July 1987; Ambler, "Hatfield-McCoy Feud."

47. Indigenous Press Network, "[Northern] Utes Demand Congress Help with Funding," *Navajo Times*, 5 May 1988; John Martin, "Referendum Election," *Ute Bulletin*, 30 March 1988; Anne MacKinnon, "State, Tribe Announce Water, Tax Agreement," *Casper Star-Tribune*, 18 February 1989.

48. Terry P. Wilson, *The Underground Reservation: Osage Oil* (Lincoln: University of Nebraska Press, 1985), 101.

49. Stephen G. Boyden and Scott C. Pugsley, *Use of Indian Water in Developing Mineral Property* (Denver: Rocky Mountain Mineral Law Foundation, 16–17 March 1978), 5–16; Frank J. Trelease, "Indian Water Rights for Mineral Development," in *Natural Resources Law on American Indian Lands*, ed. Peter C. Maxfield (Denver: Rocky Mountain Mineral Law Foundation, 1977), 230–233.

50. Alvin M. Josephy, Jr., "The Murder of the Southwest," *Audubon*, July 1971, 62; Atkins interview (Utah International also provided water to the San Juan Power Plant); Philip Reno, *Mother Earth, Father Sky and Economic Development: Navajo Resources and Their Use* (Albuquerque: University of New Mexico Press, 1981), 105–108.

51. Josephy, "Murder of the Southwest," 63.

52. Ibid.; Peter Iverson, *The Navajo Nation* (Albuquerque: University of New Mexico Press, 1981, 1983), 109–112; Reno, *Mother Earth, Father Sky*, 162, note 35. Reno's figures are based upon 34,000 acre-feet for the Navajo Generating Station at Page.

53. Josephy, "Murder of the Southwest," 63; Iverson, *Navajo Nation*, 111-112; Price and Weatherford, "Navajo Experience," 115, 119–131; Reno, *Mother Earth, Father Sky*, 52; Christopher McLeod, "Navajos Resist Relocation," *High Country News*, 12 May 1986, 13.

54. Josephy, "Murder of the Southwest," 65; Iverson, *Navajo Nation*, 110.

55. "Black Mesa, a Deal for Slurry's Future; Williams Pays $36 Million," *Coal Week*, 16 May 1988; "Renegotiated Water and Coal Leases by the Navajo Nation and the Hopi Tribe," *Water Market Update*, October 1987.

56. U.S. Department of Interior (DOI), Office of Surface Mining Regulation and Enforcement (OSMRE), *Cumulative Hydrological Impact Assessment of the Peabody Coal Company Black Mesa/Kayenta Mine* (Denver: OSMRE, January 1988), Chapter 6; Mike Foley (hydrologist with Navajo Department of Surface Mining), telephone interview, March 1988; Michael O'Connell (Hopi attorney), telephone interview, April 1988; Mike Nelson (special counsel to Navajo Chairman Peterson Zah), telephone interview, October 1985.

57. Atkins interview.

58. Navajo Justice Department attorneys, telephone interviews, February 1988; Joseph Shields, "600 Navajos Due $1,700 Each in PNM Job Action," *Navajo Times*, 7 April 1988.

59. "Historic Moment for Tribe: Minerals Restoration Bill Signed into Law," *Ahead of the Herd Fort Berthold Press*, 2 November 1984; Rich Schilf (non-Indian natural resources director at Fort Berthold), interview in Denver, Colo., November 1986; Lawrin "Hugh" Baker (Three Affiliated Tribes councilman), telephone interview, March 1988.

60. Steven J. Shupe, "Water Management in Indian Country," *Journal of the American Waterworks Association*, October 1986, 57; *Federal Indian Law II*, 706–715; Folk-Williams, *What Indian Water Means*, "Position Paper by Montana Governor Ted Schwinden," 119, and "Memorandum to Chairman Earl Old Person, Blackfeet Indian Nation from William H. Veeder," 120–127.

61. Marjane Ambler, "Fort Peck Water Code Approved," *Lakota Times*, 22 October 1986; Shupe, "Water Management," 57–58.

62. *Colville Confederated Tribes v. Walton*, 647 F.2d 42 (9th Cir.), cert. denied, 454 U.S. 1092 (1981); *United States v. Anderson*, 736 F.2d 1358 (9th Cir. 1984); *Federal Indian Law II*, 680, 706–711; David H. Getches, "Water Rights on Indian Allotments," *South Dakota Law Review* 26 (Summer 1981): 405–433; Glen A. Wilkinson, *Indian Control and Use of Water for Mineral Development* (Denver: Rocky Mountain Mineral Law Foundation, 2 April 1976), 9–12; Trelease, "Indian Water Rights for Mineral Development," 231; Richard B. Collins, "Indian Allotment Water Rights," *Land and Water Law Review* 20 (1985), 436.

63. Reid Peyton Chambers, telephone interview, September 1986; *Federal Indian Law II*, 715; Ambler, "Fort Peck Code Approved"; Assiniboine and Sioux Tribes, "Fort Peck Tribal Water Code," adopted and approved 9 October 1986; Bob Delk (chief of water services, BIA Billings area office), telephone interview, March 1988; Shupe, "Water Management," 57–58; U.S. Department of Interior press release, "Interior Approves First Tribal Code since 1975," 9 October 1986.

64. Marjane Ambler, "Ruling May Change Wind River Tribes," *Denver Post*, 13 February 1983; Folk-Williams and Cannon, *Water for Energy*, 23, 59.

65. Folk-Williams and Cannon, *Water for Energy*, 33, 78, 80; Folk-Williams,

What Indian Water Means, 70; Navajo Nation, "Navajo Nation Water Code," Title 22, Navajo Tribal Code, Chapter 7 (approved August 1984); Shupe, "Water Management."

66. Shupe, "Water Management"; Nelson interview; Lawrence Lano (Navajo administrator for the tribal water code), telephone interview, March 1988.

67. Folk-Williams and Cannon, *Water for Energy*, 88; Greg Leisse (Peabody Coal Company, Flagstaff, Ariz.) and Fred Meurer (Pittsburgh-Midway, Denver), telephone interviews, October 1985; Lano interview.

68. Ambler, "Paper Rights."

69. Wyoming Assistant Attorney General R. T. Cox, telephone interview, August 1985.

70. Nelson interview; Assiniboine and Sioux Tribes, "Fort Peck Water Code," Section 801 (approved October 1986).

71. Douglas R. Nash, "Water Code Development," in Miklas and Shupe, *Indian Water 1985, Collected Essays*, 117–137.

72. Marjane Ambler, "Use It or Lose It, Officials Say," *Lakota Times*, 16 October 1985; Presidential Commission, *Report and Recommendations*, II-48 and 129–133; Robert Pelcyger, speech given at Natural Resources Law Center Ninth Annual Summer Program, in Boulder, Colo., 10 June 1988.

73. Folk-Williams, *What Indian Water Means*, 13; Price and Weatherford, "Navajo Experience," 126; Collins, "Future Course of *Winters*," 93; Wyoming State Supreme Court decision, 46.

74. Reno, *Mother Earth, Father Sky*, 62–63; Collins, "Future Course of *Winters*," 93; "Navajo Nation Water Code," Sec. 1301; "Fort Peck Water Code," Sec. 107 (o).

75. Ambler, "Ruling May Change Wind River."

76. Collins, "Indian Allotment Water," 430; *Federal Indian Law I*, 1983 supplement, 166, quoting Ninth Circuit Court in *Colville Confederated Tribes v. Walton*.

77. Northern Cheyenne Tribal President Robert Bailey, interview in Denver, Colo., November 1987; Crow Tribe, "Hydropower Memorandum of Understanding Delivered by Chairman Richard Real Bird" (mimeographed press release), 7 October 1986.

78. Janelle Conaway, "Navajos, PNM Join Forces to Build Power Factory," *Albuquerque Journal*, 28 May 1987.

79. Tucson Myers and Associates, "Water for Western Energy Development Update 1977," for the Western States Water Council, Salt Lake City, Utah, September 1977, 12, 17 (photocopy).

80. McCool, *Command of the Waters*, 243–245; Marjane Ambler, "A Tale of Two Irrigation Districts," in *Western Water Made Simple*, ed. *High Country News* (Washington, D.C.: Island Press, 1987), 148–149.

81. Based upon interviews with tribal representatives and attorneys.

82. Steven J. Shupe (coeditor of *Water Market Update*), telephone interview, October 1988.

83. Ambler, "Paper Rights"; Senate Committee on Interior, "Missouri Water Sale"; Folk-Williams, *What Indian Water Means*, 25.

84. Steven J. Shupe, "Issues and Trends in Western Water Marketing," *Resource Law Notes* (Boulder: University of Colorado Natural Resources Law Center, May 1988); Richard B. Collins, "Indian Reservation Water Rights," *Journal of the American Waterworks Association*, October 1986, 53; Ambler, "Paper Rights"; Wyoming Governor Ed Herschler to Wyoming congressional delegation, 28 October

1985, on proposed water rights settlement, author's files; *Federal Indian Law II,* 705.

85. *Federal Indian Law II,* 680, 705.

86. McCool, *Command of the Waters,* 224.

87. Ambler, "Nation Molding Indian Water Policy"; *Federal Indian Law II,* 705; Collins, "Indian Reservation Water Rights," 56-57; Scott McElroy (Southern Ute tribal attorney), telephone interview, October 1988; "Opposition to Colorado Ute Settlement Marketing Provisions," *Water Market Update,* November 1987; Interior Secretary Donald Paul Hodel to Senator Pete Domenici, 28 April 1988, author's files.

88. Richard Trudell and Joseph Myers, "How Indian Water Rights Are Resolved May Determine Future of Western United States," *CERT Report,* 23 July 1982.

89. Marjane Ambler, "Cheyenne Tribe Wins First Round," *Lakota Times,* 14 August 1985; Associated Press (AP), "Indian Tribe Wins Partial Victory in Coal Leasing Case," *Casper Star-Tribune,* 16 March 1988.

90. "Northern Tier," *CERT Report,* 1 November 1979, 3-4; Joseph G. Jorgensen, "The Political Economy of the Native American Energy Business," in *Native Americans and Energy Development II,* ed. Joseph G. Jorgensen (Boston: Anthropology Resource Center, 1984), 41-45; Dean Suagee, "American Indian Religious Freedom and Cultural Resources Management: Protecting Mother Earth's Caretakers," *American Indian Law Review* 10 (1982): 48; the *Peshlakai* lawsuit challenging the Dalton Pass EIS is discussed in Chapter 6.

91. Charles Cameron (non-Indian former geologist for Northern Ute Tribe), interview in Golden, Colo., November 1987; U.S. Department of Interior (DOI), Bureau of Land Management (BLM), *Final Socioeconomic Technical Report for the Uintah Basin Synfuels Development Environmental Impact Statement* (Salt Lake City: BLM, February 1983).

92. U.S. Department of Interior (DOI), Bureau of Land Management (BLM), *Final San Juan River Regional Coal Environmental Impact Statement,* (Albuquerque, N.Mex.: BLM, March 1984); Folk-Williams and Cannon, *Water for Energy,* 80; the allottees' case, *Eicilly v. United States,* is discussed in Chapter 6.

93. CERT Resolution No. 85-12, adopted 26 September 1985, Denver, Colo., CERT files; Folk-Williams and Cannon, *Water for Energy,* 80; Edwin Dahl (Northern Cheyenne representative on regional coal team), telephone interview, April 1988.

94. Nuclear Waste Policy Act, 7 January 1983, P.L. 97-425, 96 Stat. 2201; Marjane Ambler, "Law Recognizes Tribal Concerns; DOE Ignores Them, Say 3 Tribes," *Navajo Times Today,* 29 April 1985; Marjane Ambler, "Russell Jim Is Pro-Safety, Not Anti-Nuclear," *High Country News,* 7 July 1986.

95. Daniel Hester, "Protection of Sacred Sites and Cultural Resources: An Obstacle to Mineral Development in Indian Country?" speech presented at Rocky Mountain Mineral Law Institute (RMMLI), Mineral Development on Indian Lands, 16 February 1989, in Albuquerque, N.Mex., 19-38, RMMLI files; Jim Robbins, "Blackfeet Fight to Preserve Sacred Land, So Legends Won't Die," *Denver Post,* 29 December 1986; Folk-Williams, *Water for Energy,* 80; Bob Tucker, "Sioux Indians Allowed to Join Uranium Lawsuit," *Rapid City Journal,* 2 July 1981.

96. Richard O. Clemmer, "Effects of the Energy Economy on Pueblo Peoples," in Jorgensen, *Native Americans and Energy Development II,* 107.

97. Charles Wilkinson, "The Future of Western Water Law and Policy," in Miklas and Shupe, *Indian Water 1985, Collected Essays,* 58.

CHAPTER NINE. INTO THE BOSS'S SEAT:
THE TRIBE AS DEVELOPER

1. Dolores Proubasta, "Turning Energy into Power on Indian Lands," *Geophysics: The Leading Edge of Exploration*, December 1983, 39. Earl Old Person was the tribal chairman of the Blackfeet Tribe.

2. Bill Richards, "Crow Indians Find Huge Coal Reserves Help Them Very Little," *Wall Street Journal*, 31 January 1984; Charlie Cameron (former Northern Ute staff geologist), interview in Golden, Colo., November 1987; Marvin Cook (Ute director of the Southern Ute Natural Resources Division), telephone interview in Ignacio, Colo., December 1987.

3. Marjane Ambler, "Mineral Law Hamstrung by BIA Delay," *Lakota Times*, 14 March 1989.

4. Federal Trade Commission (FTC), *Staff Report on Mineral Leasing on Indian Lands* (Washington, D.C.: GPO, October 1975), 20, 59; Charles Lipton, "The Pros and Cons of Petroleum Agreements," *American Indian Journal*, February 1980, 2–10; U.S. Congress, Senate, Select Committee on Indian Affairs, *Hearings on S. 1894* [Indian Mineral Development Act] *to Permit Indian Tribes to Enter into Certain Agreements for the Disposition of Tribal Mineral Resources*, 97th Cong., 2d sess. (Washington, D.C.: GPO, March 1982), 128 [hereafter *Hearings on S. 1894 (IMDA)*].

5. Senate Select Committee, *Hearings on S. 1894 (IMDA)*, 128 (Marvin Sonosky testimony).

6. Ibid., 1–199.

7. Indian Mineral Development Act, 22 December 1982, P.L. 97-382, 96 Stat. 1938; David H. Getches and Charles F. Wilkinson, eds., *Federal Indian Law Cases and Materials* 2d ed. (St. Paul, Minn.: West Publishing Co., 1986), 635 [hereafter *Federal Indian Law II*].

8. David Allison (acting chief of BIA Energy and Minerals Division), interview in Boulder, Colo., June 1988; B. Reid Haltom, "Mineral Development Alternatives on Indian Lands," speech presented at Rocky Mountain Mineral Law Institute, Albuquerque, N.Mex., 16 February 1989.

9. Reid Peyton Chambers, "Mineral Leasing in Indian Country," speech presented at Natural Resources Law Center Ninth Annual Summer Program, Boulder, Colo., 9 June 1988; Council of Energy Resource Tribes (CERT), *Environmental Assessment of Oil and Gas Development on the Ute Mountain Ute Reservation* (Denver: CERT, May 1981), 151; Thomas E. Luebben, "Mining Agreements with Indian Tribes," *American Indian Journal*, May 1976, 2–8; Lipton, "Pros and Cons."

10. Thurman Velarde (Jicarilla Apache oil and gas administrator), interview in Dulce, N.Mex., October 1987.

11. Pete Aguilar (BIA Energy and Minerals Division staff), telephone interview, May 1988; Senate Select Committee, *Hearings on S. 1894 (IMDA)*, 169; "Tribal Council Statement regarding Allottee Oil and Gas Development," *Southern Ute Drum*, 30 October 1987; Rich Schilf (director of Three Affiliated Tribes Natural Resources Department at Fort Berthold), interview in Denver, Colo., November 1986; U.S Department of Interior (DOI), BIA Energy and Minerals Division, "Review of Oil and Gas Exploration Joint Venture Agreement between Blackfeet Tribe and Quantum Exploration, Inc.," Golden, Colo., no date, author's files.

12. Larry Monson (Fort Peck tribal geologist), telephone interview, May 1988; Thomas J. Vogenthaler (president of Cooper Petroleum), telephone interview, April

1984 (Cooper was a subsidiary of U.S. Energy and the operator for the Fort Peck–U.S. Energy agreement).

13. Lorraine Turner Ruffing, "Navajo Mineral Development," *American Indian Journal*, September 1978, 11; Stephen Zorn, "Getting a Fair Deal in Mining Projects," in *You Don't Have to Be Poor to Be Indian*, ed. Maggie Gover (Washington, D.C.: Americans for Indian Opportunity, n.d., 53–63.

14. Thomas E. Luebben, *American Indian Natural Resources: Oil and Gas* (Vienna, Va.: Institute for the Development of Indian Law, 1980), 5, 28; Charles Lipton, "Joint Venture Rip-Offs," paper delivered at the CERT Tribal Resource Development Workshop in Billings, Mont., 13 July 1982.

15. Based upon interviews with these tribes' staffs and attorneys; "Northern Cheyenne/ARCO Exploration and Drilling Agreement," and "Crow/Shell Coal Mining Agreement" in *CERT Tribal Resource Development Notebook* (Denver: CERT, 13–14 July 1982).

16. Lipton, "Pros and Cons," 9.

17. Aguilar interview; Marjane Ambler, "Indians Have New Avenues to Deal Oil," *Western Business*, May 1984; Ruffing, "Navajo Mineral Development," 11; "Andrus Calls for Renegotiation of Navajo Coal Lease," *Indian Law Reporter*, 20 July 1977.

18. Based upon interviews with tribal and BIA officials. CERT economist Ahmed Kooros said in a March 1984 telephone interview that he knew of only one negotiated oil and gas deal that would have provided more return if it had been opened to competition.

19. Patrick Dawson, "Tribe's Oil Self-Development: Why Did the Proposal Fall Apart?" *Billings Gazette*, 19 January 1981.

20. James P. Boggs, "The Challenge of Reservation Resource Development: A Northern Cheyenne Instance," in *Native Americans and Energy Development II*, ed. Joseph G. Jorgensen (Boston: Anthropology Resource Center, 1984), 223; U.S. Department of Interior (DOI), Bureau of Indian Affairs (BIA), *Environmental Assessment for the Northern Cheyenne Petroleum Development Project* (Billings, Mont.: BIA, August 1980).

21. Tom Acevedo (former Interior Department attorney), telephone interview, May 1988; David L. Baldwin (Osage; former chief of BIA Energy and Minerals Division), telephone interview, July 1988; Tom Fredericks, telephone interview, January 1987. The ARCO contract became the focal point for the Interior Department's reinterpretation of the competition requirements of the 1938 Mineral Leasing Act (see Chapter 3).

22. "Wenona Well Strikes Oil," *Wotanin Wowapi*, 9 March 1984; Velarde interview.

23. "Oil and Gas Office Has Many Income-Making Functions," *Wotanin Wowapi*, 9 March 1984; Vogenthaler interview; Larry Wetsit, telephone interview, July 1988.

24. Velarde interview; Robert Nordhaus, attorney for the Apaches, to the author, 30 December 1987; Reid Haltom (attorney for the Apaches), telephone interview, June 1988; "Jicarilla Apache/Odessa Natural Corp. Joint Exploration and Development Agreement," 21 March 1977, Jicarilla Apache Tribe's files.

25. James Cook, "New Hope on the Reservations," *Forbes*, 9 November 1981, 114; Barrie Damson (president of Damson Oil Company), telephone interview, November 1981; "Apache/Odessa Agreement," 2, 15.

26. Aguilar interview, referring to Crow-Raven agreement; Lipton, "Pros and Cons"; Ruffing, "Navajo Mineral Development"; David Harrison, "Negotiations: A Paradigm Shift of Tribal Self-Determination," paper presented at CERT annual

meeting, Denver, Colo., November 1986); Haltom, "Mineral Development Alternatives"; Chambers, "Mineral Leasing in Indian Country."

27. Daniel H. Israel, "The Reemergence of Tribal Nationalism and Its Impact on Reservation Resource Development," *Colorado Law Review* 47 (Summer 1976): 646; Cook, "New Hope," 114; Marjane Ambler, "Oil on the Reservation," *Western Energy*, March 1982, 41-44; Lipton, "Pros and Cons"; Luebben, *Oil and Gas*, 28.

28. Ruffing, "Navajo Mineral Development," 12; Lipton, "Joint Venture Rip-Offs"; *Western Nuclear Inc. v. Hodel*, Civil Action No. 87-618-JLQ, U.S. District Court, Spokane, Wash., filed 22 October 1987, 2-4, exhibit B.

29. Nordhaus letter; Chuska Energy Company, "Wind River Tribes/Chuska Proposed Oil and Gas Operating Agreement—Undeveloped Land," no date, Wind River Tribes' files.

30. "Wind River Tribes/Chuska Proposed Agreement," 3, 4; *Western Nuclear v. Hodel*, exhibit A ("Spokane Tribe of Indians/Western Nuclear Mining Lease").

31. Velarde interview; "Apache/Odessa Agreement."

32. FTC, *Staff Report*, 31, 32.

33. Haltom, "Mineral Development Alternatives"; Chambers, "Mineral Leasing in Indian Country."

34. Dianna Troyer, "Watt Proposal Filled with Problems, No New Ideas," *Wyoming State Journal*, 18 March 1985; "Wind River/Chuska Agreement."

35. Marjane Ambler, "Watt's Client Loses to Conoco in Contest for Wind River Tribal Contracts," *Inside Energy*, 29 July 1985; Marjane Ambler, "Arapahoe Attorney Accuses Watt of Influence Peddling," *Lakota Times*, 31 July 1985.

36. Marjane Ambler, "Crow Council Approves Contract," *Inside Energy*, 18 April 1983; David L. Baldwin, BIA Division of Energy and Minerals, "Review of O'Hare Energy Corporation Proposal to Crow Tribe," 6 December 1982, author's files; "Raven Oil Company Partnership Agreement," 18 March 1983, author's files; Joe Rawlins (Crow consulting geologist), telephone interview, June 1988; BIA Energy and Minerals Division, "Review of Oil and Gas Partnership Agreement between Crow Tribe and Buffalo Exploration Company," 29 September 1983, author's files; Aguilar interview;

37. Rawlins interview; Alan O'Hare (president of O'Hare Energy Corporation), telephone interview, July 1988.

38. Associated Press (AP), "ARCO Exploration on Reservation Ends," *Char-Koosta*, 21 August 1984; DOI, *Northern Cheyenne Petroleum Development Project*, A-6, A-15, 3-8.

39. Charles D. Williams (senior vice president of exploration, Wintershall Corporation), interview in Denver, Colo., November 1987; CERT, "Oil and Gas on the Ute Mountain Ute Reservation."

40. Judy Knight (vice chairman of Ute Mountain Ute Tribe), telephone interview, December 1987; Williams interview.

41. Claude Neely (former Amoco Production Company regional land manager), interview in Denver, Colo., November 1987; "Crow/Shell Coal Mining Agreement."

42. Perry Baker (Mandan-Hidatsa; Uintah and Ouray Agency BIA superintendent), interview in Fort Duchesne, Utah, May 1988; Lester Chapoose (Northern Ute tribal chairman), interview in Fort Duchesne, Utah, May 1987; Georgia Wyasket (Santa Clara Pueblo; Northern Ute Energy and Minerals Department staff), interview in Fort Duchesne, Utah, May 1987; Charles Cameron (former Northern Ute staff geologist), interview in Golden, Colo., November 1987; Charging Ute Corporation (CUC), "Final Environmental Assessment of the Ute Oil Refinery," December 1986, 4, Ute Tribe's files.

43. Maxine Natchees, "Charging Ute Corporation," *Ute Bulletin*, June 1987; John Martin, "Promises from Charging Ute," *Ute Bulletin*, March 1987; George M. Elliott (president of CUC), "CUC Responds to Maxine Natchees' Letter," *Ute Bulletin*, 2 September 1987.

44. Interviews with Chapoose, Baker, and Cameron.

45. Boggs, "Northern Cheyenne Instance," 221–223; Mike Nelson (former special counsel to Navajo Chairman Peterson Zah), interview in Boulder, Colo., June 1988; Marvin Cook (Ute; director of the Southern Ute Natural Resources Division), telephone interview in Ignacio, Colo., December 1987.

46. Based upon interviews with tribal and BIA officials.

47. Council of Energy Resource Tribes (CERT), "Tribal Self-Government: A Self-Sufficiency Analysis," by Ahmed Kooros and Anne C. Seip, Denver, Colo., October 1983, 41 (photocopy).

48. Velarde and Cook interviews.

49. Based upon conversations with CERT, BIA, and tribal officials and upon Luebben, "Mining Agreements," and Lorraine Turner Ruffing, "The Role of Policy in American Indian Mineral Development," paper for Native American Studies Center at the University of New Mexico, Albuquerque, 1980 (photocopy).

50. "Tribal Council Weighs $78.6 Million Budget," *Navajo Times*, 8 October 1987; Chambers, "Mineral Leasing in Indian Country," William A. White (attorney for Westmoreland Resources), "The Industry Perspective: The Pros and Cons of Mineral Development in Indian Country," and Donald R. Wharton, "Resources Development on Navajo: The Dineh Power Project," papers presented at Natural Resources Law Center Ninth Annual Summer Program, Boulder, Colo., 9 June 1988.

51. Harrison, "Negotiations: Paradigm Shift," 5; "Blackfeet/Damson Exploration and Operations Service Agreement" and "Northern Cheyenne/ARCO Exploration and Drilling Agreement" in *CERT Tribal Resource Development Notebook*, 13–14 July 1982.

52. Wharton, "Dineh Power Project," 5–7, 12–14.

53. Ibid., 21–22.

54. Daniel Israel, "Industry Perspective on Tribal Sovereignty and Reservation Taxation," speech presented at Energy Bureau Forum, Denver, Colo., 18 May 1982; "Blackfeet/Damson Agreement."

55. "Crow/Shell Coal Mining Agreement." The U.S. Supreme Court later ruled that the Montana severance tax could not be applied to Crow coal, but by then Shell had withdrawn from the contract.

56. Williams interview.

57. Patrick Dawson, "Reach Coal Accord, U.S. Warns Crows," *Billings Gazette*, 29 October 1977.

58. Boggs, "Northern Cheyenne Instance," 221; James G. Watt to Northern Arapahoe Business Council Chairman Chester Armajo, 19 March 1985, author's files; Northern Arapahoe Business Council Chairman Chester Armajo to Shoshone Business Council Chairman Robert Harris, 17 May 1985 (printed in *Wind River News*, 21 May 1985); Chester Armajo, interview in Fort Washakie, Wyo., March 1985.

59. "Blackfeet/Quantum Exploration, Inc., Oil and Gas Joint Venture," *Wotanin Wowapi*, 9 September 1983; "Tribe Rescinds Oil and Gas Agreement with Quantum Exploration, Inc.," *Blackfeet Tribal News*, October 1984; *Quantum Exploration, Inc., v. Clark*, 780 F.2d 1457 (9th Cir. 1986); telephone interviews in July 1988 with Reid Haltom (attorney for the Blackfeet), Roger Evans (attorney for Quantum), and Robert "Smokey" Doore (petroleum consultant and former Blackfeet tribal councilman).

60. Natchees, "Charging Ute Corporation"; Elliott, "CUC Responds"; Baker interview; Council of Energy Resource Tribes (CERT), "Summary of Oil and Gas Laws applicable to the Blackfeet Indian Reservation," by Vicky Santana, Denver, Colo., no date, 15 (photocopy); "Raven Oil Company Partnership Agreement," 18 March 1983, author's files; Ruffing, "Role of Policy"; Ahmed Kooros, interview in Denver, Colo., November 1987; Luebben, "Mining Agreements"; Wharton, "Dineh Power Project."

61. Velarde interview; Senate Select Committee, *Hearings on S. 1894 (IMDA)*, 40–42; Marjane Ambler, "Navajo," *CERT Report*, 20 June 1980; Ruffing, "Role of Policy," 15; Frank Long, "Alternate Development Agreements," paper presented at CERT Tribal Resource Development Workshop, Billings, Mont., 13 July 1982; Nelson interview; Marjane Ambler, "Tribe Creates Development Corporation," *High Country News*, 3 September 1982.

62. Caleb Shields, telephone interview, July 1988; Doore interview; Haltom (attorney for Jicarilla Apache Tribe and Laguna Pueblo), telephone interview; Chambers, "Mineral Leasing in Indian Country."

63. Tom Fredericks (Mandan-Hidatsa attorney representing several tribes and former assistant secretary of Interior), telephone interview, July 1988.

APPENDIX FOUR. SUGGESTIONS FOR INDUSTRY

1. Based upon interviews with industry and tribal representatives and upon Dolores Proubasta, "Turning Energy into Power on Indian Lands," *Geophysics: The Leading Edge of Exploration*, December 1983; Daniel Israel, "Industry Perspective on Tribal Sovereignty and Reservation Taxation," speech presented at Energy Bureau Forum, Denver, Colo., 18 May 1982; B. Reid Haltom, "Mineral Development Alternatives on Indian Lands," speech presented at Rocky Mountain Mineral Law Institute, Albuquerque, N.Mex., 16 February 1989.

SELECTED BIBLIOGRAPHY

When documents can be obtained from the National Indian Law Library in Boulder, Colorado, they are marked "NILL." GPO refers to the Government Printing Office.

BOOKS AND ARTICLES

Alverez, Frank H., and J. Kevin Poorman. "Real Property: Congressional Control of Allotted Mineral Interests." *American Indian Law Review* 3 (1975): 159-167. [Document can be obtained from the National Indian Law Library, Boulder, Colo. (hereafter NILL)].

Ambler, Marjane. "Indians Have New Avenues to Deal Oil." *Western Business* (Billings, Mont.), May 1984.

_____. "Oil on the Reservation." *Western Energy* (Casper, Wyo.,), March 1982, 41-44.

_____. "Oil Rich and Penny Poor." *Denver Post Empire Magazine*, 13 October 1985.

_____. "Paper Rights and Wet Water: The Question of Indians." *Northern Lights* (Missoula, Mont.), November/December 1985.

_____. "Reading Fine Print on Indian Contracts." *Western Energy*, October 1982, 3.

_____. "The Real Water Lawyers" and "A Tale of Two Irrigation Districts." In *Western Water Made Simple*, edited by *High Country News*. Washington, D.C.: Island Press, 1987.

_____. "Tribes Refining Goals for Resources." *Western Business* (Billings, Mont.), February 1985.

_____. "Who Controls the Oil?" *Nations, the Native American Magazine* (Seattle, Wash.), August 1981.

Americans for Indian Opportunity (AIO). *Messing with Mother Nature Can Be Hazardous to Your Health!* Washington, D.C.: AIO, no date.

_____. *Survey of American Indian Environmental Protection Needs on Reservation Lands*, written for the Environmental Protection Agency. Washington, D.C.: GPO, September 1986.

_____. *You Don't Have to Be Poor to Be Indian*, edited by Maggie Gover. Washington, D.C.: AIO, no date.

Anderson, David H. "Strip Mining on Reservation Lands: Protecting the Environment and the Rights of Indian Allotment Owners." *Montana Law Review* 35 (1974): 209-226. NILL.

Archer, Victor E., J. Dean Gillam, and Joseph K. Wagoner. "Respiratory Disease Mortality among Uranium Miners." *Annals of the New York Academy of Sciences* 271 (28 May 1976): 280-293.

Ashabranner, Brent. *Morning Star, Black Sun: The Northern Cheyenne Indians and America's Energy Crisis*. New York: Dodd, Mead & Company, 1982.

Barry, Tom. "Navajo Legal Services and Friends of the Earth Sue Six Federal Agencies over Alleged Careless Uranium Mining Policies." *American Indian Journal* (Vienna, Va.), February 1979, 3–7.

Bomberry, Victoria. "*Native Self-Sufficiency* Interviews Richard Trudell." *Native Self-Sufficiency* (Forestville, Calif.) 6: 4–5.

Brophy, William A., and Sophie D. Aberle. *The Indian: America's Unfinished Business—Report of the Commission on the Rights, Liberties, and Responsibilities of the American Indian*. Norman: University of Oklahoma Press, 1966.

Bryan, William L., Jr. *Montana's Indians Yesterday and Today*. Helena: *Montana Magazine*, 1985.

Burness, H. S., R. G. Cummings, W. D. Gorman, and R. R. Lansford. "United States Reclamation Policy and Indian Water Rights." *Natural Resources Journal* 20 (October 1980): 824.

Burton, Lloyd. "The American Indian Water Rights Dilemma: Historical Perspective and Dispute-Settling Policy Recommendations." *UCLA Journal of Environmental Law and Policy* 7 (Spring 1988).

Canard, Curtis. "A Petroleum Geologist Branches Out in a Burgeoning Field." *American Indian Journal* (Vienna, Va.), Feb. 1980, 21–24.

Cannon, James. "Leased and Lost." *Economic Priorities Report* (New York: Council on Economic Priorities) 5, no. 2 (1974): 1–48.

Chambers, Reid Peyton. "Judicial Enforcement of the Federal Trust Responsibility to Indians." *Stanford Law Review* 27 (1975): 1213–1230. NILL.

Chambers, Reid Peyton, and Monroe E. Price. "Regulating Sovereignty: Secretarial Discretion and the Leasing of Indian Lands." *Stanford Law Review* 26 (1974). NILL.

Charging Ute Corporation (CUC). *Final Environmental Assessment of the Ute Oil Refinery*. Fort Duchesne, Utah: CUC, December 1986.

Cohen, Felix. *Felix Cohen's Handbook of Federal Indian Law*. 2d ed. Charlottesville, Va.: Michie/Bobbs-Merrill, 1982.

_____. *Handbook of Federal Indian Law*. Albuquerque: University of New Mexico Press, 1972.

Collier, John. *Indians of the Americas: The Long Hope*. New York: New American Library, 1947.

Collins, Richard B. "Indian Allotment Water Rights." *Land and Water Law Review* 20 (1985): 422–457. NILL.

_____. "Indian Reservation Water Rights." *Journal of the American Waterworks Association*, October 1986, 48–54.

_____. "Memo to Council of Energy Resource Tribes re the Secretary of the Interior's Authority to Approve Joint Venture Mineral Development Agreements between Tribes and Mining Companies." 13 October 1981. NILL.

Cook, Chuck, Mike Masterson, and Mark N. Trahant. "Fraud in Indian Country." *Arizona Republic*, 4–11 October 1987 (reprint).

Cook, James. "New Hope on the Reservations." *Forbes*, 9 November 1981, 108–115.

Corke, Charles Philip. "Corke Testimony Exposing Interior's Failure to Protect Indian Water Rights, April 4, 1978." *Wassaja* (San Francisco: American Indian Historical Society), May 1979.

Council of Energy Resource Tribes (CERT). *The Control and Reclamation of Sur-*

face Mining on Indian Lands, by Douglas Richardson, for the Office of Surface Mining. Denver: CERT, September 1979.

————. *Crow Power Plant Feasibility Study*, for the Department of Energy. Denver: CERT, 1982.

————. *Environmental Assessment of Oil and Gas Development on the Ute Mountain Ute Reservation*. Denver: CERT, May 1981.

————. *Environmental Laws and Policies Reference Manual*, by Mervyn Tano. Denver: CERT, November 1986.

————. *First Decade*. Denver: CERT, November 1986.

————. *Inventory of Hazardous Waste Generators and Sites on Selected Indian Reservations*, by Warner Reeser, for the Environmental Protection Agency. Denver: CERT, July 1985.

————. "Negotiations: A Paradigm Shift of Tribal Self-Determination," by David Harrison. Presented at CERT annual meeting, Denver, Colo., November 1986. Photocopy.

————. "Prefeasibility Analysis of Coal Development on the Southern Ute Reservation: Six Options." Denver, Colo., June 1981. Photocopy.

————. *Proceedings of the Energy Forum—Strategic Planning for Marketing Indian Gas*. Denver: CERT, April 1986.

————. "Recommendations Submitted to FEA regarding Energy Resource Development on Indian Reservations." Washington, D.C., September 1975. Photocopy. NILL.

————. "Summary of Oil and Gas Law Applicable to the Blackfeet Indian Reservation," by Vicky Santana. Denver, Colo., no date. Photocopy. NILL.

————. "Tribal Self-Government: A Self-Sufficiency Analysis," by Ahmed Kooros and Anne C. Seip. Denver, Colo., October 1983. Photocopy.

Debo, Angie. *And Still the Waters Run: The Betrayal of the Five Civilized Tribes*. Princeton, N.J.: Princeton University Press, 1940: reprint, Norman: University of Oklahoma Press, 1984.

Deloria, Philip Sam. "CERT: It's Time for an Evaluation." *American Indian Law Newsletter* (Albuquerque, N.Mex.), September/October 1982.

DuBey, Richard A., Mervyn T. Tano, and Grant D. Parker. "Protection of the Reservation Environment: Hazardous Waste Management on Indian Lands." *Environmental Law* (Portland, Oreg.: Lewis and Clark College) 18 (1988): 449–504.

DuMars, Charles, and Helen Ingram. "Congressional Quantification of Indian Reserved Water Rights: A Definitive Solution or a Mirage?" *Natural Resources Journal* 20, no. 17 (1980): 17–43.

Energy Resource Development: Implications for Women and Minorities in the Intermountain West, State Advisory Committees to the U.S. Commission on Civil Rights, symposium, Denver, Colo., 2–3 November 1978. Washington, D.C.: GPO, 1978

Farrell, John Aloysius. "The New Indian Wars." *Denver Post*, 20–27 November 1983 (reprint).

Fluor Engineers and Constructors. *Crow Synfuels Feasibility Study*. Prepared for the Department of Energy. Washington, D.C.: DOE, August 1982.

Folk-Williams, John A. *What Indian Water Means to the West*. Santa Fe, N.Mex.: Western Network, 1982.

Folk-Williams, John A., and James S. Cannon. *Water for the Energy Market*. Santa Fe, N.Mex.: Western Network, 1983.

Ford, Bacon & Davis Utah. *A Summary of the Phase II–Title I Engineering Assessment of Inactive Uranium Mill Tailings, Shiprock Site, Shiprock, N.M.,* prepared for the U.S. Energy Research and Development Administration. Salt Lake City, Utah: Ford, Bacon & Davis Utah, March 1977.

Fowler, Loretta. *Arapahoe Politics, 1851–1978: Symbols in Crises of Authority*. Lincoln: University of Nebraska Press, 1982.

_____. *Shared Symbols, Contested Meanings: Gros Ventre Culture and History, 1778–1984*. Ithaca, N.Y.: Cornell University Press, 1987.

Fredericks, John, III. "Financing Indian Agriculture: Mortgaged Indian Lands and the Federal Trust Responsibility." *American Indian Law Review* 14 (1989).

Galuszka, Peter. "1911–1986: Coal's Rise and Fall and Rise." *Coal Age* (New York: McGraw-Hill–World News), June 1986, 47–55.

Gates, Paul W. *History of Public Land Law Development*, written for the Public Land Law Review Commission. Washington, D.C.: GPO, November 1968.

Getches, David H. "Water Rights on Indian Allotments." *South Dakota Law Review* 26 (Summer 1981): 405–433.

Getches, David H., Daniel M. Rosenfelt, and Charles F. Wilkinson, eds. *Federal Indian Law Cases and Materials*. St. Paul, Minn.: West Publishing Co., 1979.

Getches, David H., and Charles F. Wilkinson, eds. *Federal Indian Law Cases and Materials*, 2d ed. St. Paul, Minn.: West Publishing Co., 1986.

Gillenkirk, Jeff, and Mark Dowie. "The Great Indian Power Grab." *Mother Jones*, January 1982.

Gover, B. Kevin. "Indian Tribal Governments Look to Take Control of Reservation Environments." *The Workbook* (Albuquerque, N.Mex.: Southwest Research and Information Center) 12 (July–September 1987): 89–96.

Gudgell, J. Dallas C. "Fort Peck Project Environmental Study for the Fort Peck Reservation." Senior paper, Fort Peck Community College, 1985.

Hagan, William T. "Tribalism Rejuvenated: The Native American since the Era of Termination." *Western Historical Quarterly*, January 1981, 5–16.

Harris, Fred, and LaDonna Harris. "Indians, Coal, and the Big Sky." *The Progressive*, 1974 (reprint).

Hertzberg, Hazel W. "Reaganomics on the Reservation." *New Republic*, 22 November 1982, 15–18.

Hill, Arthur. "CERT Blasts DOI, BIA on Funding Commitments." *Western Oil Reporter*, October 1980, 68–70.

Hoxie, Frederick E. "Building a Future on the Past: Crow Indian Leadership in an Era of Division and Reunion." In *Indian Leadership*, edited by Walter L. Williams, 76–84. Manhattan, Kans.: Sunflower University Press, 1984.

_____. *A Final Promise: The Campaign to Assimilate the Indians, 1880–1920*. Lincoln: University of Nebraska Press, 1984.

Hundley, Norris, Jr. "The *Winters* Decision and Indian Water Rights: A Mystery Reexamined." In *The Plains Indians of the Twentieth Century*, edited by Peter Iverson, 77–106. Norman: University of Oklahoma Press, 1985.

Institute for the Development of Indian Law. *American Indian Natural Resources: Oil and Gas*, by Thomas E. Luebben. Vienna, Va.: Institute for the Development of Indian Law, 1980.

_____. *The Federal-Indian Trust Relationship*, by Gilbert L. Hall. Vienna, Va.: Institute for the Development of Indian Law, 1979.

Israel, Daniel H. "New Opportunities for Energy Development on Indian Reservations." *Mining Engineering* (publication of the Society of Mining Engineers), June 1980.

_____. "The Reemergence of Tribal Nationalism and Its Impact on Reservation Resource Development." *University of Colorado Law Review* 47 (Summer 1976): 617–652. NILL.

Iverson, Peter. "History of the Wind River Reservation Water Rights." University of Wyoming, no date. Photocopy.

———. "Knowing the Land, Leaving the Land: Navajos, Hopis, and Relocation in the American West." *Montana: The Magazine of Western History* 38, no. 1 (Winter 1988): 67–70.

———. *The Navajo Nation*. Albuquerque: University of New Mexico Press, 1981.

Jackson, Dan. "Mine Development on U.S. Indian Lands." *Engineering and Mining Journal*, January 1980.

Jacobs, Mike. *One Time Harvest: Reflections on Coal and Our Future*. Mandan: North Dakota Farmers Union, 1975.

Jefferson, James, Robert W. Delaney, and Gregory C. Thompson. *The Southern Utes: A Tribal History*, edited by Floyd A. O'Neil. Ignacio, Colo.: Southern Ute Tribe, 1972.

Johnson, Norman K. "Indian Water Rights in the West," prepared for the Western States Water Council, Salt Lake City, Utah, December 1983. NILL.

Jorgensen, Joseph G., ed. *Native Americans and Energy Development I*. Boston: Anthropology Resource Center, 1978.

———. *Native Americans and Energy Development II*. Boston: Anthropology Resource Center, 1984.

Josephy, Alvin M., Jr. "The Murder of the Southwest." *Audubon*, July 1971.

———. "The Agony of the Northern Plains." *Audubon*, July 1973.

Keene-Osborn, Sherry, and Jeff Rundles. "Wyoming Oil Thefts before Grand Jury." *Rocky Mountain Journal* (Denver), 29 October 1980.

Kelly, Lawrence C. *The Navajo Indians and Federal Indian Policy*. Tucson: University of Arizona Press, 1968.

Kickingbird, Kirke, and Karen Ducheneaux. *One Hundred Million Acres*. New York: Macmillan Publishing Co., 1973.

Kinney, J. P. *A Continent Lost—A Civilization Won: Indian Land Tenure in America*. Baltimore: Johns Hopkins Press, 1937; reprint, New York: Arno Press, 1975.

Kohn, Howard. "The Lawman Who Corralled the Oil Rustlers." *Reader's Digest*, August 1984, 43–48.

LaCourse, Richard, and John Butler. "45 Indian Tribes in a Dozen States Form Heartland of Indian Oil Production." *CERT Report*, 13 September 1982.

LaDuke, Winona. "CERT—An Outsider's View In." *Akwesasne Notes* (Rooseveltown, N.Y.), Summer 1980, 20–22.

———. "Energy Mission, Doom for Indians." *Mine Talk* (Albuquerque, N.Mex.: Southwest Research and Information Center), July/August 1981, 11–13.

Lamb, Terrence J. "Some Considerations regarding Blackfeet Indian Water and Land Use in the Early Twentieth Century." Paper delivered at the Western History Association meeting, Billings, Mont., October 1986.

Lewis, Jack. "An Indian Policy at EPA." *EPA Journal*, January/February 1986, 23–26.

Lipton, Charles. "Indian Mineral Leases." *American Indian Journal* (Vienna, Va.), May 1976, 9–10.

———. "The Pros and Cons of Petroleum Agreements." *American Indian Journal* (Vienna, Va.), February 1980, 2–10.

Luebben, Thomas E. "Mining Agreements with Indian Tribes." *American Indian Journal* (Vienna, Va.), May 1976, 2–8.

MacDonald, Peter. Letter to President Jimmy Carter, 20 July 1979. NILL.

Maddox, Sam. "Can 17th Street Make Peace with the Energy Tribes?" *Denver Magazine*, May 1980, 32–37.

Massie, Michael A. "The Cultural Roots of Indian Water Rights." *Annals of Wyoming* 59, no. 1 (Spring 1987): 15–28.

Maxfield, Peter C. "Tribal Control of Indian Mineral Development." *Oregon Law Review* 62 (1983): 49–72. NILL.

McCool, Daniel. *Command of the Waters: Iron Triangles, Federal Water Development, and Indian Water.* Berkeley: University of California Press, 1987.

McMahan, Robert L., and Porter B. Bennett (Abt/West). *Energy Activity in the West,* written for the Western Governors Policy Office, in cooperation with Four Corners Regional Commission. Denver: Western Governors Office, March 1981.

McNickle, D'Arcy. *Native American Tribalism: Indian Survivals and Renewals.* New York: Oxford University Press, 1973.

Miklas, Christine L., and Steven J. Shupe, eds. *Indian Water 1985, Collected Essays.* Oakland, Calif.: American Indian Resources Institute, 1986.

Miner, H. Craig. *The Corporation and the Indian: Tribal Sovereignty and Industrial Civilization in Indian Territory, 1865–1907.* Columbia: University of Missouri Press, 1976.

"Missouri: River of Promise or River of Peril?" *Missouri River Brief Series.* Missoula, Mont.: Northern Lights Institute, February 1988.

National Congress of American Indians (NCAI). *Environmental Protection in Indian Country: A Handbook for Tribal Leaders.* Washington, D.C.: NCAI, 1987.

Native American Natural Resources Development Federation (NANRDF) of the Northern Great Plains. "Declaration of Indian Rights to the Natural Resources in the Northern Great Plains States," prepared by member tribes, in conjunction with Native American Rights Fund, Bureau of Indian Affairs, and private consultants. Denver, Colo., June 1974. Photocopy. NILL.

Natural Resource Development in Indian County. Boulder: University of Colorado Natural Resources Law Center, June 1988.

Newmont Mining Corporation. "1987 Annual Report." New York, N.Y., April 1988.

O'Neil, Floyd A., and Kathryn L. MacKay. *A History of the Uintah-Ouray Ute Lands.* Salt Lake City: American West Center, 1982.

Ortiz, Roxanne Dunbar, ed. *Economic Development in American Indian Reservations.* Albuquerque: University of New Mexico Native American Studies Development Series, 1979.

Parlow, Anita. *Cry, Sacred Ground.* Washington, D.C.: Christic Institute, 1988.

Pearson, Jessica S. *A Sociological Analysis of the Reduction of Hazardous Radiation in Uranium Mines,* prepared for the National Institute for Occupational Safety and Health, Public Health Service. Washington, D.C.: GPO, 1975.

Peres, Ken, and Fran Swan. "The New Indian Elite: Bureaucratic Entrepreneurs." *Akwesasne Notes* (Rooseveltown, N.Y.), Late Spring 1980, 18.

Philp, Kenneth R. *Indian Self-Rule: First-Hand Accounts of Indian-White Relations from Roosevelt to Reagan.* Salt Lake City: Howe Brothers Press, 1986.

———. "Termination: A Legacy of the Indian New Deal." *Western Historical Quarterly,* April 1983, 165–180.

Popkin, Roy. "Indians Act for a Cleaner Environment." *EPA Journal,* April 1987, 28–31.

Price, Monroe E., and Gary D. Weatherford. "Indian Water Rights in Theory and Practice: Navajo Experience in the Colorado River Basin." In *American Indians and the Law,* edited by Lawrence Rosen, 97–131. New Brunswick, N.J.: Transaction Books, 1976. NILL.

Proubasta, Dolores. "Turning Energy into Power on Indian Lands." *Geophysics: The Leading Edge of Exploration,* December 1983, 32–39. NILL.

Prucha, Francis Paul. *The Great Father: The United States Government and the American Indians.* Abridged ed. Lincoln: University of Nebraska Press, 1986.

Redhorse, David, and Theodore Reynolds Smith. "American Indian Tribal Taxation of Energy Resources." *Natural Resources Journal* 22 (July 1982): 659–671.

Reiniger, Clair. *The Directory of Indian Appropriate Technology Projects in the U.S.A.* Santa Fe, N.Mex.: Designwright's Collaborative, 1981.

Reno, Philip. "High, Dry, and Penniless." *The Nation*, 29 March 1975, 359–363.

———. *Mother Earth, Father Sky, and Economic Development: Navajo Resources and Their Use.* Albuquerque: University of New Mexico Press, 1981.

Rocky Mountain Mineral Law Institute. *The Acquisition of Rights to Prospect for and Mine Coal from Tribal and Allotted Indian Lands,* by Jerome H. Simonds. Denver: Rocky Mountain Mineral Law Foundation, 1975.

———. *Federal Valuation of Indian Oil and Gas for Royalty Purposes,* by Patricia L. Brown. Denver: Rocky Mountain Mineral Law Foundation, 1988.

———. *Indian Self-Determination: Patterns for Mineral Development,* by Alvin J. Ziontz. Denver: Rocky Mountain Mineral Law Foundation, 1976.

———. *An Indian View of Mineral Development on Indian Land,* by Peter Mac-Donald. Denver: Rocky Mountain Mineral Law Foundation, 1976. NILL.

———. "Indian Water Rights for Mineral Development," by Frank J. Trelease. In *Natural Resources Law on American Indian Lands,* edited by Peter C. Maxfield, 230–233. Denver: Rocky Mountain Mineral Law Foundation, 1977.

———. *Mineral Development on Indian Lands.* Denver: Rocky Mountain Mineral Law Foundation, 1989.

———. *Mineral Development on Indian Lands—Cooperation and Conflict,* by Louis R. Moore. Denver: Rocky Mountain Mineral Law Foundation, 1983.

———. *Negotiations for Acquiring Exploration Rights on Indian Lands,* by Edward B. Berger. Denver: Rocky Moutain Mineral Law Foundation, 1973.

———. *Title Examination of Indian Lands,* by Louis R. Moore. Denver: Rocky Mountain Mineral Law Foundation, 1976.

———. *Use of Indian Water in Developing Mineral Property,* by Stephen G. Boyden and Scott C. Pugsley. Denver: Rocky Mountain Mineral Law Foundation, 1978.

Ruffing, Lorraine Turner. "Navajo Mineral Development." *American Indian Journal* (Vienna, Va.), September 1978, 2–15. NILL.

———. "The Role of Policy in American Indian Mineral Development," prepared for University of New Mexico Native American Studies Center. Albuquerque, N.Mex., 1980. Photocopy.

Sando, Joe S. *The Pueblo Indians.* San Francisco: Indian Historian Press, 1976.

Shanks, Bernard. "The American Indian and Missouri River Water Developments." *Water Resources Bulletin* 10 (June 1974): 573–579.

Shuey, Chris. "Accident Left Long-Term Contamination of Rio Puerco, But Seepage Problem Consumes New Mexico's Response." *Mine Talk* (Albuquerque, N.Mex.: Southwest Research and Information Center), Summer/Fall 1982, 10–26.

———. "The Puerco River: Where Did the Water Go?" *The Workbook* (Albuquerque, N.Mex.: Southwest Research and Information Center) 11 (January-March 1986): 1–10.

———. "Uranium Mines and Their Problems." *The Workbook* (Albuquerque, N.Mex.: Southwest Research and Information Center) 10 (July-September 1985): 113–115.

Shupe, Steven J. "Water in Indian Country: From Paper Rights to a Managed Resource." *University of Colorado Law Review* 57 (Spring 1986): 561–592.

Siskind, Janet. "A Beautiful River That Turned Sour." *Mine Talk* (Albuquerque, N.Mex.: Southwest Research and Information Center), Summer/Fall 1982, 37–59.

Smillie, John D. "Whatever Happened to the Energy Crisis?" *The Plains Truth* (Billings, Mont.: Northern Plains Resource Council), April 1986.

Smith, Duane A. *Mining America*. Lawrence: University Press of Kansas, 1987.

Smith, Theodore Reynolds. "Overview of the Council of Energy Resource Tribes." *Mineral & Energy Resources* (Golden: Colorado School of Mines), 1982.

Smith, Patrick, and Jerry D. Guenther. "Environmental Law—Protecting Clean Air: The Authority of Indian Governments to Regulate Reservation Airsheds." *American Indian Law Review* 9 (1981): 83–119.

Sonosky, Marvin J. "Oil, Gas, and Other Minerals on Indian Reservations." *Federal Bar Journal* 20 (1960): 230–234.

St. Aubin, Kim, and Sue Massie. *The Abandoned Mine Land Program, 1977–1987*. Springfield, Ill.: Association of Abandoned Mined Land Programs, May 1987.

Suagee, Dean. "American Indian Religious Freedom and Cultural Resources Management: Protecting Mother Earth's Caretakers." *American Indian Law Review* 10 (1982): 1–58.

Swenson, Robert W. "Legal Aspects of Mineral Resources Exploitation." In *History of Public Land Law Development*, edited by Paul W. Gates, 699–764. Washington, D.C.: GPO, November 1968.

Taylor, Lynda. "The Health Effects of Radiation." *The Workbook* (Albuquerque, N.Mex.: Southwest Research and Information Center) 10 (October-December 1985): 147–159.

————. "Uranium Legacy." *The Workbook* (Albuquerque, N.Mex.: Southwest Research and Information Center) 8 (November/December 1983): 192–207.

Temple, John B. "A Conversation with CERT Executive Director David Lester." *The Landman*, November 1986.

Tiller, Veronica E. Velarde. *The Jicarilla Apache Tribe: A History, 1846–1970*. Lincoln: University of Nebraska Press, 1983.

Tolan, Sandy. "Uranium Plagues the Navajos." *Sierra* (San Francisco: Sierra Club), November/December 1983, 55–60.

Toole, K. Ross. *The Rape of the Great Plains: Northwest America, Cattle and Coal*. Boston: Atlantic Monthly Press Book, Little, Brown and Company, 1976.

Touche Ross & Company. Letter to CERT board of directors, 11 January 1980. Author's files.

"U.S. Indians Demand a Better Energy Deal," *Business Week*, 19 December 1977, 53.

Veeder, William H. "Indian Water Rights and Reservation Development." In *Red Power*, edited by Alvin M. Josephy, Jr., 177–184. New York: McGraw-Hill, 1971.

Watt, James G. "Conservative Counterpoint." Satellite Program Network, 19 January 1983.

Weatherford, Gary D., and Gordon C. Jacoby. "Impact of Energy Development on the Law of the Colorado River." *Natural Resources Journal* 15 (January 1975).

Weist, Tom. *A History of the Cheyenne People*. Billings: Montana Council for Indian Education, 1977.

Western States Water Council. "Water for Western Energy Development Update 1977," by Tucson Myers and Associates, Salt Lake City, Utah, September 1977.

Western States Water Council. "Western States Water Requirements for Energy Development to 1990." Salt Lake City, Utah, November 1974.

Wetsit, Lawrence. "From the Heart." *EPA Journal*, April 1977, 30.

Wiener, Daniel Philip. "Indian Coal—Resource for the Future." *The Exchange* (Washington, D.C.: Phelps-Stokes Fund) 3, no. 4 (1981): 3–6.

Wilkinson, Charles. "Perspectives on Water and Energy in the American West and in Indian Country." *South Dakota Law Review* 26 (Summer 1981).

_____. "Western Water Law in Transition." *University of Colorado Law Review* 56 (Spring 1985): 317–345.

Williams, Ethel J. "Too Little Land, Too Many Heirs—The Indian Heirship Land Problem." *Washington Law Review* 46 (1971): 709–744.

Williams, Grant, Mary Hargrove, and Edward M. Eveld. "A Vanishing Trust." *Tulsa Tribune*, 9–18 November 1987.

Williams, Susan M., and David C. Cole. "Resource, Revenue, and Rights Reclamation: The Tax Program for the Navajo Nation." 1 August 1977. Photocopy. Author's files.

Wilson, Terry P. *The Underground Reservation: Osage Oil*. Lincoln: University of Nebraska Press, 1985.

Zeilig, Nancy. "Face to Face: Interview with Susan M. Williams on Indian Water Rights." *Journal of the American Water Works Association*, March 1985.

GOVERNMENT DOCUMENTS

The following list omits court decisions and all but the most important Department of Interior memorandums. Interior agencies are abbreviated as follows: Bureau of Indian Affairs (BIA); Bureau of Land Management (BLM); Bureau of Reclamation (BuRec); Minerals Management Service (MMS); Office of Surface Mining and Reclamation (OSMRE).

American Indian Policy Review Commission. *Final Report*. 2 vols. Washington, D.C.: GPO, 1977.

_____. *Task Force 7 Final Report: Reservation and Resource Development and Protection*. "Indian Minerals," by Ronald L. Trosper, Appendix. Washington, D.C.: GPO, 1977.

Assiniboine and Sioux Tribes. "Fort Peck Tribal Water Code." Poplar, Mont., 1986. NILL.

Collier, John. Indian commissioner to the secretary of the Interior. "Restoration of Lands Formerly Indian to Tribal Ownership." 10 August 1934. NILL.

Crow Tribe. "Section 710 Crow Indian Lands Study Final Report." Prepared for the Office of Surface Mining. Crow Agency, Mont., 1979.

Jicarilla Apache Tribe. *Jicarilla Apache Tribal Government Operations Handbook*, with the Institute for Development of Indian Law. Dulce, N.Mex.: Jicarilla Apache Tribe, August 1983.

Missouri River Basin Commission. "Draft Water Assessment of the Great Plains Gasification Associates Project," prepared for the U.S. Water Resources Council. Omaha, April 1980. Photocopy.

Navajo Nation. "Navajo Nation Water Code." Window Rock, Ariz., 1984. NILL.

Presidential Commission on Indian Reservation Economies. *Report and Recommendations to the President of the United States*. Washington, D.C.: GPO, November 1984.

Reagan, Ronald. *Statement by the President: Indian Policy*. Washington, D.C.: White House, January 1983. Photocopy.

Shoshone and Arapahoe Tribes. *Wind River Reservation Tax Regulations*. Fort Washakie, Wyo., 1988. NILL.

U.S. Congress. House. Committee on Government Operations. *Hearing on Problems Associated with the Department of Interior's Distribution of Oil and Gas Royalty Payments to Indians*. 99th Cong., 1st sess. Washington, D.C.: GPO, 1985.

_____. *Indian Oil and Gas Royalty Payments: Problems Persist.* 99th Cong., 1st sess. Washington, D.C.: GPO, 1985.

_____. Committee on Interior and Insular Affairs. *Federal Minerals Royalty Management*, by the committee staff, with the assistance of the General Accounting Office staff. 98th Cong., 2d sess. Washington, D.C.: GPO, 1985.

U.S. Congress. Senate. Committee on Interior and Insular Affairs, Subcommittee on Energy Resources and Water Resources. *Hearing on the Sale of Water from the Upper Missouri River Basin by the Federal Government for the Development of Energy.* 94th Cong., 1st sess. Washington, D.C.: GPO, 1975.

_____. Select Committee on Indian Affairs. *Hearing on Federal Supervision of Oil and Gas Leases on Indian Lands.* Parts 1, 2, and 3. 97th Cong., 1st sess. Washington, D.C.: GPO, 1981.

_____. *Hearing on S. 1894 to Permit Indian Tribes to Enter into Certain Agreements for the Disposition of Tribal Mineral Resources.* 97th Cong., 2d sess. Washington, D.C.: GPO, 1982.

_____. *Hearing on S. 1039 to Review and Determine the Impact of Indian Tribal Taxation on Indian Reservations and Residents, Nov. 12, 1987.* 100th Cong., 1st sess. Washington, D.C.: GPO, 1988.

_____. *Montana Water Rights: First Session on Oversight on Litigation Involving Water Rights in Montana.* 96th Cong., 1st sess. Washington, D.C.: GPO, 1979.

U.S. Department of Energy (DOE). *Environmental Assessment of Remedial Action at the Riverton Uranium Mill Tailings Site, Riverton, Wyo.* Albuquerque, N.Mex.: DOE, June 1987.

_____. *Environmental Assessment of Remedial Action at the Shiprock Uranium Mill Tailings Site, Shiprock, N.Mex.* Albuquerque, N.Mex.: DOE, May 1984.

U.S. Department of Health and Human Services. Public Health Service. *Health Hazards related to Nuclear Resource Development on Indian Land.* Washington, D.C.: Public Health Service, November 1982.

U.S. Department of Interior (DOI). *American Indians*, by Vince Lovett and Larry Rummel. Washington, D.C.: BIA, 1984.

_____. *Cumulative Hydrological Impact Assessment of the Peabody Coal Company Black Mesa/Kayenta Mine.* Denver: OSMRE, 1988.

_____. *Environmental Assessment for the Northern Cheyenne Petroleum Development Project.* Billings, Mont.: BIA, August 1980.

_____. *Final San Juan River Regional Coal Environmental Impact Statement.* Albuquerque, N.Mex.: BLM, March 1984.

_____. *Final Socioeconomic Technical Report for the Uintah Basin Synfuels Development Environmental Impact Statement.* Salt Lake City: BLM, February 1983.

_____. *Guidelines for Review of Tribal Ordinances Imposing Taxes on Mineral Activities.* Washington, D.C.: DOI, January 1983. NILL.

_____. *Indian Natural Resources—Part II: Coal, Oil, and Gas. Better Management Can Improve Development and Increase Indian Income and Employment*, by the comptroller general of the United States. Washington, D.C.: GPO, 1976.

_____. *Indian Service Population and Labor Force Estimates.* Washington, D.C.: BIA, January 1983.

_____. *Jackpile-Paguate Uranium Mine Reclamation Project Environmental Impact Statement.* 2 vols. Albuquerque, N.Mex.: BLM, October 1986.

_____. *Mineral Revenues: The 1986 Report on Receipts from Federal and Indian Leases.* Washington, D.C.: GPO, 1987.

_____. "Nature and Extent of the Secretary's Trust Responsibility When Indian Interests Conflict with Other National or Departmental Interests," memorandum prepared by Associate Solicitor for Energy and Resources Alexander Good for the undersecretary of the Interior. Washington, D.C., 28 October 1981. Photocopy. NILL.

_____. *North Central Power Study*. Billings, Mont.: BuRec, October 1971.

_____. *Office of Inspector General Audit Report: Review of Bureau of Land Management's Inspection and Enforcement Program*. Washington, D.C.: DOI, June 1986.

_____. *Office of Inspector General Audit Report: Review of the Distribution of Oil and Gas Revenues Generated from Indian Lands*. Washington, D.C.: DOI, March 1985.

_____. "Procedures for Evaluating the Effects of Federal Water Projects That Impact upon Indian Reservations," approved by Interior Secretary Cecil Andrus. Washington, D.C., 20 May 1980. Photocopy. NILL.

_____. *Report of the Commission on Fiscal Accountability of the Nation's Energy Resources*, by David F. Linowes, Michel T. Halbouty, Mary Gardiner Jones, Charles J. Mankin, and Elmer B. Staats. Washington, D.C.: DOI, January 1982.

_____. *Report of the Task Force on Indian Economic Development*. Washington, D.C.: GPO, July 1986.

_____. *Report to the U.S. Congress: Proposed Legislation Designed to Allow Indian Tribes to Elect to Regulate Surface Mining of Coal on Indian Lands*. Washington, D.C.: OSM, 1984.

_____. *Sherwood Uranium Project, Spokane Indian Reservation, Final Environmental Statement*. Portland, Oreg.: BIA, August 1976.

_____. *Study of Tribal Capability to Assume Regulatory Primacy*. Washington, D.C.: OSMRE, August 1987.

_____. *Uranium Development in the San Juan Basin Region, Final Report*. Albuquerque, N.Mex.: BLM, 1980.

_____. *Water for Energy: Missouri River Reservoirs, Pick-Sloan Missouri Basin Program, Final Environmental Impact Statement*. Billings, Mont.: BuRec, December 1977.

U.S. Environmental Protection Agency. "EPA Policy for the Administration of Environmental Programs on Indian Reservations." 8 November 1984. Photocopy.

_____. *Indian Drinking Water Supply Study*. Washington, D.C.: GPO, 1988.

_____. *Potential Health and Environmental Hazards of Uranium Mine Wastes*. 3 vols. Washington, D.C.: EPA, June 1983.

_____. *Radiological Quality of the Environment in the United States, 1977*. Washington, D.C.: GPO, September 1977.

_____. *Report to Congress: Indian Wastewater Treatment and Assistance*. Washington, D.C.: GPO, 1989.

U.S. Federal Trade Commission (FTC). *Staff Report on Mineral Leasing on Indian Lands*. Washington, D.C.: GPO, October 1975.

U.S. General Accounting Office (GAO). *Administration of Regulations for Surface Explorations, Mining and Reclamation of Public and Indian Coal Lands*. Washington, D.C.: GAO, August 1972.

_____. *Farmers Home Administration Information on Agricultural Credit Provided to Indians on 14 Reservations*. Gaithersburg, Md.: GAO, March 1987.

_____. *Indian Natural Resources, Opportunities for Improving Management: Part II, Coal, Oil and Gas*. Washington, D.C.: GAO, 31 March 1976.

_____. *Indian Royalties: Interior Has Not Solved Indian Oil and Gas Royalty Payment Problems*. Washington, D.C.: GAO, March 1986.

_____. *Oil and Gas Royalty Collections—Serious Financial Management Problems Need Congressional Attention*. Washington, D.C.: GAO, April 1979.

———. *Reserved Water Rights for Federal and Indian Reservations: A Growing Controversy in Need of Resolution.* Washington, D.C.: GAO, November 1978.

———. *Surface Mining: Issues Associated with Indian Assumption of Regulatory Authority.* Washington, D.C.: GAO, May 1986.

Wyoming. Office of the Attorney General. "Confidential Memorandum re Severance Taxes on the Wind River Indian Reservation," by Assistant Attorney General Robert Nicholas. Cheyenne, Wyo., 5 October 1987. Author's files.

INTERVIEWS WITH AUTHOR

Acevedo, Tom (Arapahoe tribal attorney). Telephone interviews, August 1986 and May 1988.

Aguilar, Pete (BIA Energy and Minerals Division staff). Interview in Golden, Colo., August 1986; telephone interview, May 1988.

Aitkin, Bob (formerly of the Southern Ute staff). Interview in Durango, Colo., October 1987.

Allison, David (acting chief of BIA Energy and Minerals Division). Interview in Boulder, Colo., June 1988.

Armajo, Chester (Northern Arapahoe Business Council chairman). Interview in Fort Washakie, Wyo., March 1985.

Atkins, John (Vice president of BHP-Utah International). Telephone interview, March 1988.

Atole, Leonard (Jicarilla Apache tribal president). Interview in Dulce, N.Mex., October 1987.

Aubertin, Don (BIA Energy and Minerals Division staff). Interview in Golden, Colo., August 1986; telephone interview, March 1988.

Bailey, Robert (Northern Cheyenne tribal president). Interview in Denver, Colo., November 1987.

Baker, Anson (BIA Billings area director and former BIA superintendent). Interview in New Town, N.Dak., April 1977, and in Billings, Mont., February 1981.

Baker, Chris (Southern Ute tribal chairman). Telephone interview, November 1986; interview in Denver, Colo., November 1987.

Baker, Lawrin "Hugh" (Fort Berthold tribal councilman). Interview in Denver, Colo., November 1987; telephone interviews, March and April 1988.

Baker, Perry (Uintah and Ouray Agency BIA superintendent). Interview in Fort Duchesne, Utah, May 1988.

Bettenberg, William (director of MMS). Interview in Denver, Colo., November 1986.

Bragg, J. W. (spokesman for Exxon). Telephone interview, October 1980.

Brewer, Del (BIA Energy and Minerals Division staff). Interview in Golden, Colo., August 1986.

Brock, John (Tesoro Petroleum landman). Telephone interview, September 1980.

Cameron, Charles (former Northern Ute staff geologist). Interview in Golden, Colo., November 1987.

Chambers, Reid Peyton (tribal attorney and former Interior associate solicitor). Telephone interviews, September 1986 and May 1988.

Chapoose, Lester (Northern Ute tribal chairman). Interview in Fort Duchesne, Utah, May 1987.

Chestnut, Steven (tribal attorney). Telephone interview, December 1987.

Chohamin, Ron (Northern Ute minerals staff). Interview in Denver, Colo., October 1981.

Comes at Night, Gary (Blackfeet Revenue Department inspector). Interview in Browning, Mont., September 1981; telephone interview, March 1988.

Connor, Carol (attorney). Telephone interviews, December 1979, January 1981, and April 1986.

Cook, Marvin (director of the Southern Ute Natural Resources Division). Interview in Ignacio, Colo., October 1987; telephone interview, December 1987.

Cox, R. T. (Wyoming assistant attorney general). Telephone interview, August 1985.

Dahl, Edwin (former Northern Cheyenne tribal councilman). Telephone interview, April 1988.

Damson, Barrie (president of Damson Oil Company). Telephone interview, November 1981.

DeGuire, Marcel (president of Dawn Mining Company). Telephone interview, April 1988.

Delk, Bob (chief of water services, BIA Billings area office). Telephone interview, March 1988.

Echohawk, John (executive director of NARF). Telephone interview, February 1988.

Epley, Paul (ex-staff member of CERT). Interview in Denver, Colo., November 1980.

Fairbanks, Mike (BIA Blackfeet Agency superintendent). Interview in Browning, Mont., September 1981.

Fernando, Chester (Laguna Pueblo governor). Interviews in Denver, Colo., November 1986 and November 1987.

Foley, Mike (hydrologist with Navajo Department of Surface Mining). Telephone interview, March 1988.

Ford, Glenn (Spokane tribal councilman). Telephone interview, April 1988.

Fredericks, Ken (BIA national director of realty). Telephone interviews, September and December 1980.

Fredericks, Tom (tribal attorney and former assistant secretary of Interior). Telephone interviews in September 1985, January 1987, February 1987, and July 1988.

Gabriel, Ed (former executive director of CERT). Telephone interview, January 1981; interview in Denver, Colo., November 1982.

Giedt, John (EPA, region 8). Telephone interview, August 1984.

Hans, Joseph M., Jr. (chief of field studies branch for EPA, Las Vegas). Telephone interview, August 1984.

Harris, LaDonna (founder of AIO). Interviews in Sun Valley, Idaho, August 1983, and in Denver, Colo., November 1986.

Harrison, David (CERT's chief operating officer and former director of trust responsibility for BIA). Interview in Denver, Colo., August and November 1986.

Holder, Berdena (founder of Oklahoma Mineral Owners Association). Interview in Anadarko, Okla., April 1985.

Ingraham, Vern (MMS chief of office of external affairs). Interview in Lakewood, Colo., August 1986; telephone interview, February 1988.

Irvin, John (Shell Oil Company spokesman). Telephone interview, January 1986.

Johnson, Ruth (attorney for Amoco). Telephone interview, February 1988.

Kipp, G. G. (Blackfeet natural resources director). Interview in Browning, Mont., September 1981.

Knight, Judy (vice chairman of Ute Mountain Ute Tribe and CERT board chairman). Interview in Denver, Colo., November 1986; telephone interview, December 1987.

Knows Gun, Ellis (former director of the Crow Coal Reclamation office and tribal secretary). Telephone interviews, March 1984 and January 1988.

Kooros, Ahmed (CERT's chief economist). Telephone interview, January 1987; interview in Denver, Colo., November 1987.

Lamb, Keith (OSM abandoned mines lands coordinator). Telephone interview, February 1988.

LeBret, Jim (BIA geologist in Portland area office). Telephone interviews, August 1987 and March 1989.

Leisse, Greg (Peabody Coal Company spokesman). Telephone interview, October 1985.

Lester, A. David (CERT executive director). Interview in Spokane, Wash., October 1983; telephone interview, November 1983.

Littlewhiteman, Wayne (Northern Cheyenne elder). Interview in Lame Deer, Mont., February 1981.

McElroy, Scott (Southern Ute tribal attorney). Telephone interview, October 1988.

Mankiller, Wilma (principal chief of the Cherokee Nation). Interview in Denver, Colo., November 1986.

Manning, Bill (former Southern Ute staff member). Interview in Durango, Colo., October 1987, and in Ignacio, Colo., November 1980.

Martel, Wes (Wind River Tax Commission chairman). Interview in Fort Washakie, Wyo., August 1987.

Martz, Clyde O. (attorney for Interior). Telephone interview, September 1980.

Meurer, Fred (Pittsburgh-Midway Coal Mining Company spokesman). Telephone interview, October 1985.

Miller, Anne (director of special programs for EPA). Telephone interview, May 1988.

Monsen, Marie (local and Indian affairs chief for DOE). Telephone interview, December 1986.

Monson, Larry (Fort Peck tribal geologist). Telephone interview, May 1988.

Moore, Bobby (GAO investigator). Telephone interview, November 1981.

Nagel, Barbara (ex-staff member of CERT). Interview in Denver, Colo., November 1980.

Nagle, Bill (ex-staff member of CERT). Interview in Denver, Colo., November 1980.

Neely, Claude (former Amoco Production Company regional land manager). Interview in Denver, Colo., November 1987.

Nelson, Marc (Jacobs Engineering staff). Telephone interview, January 1987.

Nelson, Mike (former special counsel to Navajo Chairman Peterson Zah). Telephone interview, October 1985; interview in Boulder, Colo., June 1988.

Nutongla, Nat (director of the Hopi Division of Mining and Reclamation Enforcement). Telephone interview, April 1988.

O'Connell, Michael (Hopi attorney). Telephone interview, April 1988.

Ortiz, Richard (director of the Wind River Tax Commission). Interview in Boulder, Colo., June 1988.

Partee, Grover (Environmental engineer with EPA, region 10). Telephone interview, April 1988.

Pettingil, Wes (director of Northern Ute Energy and Minerals Resource Department). Telephone interview, February 1988.

Plummer, Ed (Crownpoint BIA superintendent). Interview in Denver, Colo., May 1982; telephone interview, April 1986.

Pollack, Stanley (Navajo tribal attorney). Telephone interview, March 1988.

Ragsdale, Pat (BIA area director). Interview in Anadarko, Okla., April 1985.

Rana, Mahmood (ex-staff member of CERT). Interview in Denver, Colo., November 1980.

Robbins, Leonard (director of the Navajo Department of Surface Mining). Telephone interview, April 1988.

Roberts, William (ex-staff member of CERT). Interview in Denver, Colo., November 1980.

Rowland, Allen (Northern Cheyenne tribal president). Telephone interview, December 1980; interview in Lame Deer, Mont., February 1981.

Rudzik, Robert (WESCO project manager, Crow synfuels project manager, and vice president of Pacific Lighting). Telephone interview, December 1981.

St. Clair, Orville (Shoshone business councilman). Interview in Fort Washakie, Wyo., August 1987; telephone interview, January 1988.

Schilf, Rich (director of the Three Affiliated Tribes Natural Resources Department). Interview in Denver, Colo., November 1986.

Schryver, Robert (former Navajo director of mineral development). Telephone interview, January 1988.

Scott, Wilfred (Nez Perce tribal chairman). Telephone interview, November 1980.

Shirley, Perry (Navajo auditor general). Telephone interview, August 1986; interview in Denver, Colo., November 1988.

Shuey, Chris (Southwest Research and Information Center staff). Telephone interview, May 1988.

Shupe, Steven J. (water marketing consultant and coeditor of *Water Market Update*). Telephone interview, October 1988.

Sledd, John (DNA-People's Legal Services director of litigation). Telephone interview, April 1986.

Solimon, Ron (Laguna Pueblo negotiating team member). Interview in Denver, Colo., November 1987.

Sonosky, Marvin J. (tribal attorney). Telephone interview, April 1987.

Stevens, Ernest (Navajo economic development director). Telephone interviews, May and August 1980.

Stone, Richard (assistant to Energy Secretary Charles Duncan). Telephone interview, December 1980.

Swimmer, Ross (assistant secretary for Interior). Interview in Denver, Colo., November 1986.

Taradash, Alan (attorney representing allottees). Telephone interviews, April 1986 and April 1988.

Taylor, Les (tribal attorney). Telephone interview, March 1988.

Tome, Harry (Navajo tribal councilman). Telephone interview, August 1984.

Trujillo, Bill (MMS ombudsman). Telephone interview, July 1988.

Vail, Dorothy (BIA Billings, Mont., area realty specialist). Telephone interview, April 1987.

Velarde, Thurman (Jicarilla Apache oil and gas administrator). Interview in Dulce, N.Mex., October 1987.

Vigil, Dale (vice president of Jicarilla Apache Tribe). Interview in Dulce, N.Mex., October 1987.

Vigil, Sherryl (Jicarilla Apache BIA Agency superintendent). Interview in Dulce, N.Mex., October 1987.

Vlassis, George (Navajo tribal attorney). Telephone interview, May 1980.

Vogenthaler, Thomas J. (president of Cooper Petroleum). Telephone interview, April 1984.

Ware, Kent (Gulf Oil spokesman). Interview in Denver, Colo., October 1980.

Whiteing, Jeanne (Blackfeet tribal attorney). Telephone interview, June 1988.

Whitesell, Richard (BIA area director). Interview in Billings, Mont., September 1983.

Wiley, Marcus (Burnham mine superintendent for Consolidation Coal). Telephone interview, October 1980.

Wilkinson, Gerald (executive director of National Indian Youth Council). Interview in Sun Valley, Idaho, August 1983.

Williams, Charles D. (senior vice president of exploration, Wintershall Corporation). Interviews in Denver, Colo., November 1987 and February 1988.

Williams, Susan (tribal attorney). Telephone interviews, April 1985 and April and May 1988.

Wong, David (Jicarilla Apache tribal accountant). Interview in Dulce, N.Mex., October 1987.

Yazzie, Kee Ike (director of the Navajo Tax Commission). Telephone interview, May 1988.

NEWSPAPERS AND NEWSLETTERS

Ahead of the Herd, Three Affiliated Tribes of the Fort Berthold Reservation, New Town, N.Dak.

Billings Gazette, Billings, Mont.

Blackfeet Tribal News, Blackfeet Tribe, Browning, Mont.

Casper Star-Tribune, Casper, Wyo.

CERT Report, Council of Energy Resource Tribes, Washington, D.C. (defunct).

Char-Koosta, Salish-Kootenai Tribes, Pablo, Mont.

Crow Office of Reclamation News, Crow Tribe, Crow Agency, Mont.

High Country News, Paonia, Colo.

Jicarilla Chieftain, Jicarilla Apache Tribe, Dulce, N.Mex.

Lakota Times, Native American Publishing, Martin, S.Dak.

McGraw Hill–World News publications in New York: *Coal Age*, *Coal Week*, *Engineering and Mining Journal*, *Engineering News-Record*, *Inside Energy*, *Nuclear Fuel*, and *Platt's Oilgram News*.

NARF Legal Review, Native American Rights Fund, Boulder, Colo.

Navajo Times and *Navajo Times Today*, Navajo Tribe, Window Rock, Ariz.

NCAI Sentinel, National Congress of American Indians, Washington, D.C.

Northern Cheyenne News, Northern Cheyenne Tribe, Lame Deer, Mont.

The Plains Truth, Northern Plains Resource Council, Billings, Mont.

Riverton Ranger, Riverton, Wyo.

Sho-Ban News, Shoshone-Bannock Tribes, Fort Hall, Idaho.

Southern Ute Drum, Southern Ute Tribe, Ignacio, Colo.

Ute Bulletin, Northern Ute Tribe, Fort Duchesne, Utah.

Water Market Update, Western Network, Santa Fe, N.Mex.

Wind River News, *Wyoming State Journal*, Lander, Wyo.

Wotanin Wowapi, Fort Peck Assiniboine and Sioux Tribes, Poplar, Mont.

Wyoming State Journal, Lander, Wyo.

INDEX

Abandoned mines, 177, 179–180, 187, 189, 192, 194

Abandoned Mines Land (AML) funds, 180, 187, 189, 192, 194

Abourezk, James, 150

Absaloka mine. *See* Westmoreland Resources

Acoma Pueblo, 92, 111, 173, 174, 200, 234

Administration for Native Americans, 111

Agriculture Department, 108

Agriculture of the West, threats to, 67, 71, 153, 207–208, 209–210

Air quality: federal standards, 54, 63, 89, 183–185, 192–193; tribal monitoring, 115, 184–185; tribal standards, 84, 104–105, 183–185, 199, 234

Ak-Chin, 219

Akwesasne Notes, 101, 218

Alaska, xiv, xvi, 118

Albuquerque, N.Mex., 213, 219

Allotment, 42, 145; fee patents for, 45, 53, 147–148; land losses caused by, 45, 53, 147–148, 272n.28, 274n.53; mineral development of, 37; policy, 10–16, 45; problems caused by, 15, 29, 186, 281n.50; reservations affected by, 4, 272n.27; trust protection for, 45, 53, 147–148. *See also* Allottees

Allottees, 145–171; alternative contracts for, 241; and CERT, 105, 106, 111, 146, 298n.4; consent of for leasing, 147, 151–153, 157–160, 170; fractionated ownership of, 146–148; mineral information of, 147–148, 170–171; mineral ownership of, xvi, 42–47, 53–54, 105–106, 145–146; oil and gas income of, 134, 135, 139, 157; organizations of, 149, 153, 166–170; royalties accounting of, 111, 118; split estates of, 42–47, 148–154; tribes' efforts for, 121, 146, 163–166, 175, 241; trust responsibility for, 155. *See also* Allotment

Alternative contracts, 72, 85–90, 237–262; accounting for, 137, 142, 247–249, 268–269; advantages of, 240–241; negotiating for, 164–165; risks of, 237–240; types of, 241–243. *See also* Tribally owned wells; *individual companies and tribes*

AMAX Coal Company, 64, 65, 69, 70, 209

American Indian Defense Association, 16

American Indian Law Center, 105, 155

American Indian Policy Review Commission, 17, 19, 22, 274n.53

American Indian Resources Institute, 227, 228

American Religious Freedom Act, 234–236

Americans for Indian Opportunity (AIO), 93, 110, 115, 192

Ami, Don, 101 (photo)

Amoco Production Company, 81–82; royalty problems of, 125–126, 129, 166–167

Anaconda Minerals Company, 58, 181–183

Andrus, Cecil, 24, 82, 85, 98

Aneth, Utah, 74

Apache, 12, 134, 136. *See also* Fort Sill Apache; Jicarilla Apache; Mescalero Apache

Aqua Caliente, xvi

Arapahoe. *See* Cheyenne Arapahoe; Wind River Reservation

Arikara, 204. *See also* Fort Berthold Reservation

Arizona: environmental standards of, 175–176, 199; and Indians, 201, 219, 230–231, 232; water in, 175–176, 206, 232

Arizona Public Service Company, 199, 224

Arizona Republic, 169

Armajo, Chester, 258–259

Army Corps of Engineers, 20, 206, 209, 311n.4

Arthur, Claudeen, 159

Ashurst, Henry, 39

Aspinall, Wayne, 222

Assimilation. *See* Termination

Assiniboine. *See* Fort Belknap Reservation; Fort Peck Reservation

Atlantic Richfield Company (ARCO), 80, 87–89, 153–154, 181–183, 209, 243, 246

Atomic Energy Commission, 178, 191

Baker, Anson, 67, 70, 71 (photo)

Baker, Perry, 253

Baldwin, David, 86

Bannock Tribe. *See* Fort Hall Reservation

Basin Electric Power Cooperative, 205, 212

Battin, James, 233

Bavarskis, Justas, 178

Bechtel Power Corporation, 229, 255–256, 259

Benally v. UNC, 175